Handbook of the Mammals of Europe

Series Editors

Klaus Hackländer, University of Natural Resources and Life Sciences, Vienna, Department of Integrative Biology and Biodiversity Research, Institute of Wildlife Biology and Game Management, Vienna, Austria

Deutsche Wildtier Stiftung (German Wildlife Foundation), Hamburg, Germany

Frank E. Zachos, Natural History Museum Vienna, Mammal Collection, Vienna, Austria

University of Vienna, Department of Evolutionary Biology, Vienna, Austria

University of the Free State, Department of Genetics, Bloemfontein, South Africa

This handbook offers a unique collection of information on all wild mammal species living in Europe and will serve as a standard reference guide for all mammalogists and readers interested in research on mammals. The introductory volume provides an overview of mammalian phylogeny, the history and current status of European mammals and their management as well as their habitats and the history of mammalogy in Europe. The remaining volumes present comprehensive species-specific chapters covering all aspects of mammalian biology, including palaeontology, physiology, genetics, reproduction and development, ecology, habitat, diet, mortality and behaviour. The economic significance and management of mammals and future challenges for research and conservation are addressed as well. Each chapter includes a distribution map, a photograph of the animal, and a list of key literature. This authoritative handbook provides current and detailed descriptions of all European mammals; it will appeal to academics and students in mammal research alike, as well as to practitioners whose work involves mammal management, control, use and conservation.

Klaus Hackländer • Paulo C. Alves
Editors

Primates and Lagomorpha

With 40 Figures and 17 Tables

Springer

Editors
Klaus Hackländer
University of Natural Resources
and Life Sciences, Vienna, Department
of Integrative Biology and Biodiversity
Research, Institute of Wildlife Biology
and Game Management
Vienna, Austria

Paulo C. Alves
Faculty of Sciences
CIBIO-InBIO/BIOPOLIS
University of Porto
Vairão, Portugal

ISSN 2730-7387 ISSN 2730-7395 (electronic)
Handbook of the Mammals of Europe
ISBN 978-3-030-34042-1 ISBN 978-3-030-34043-8 (eBook)
https://doi.org/10.1007/978-3-030-34043-8

Cover photos with kind permission from © Ingo Arndt (young European hares), Bonaventura Majolo (Barbary macaque), Rolf Giger (European wild rabbit)

This Springer imprint is published by the registered company Springer Nature Switzerland AG.
The registered company address is: Gewerbestrasse 11, 6330 Cham, Switzerland

Acknowledgments

We are grateful to the editorial team of Springer Nature, especially to Barbara Wolf, Veronika Mang, Verena Penning, and Lars Koerner. In addition, we thank our volume editors and all species chapter authors for their support and contributions.

Klaus Hackländer
Frank Zachos
Series Editors

Contents

About the Series Editors

Klaus Hackländer studied Zoology at Philipps-University Marburg/Lahn and the University of Vienna. He is currently a Full Professor of Wildlife Biology and Wildlife Management at the University of Natural Resources and Life Sciences, Vienna (BOKU), where he also coordinates the master program on Wildlife Ecology and Wildlife Management. Prof. Hackländer is a member of numerous mammalogical societies and served as President of the German Society for Mammalian Biology from 2009 to 2016. He was Editor-in-Chief of the journal *Mammal Review* from 2009 to 2015. Prof. Hackländer's research covers all aspects of mammalian biology, including ecology, behavior, physiology, diseases, and management. Since 2021, he is Head of the German Wildlife Foundation (Deutsche Wildtier Stiftung), besides being Head of the Institute of Wildlife Biology and Game Management at BOKU.

Frank E. Zachos is Head of the Mammal Collection at the Natural History Museum in Vienna, Austria, an Affiliated Lecturer at the University of Vienna, and an Affiliated Professor at the Department of Genetics, University of the Free State in Bloemfontein, South Africa. He studied biology, the history of science, and philosophy in Kiel and Jena, Germany, and received his Ph.D. in Zoology in 2005. In 2009, he completed his habilitation (postdoctoral degree) in Zoology and Evolutionary Biology. His main research interests are intraspecific biodiversity, phylogeography, population genetics, and conservation of mammals and birds, with a particular focus on red deer and other ungulates. He also has a long-standing interest in species concepts and the species problem, as well as other theoretical and philosophical issues in evolutionary biology and systematics. Until 2021, he was Editor-in-Chief of the journal *Mammalian Biology*, and he is currently the president of the German Society for Mammalian Biology and the Editor of the Mammalia series within the *Handbook of Zoology*.

About the Volume Editors

Klaus Hackländer's favorite mammalian taxon is Lagomorpha. He studied leporids in Europe, North America, and Africa. Together with the co-editor of this volume, he founded the World Lagomorph Society. Prof. Hackländer serves in the Lagomorph Specialist Group of the IUCN Species Survival Commission and contributed to the global Red List Assessment of *Lepus europaeus*. In addition, he is responsible for the European Red List Assessment of leporids. His publications include the leporid chapter of the *Handbook of Mammals of the World*, two books on lagomorphs, and numerous original work, mainly on *L. europaeus* and *L. timidus*.

Paulo Célio Alves is an Associate Professor of Biology at the Faculty of Sciences, University of Porto and the Head of the Conservation Genetics and Wildlife Management research group at CIBIO (Research Center in Biodiversity and Genetics Resources/BIOPOLIS). Editor of the *European Journal of Wildlife Research*, author of more than 200 publications, including papers, chapters, and books, namely two co-edited books on lagomorphs. Cofounder and current President of the World Lagomorph Society. Currently is the Director of recently created Biological Station of Mértola, Portugal. His main research topics cover animal ecology, wildlife management, conservation biology, and conservation genetics.

Contributors

Pelayo Acevedo Instituto de Investigación en Recursos Cinegéticos IREC (CSIC-UCLM-JCCM), Ciudad Real, Spain

Paulo C. Alves CIBIO, Centro de Investigação em Biodiversidade e Recursos Genéticos, InBIO, Universidade do Porto, Vairão, Portugal

BIOPOLIS Program in Genomics, Biodiversity and Land Planning, CIBIO, Vairão, Portugal

Departamento de Biologia, Faculdade de Ciências, Universidade do Porto, Porto, Portugal

Estação Biológica de Mértola (EBM), Mértola, Portugal

Francesco Maria Angelici Fondazione Italiana per la Zoologia dei Vertebrati (FIZV), Rome, Italy

Anders Angerbjörn Department of Zoology, Stockholm University, Stockholm, Sweden

Fernando Ballesteros Fundación Oso Pardo (Brown Bear Foundation), Santander, Spain

Sandro Bertolino Department of Life Sciences and Systems Biology, University of Turin, Torino, Italy

David E. Brown Natural History Collections, Arizona State University, Tempe, AZ, USA

Antonio Canu Department of Veterinary Medicine, University of Sassari, Sassari, Italy

Miguel Carneiro CIBIO, Centro de Investigação em Biodiversidade e Recursos Genéticos, InBIO, Universidade do Porto, Vairão, Portugal

BIOPOLIS Program in Genomics, Biodiversity and Land Planning, CIBIO, Vairão, Portugal

Jacopo Cerri Institute for Environmental Protection and Research, Livorno, Italy

Anna Maria De Marinis Italian Institute for Environmental Protection and Research (ISPRA), Ozzano dell'Emilia, Italy

Miguel Delibes-Mateos Instituto de Estudios Sociales Avanzados (IESA-CSIC), Córdoba, Spain

Klaus Hackländer University of Natural Resources and Life Sciences, Vienna, Department of Integrative Biology and Biodiversity Research, Institute of Wildlife Biology and Game Management, Vienna, Austria

Deutsche Wildtier Stiftung (German Wildlife Foundation), Hamburg, Germany

John L. Koprowski Haub School of Environment & Natural Resources, University of Wyoming, Laramie, WY, USA

Bonaventura Majolo School of Psychology, University of Lincoln, Lincoln, UK

Laëtitia Maréchal School of Psychology, University of Lincoln, Lincoln, UK

José Melo-Ferreira Departamento de Biologia, Faculdade de Ciências da Universidade do Porto, Porto, Portugal

CIBIO/InBIO, Centro de Investigação em Biodiversidade e Recursos Genéticos, Universidade do Porto, Vila do Conde, Portugal

BIOPOLIS Program in Genomics, Biodiversity and Land Planning, CIBIO, Vairão, Portugal

Heiko G. Rödel Laboratoire d'Ethologie Expérimentale et Comparée (LEEC), UR4443, Université Sorbonne Paris Nord, Villetaneuse, France

Carlos Rouco Área de Ecología, Departamento de Botánica, Ecología y Fisiología Vegetal, Universidad de Córdoba, Córdoba, Spain

Sociedad, Ecología y Gestión del Medio Ambiente, UCO-IESA, Unidad Asociada al CSIC, Córdoba, Spain

Massimo Scandura Department of Veterinary Medicine, University of Sassari, Sassari, Italy

Stéphanie Claire Schai-Braun Department of Integrative Biology and Biodiversity Research, Institute of Wildlife Biology and Game Management, University of Natural Resources and Life Sciences, Vienna, Vienna, Austria

Rafael Villafuerte Instituto de Estudios Sociales Avanzados (IESA-CSIC), Córdoba, Spain

Barbary Macaque *Macaca sylvanus* (Linnaeus, 1758)

1

Bonaventura Majolo and Laëtitia Maréchal

Contents

B. Majolo (✉) · L. Maréchal
School of Psychology, University of Lincoln, Lincoln, UK
e-mail: bmajolo@lincoln.ac.uk; lmarechal@lincoln.ac.uk

K. Hackländer, P. C. Alves (eds.), *Primates and Lagomorpha*, Handbook of the Mammals of Europe,
https://doi.org/10.1007/978-3-030-34043-8_15

Common Names

English	Barbary macaque (also wrongly called Barbary ape)
German	Berberaffe, Magot
French	Magot, singe magot, macaque Berbère
Spanish	Mona de Berberíc, mona/mono de Gibraltar
Italian	Bertuccia, macaco Berbero
Russian	Берберийская макака

Taxonomy and Systematics

The Barbary macaque belongs to the genus *Macaca* comprising approximately 25 species (Li et al. 2009). It is the only nonhuman primate in Europe, the northernmost African nonhuman primate and the only extant species of the genus *Macaca* outside East-Asia. The genus *Macaca* probably originated in North Africa around 5.5 million years ago (Modolo et al. 2005), and then radiated into Europe and Asia (Li et al. 2009). However, extant macaque species all live in East Asia, with the exception of the Barbary macaque, between Afghanistan and Pakistan to the East and Japan to the West, and between China to the North and Indonesia to the South. By the early Pleistocene the Barbary macaque lineage disappeared in most of Europe and in the Middle East and as a result became separated from the other macaque lineages. The reduction in the distribution of the Barbary macaque was probably due to climate change and associated changes in vegetation and availability of resources (Elton and O'Regan 2014). Due to the geographic distance between the Barbary macaque and other macaque species, there is no record of hybridization with related species in the wild.

The Barbary macaque population living in Gibraltar has both Algerian and Moroccan origins (Modolo et al. 2005). The origins of the Gibraltar population are unclear: it may have been originated from a remnant European population or it may have been established through the translocation of some animals from Africa (Modolo et al. 2005). In any case, the current Gibraltar population might not include any remnants of the original European population, since there are no fossil records of Barbary macaques in Gibraltar after the last glaciation (Shaw and Cortes 2006).

Paleontology

The fossil record and the modeled reconstruction of their geographic distribution suggest that the Barbary macaque inhabited vast parts of Europe, North-Africa, and the Middle East (Syria and the Caucasian region) between the Miocene and Pleistocene (Fig. 1; Elton and O'Regan 2014). Humans have probably affected the geographic range of the Barbary macaques since at least the last 4,000 years (Modolo et al. 2005). In historic times, the Barbary macaque faced a dramatic decline in its distribution. In Europe, this species was restricted to Spain by the end of the nineteenth century and it was only found in Gibraltar by the beginning of the twentieth century (Taub 1984).

Current Distribution

The current distribution of the Barbary macaque is limited to Gibraltar (36°09'N, 05°21'W) and to small relic patches of forest in Morocco and Algeria (Fig. 2). In Gibraltar, around six groups of macaques, for a total of approximately 200 individuals, inhabit the Upper Rock Nature Reserve; some of these groups rely extensively of food provisioning from humans.

Description

The Barbary macaque is often wrongly called the tailless monkey or Barbary ape. The Barbary macaque is not an ape (i.e., anthropoid primate) and it has a vestigial tail, which varies in length from 4 mm to 22 mm (Fooden 2007). Sexual dimorphism can be observed in relation to body length (males: 550–600 mm, females: 450 mm)

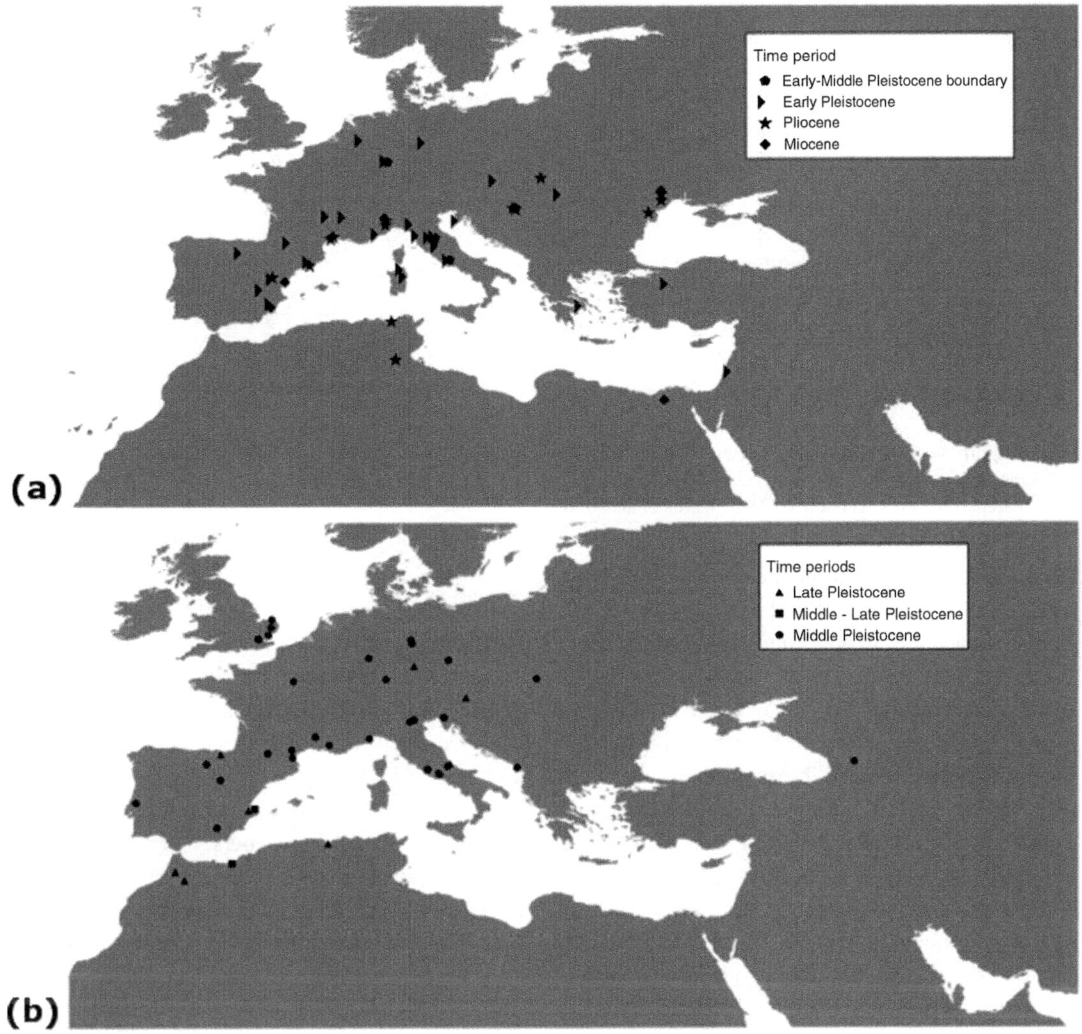

Fig. 1 Location of *Macaca* fossils in Europe, North Africa and Caucasus from (**a**) the Miocene to the Early-Middle Pleistocene boundary and (**b**) in the Middle and Late Pleistocene. Note that not all Miocene *Macaca* sites in North Africa are shown in the figure. (Redrawn from Elton and O'Regan (2014) with permission from Elsevier)

and weight (males: 15–17 kg, females: 10–11 kg; Fa 1989). The body size and morphology of Barbary macaques varies depending on age and on whether measurements are taken from provisioned or non-provisioned individuals (Fa 1984; Fooden 2007; Borg et al. 2014; Table 1 and Fig. 3). Provisioned macaques have a significantly larger body size than non-provisioned animals (Borg et al. 2014; Maréchal et al. 2016a).

The dental formula of a Barbary macaque is I2/2, C1/1, P2/2, M3/3. Sub-adult and adult males have long canine teeth, surpassing their incisors, which play an important role in dominance-related behaviors, while females do not display long canine teeth.

During the mating period (see section "Genetics" below), female sexual displays are characterized by a sexual swelling (Fig. 4). This swelling varies during the reproductive cycle of a female (Brauch et al. 2007; Young et al. 2013a) but, in provisioned conditions, a swelling might be still visible outside of the mating season despite the

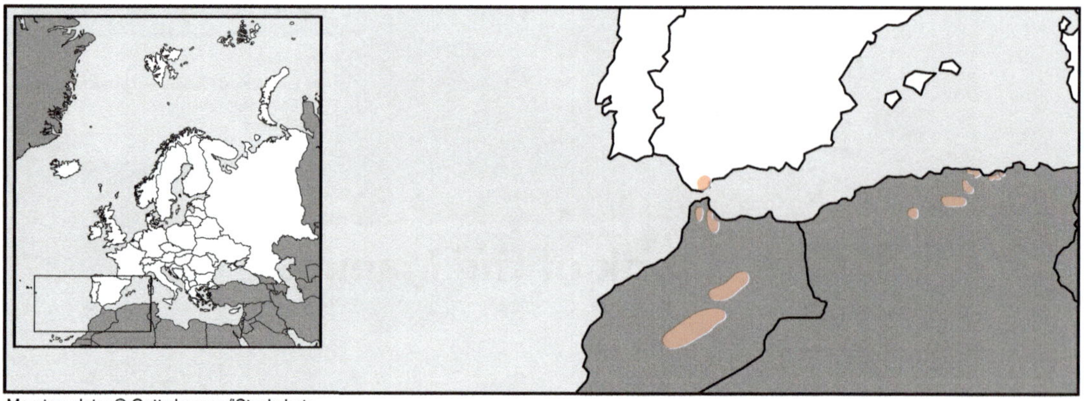

Map template: © Getty Images/iStockphoto

Fig. 2 Current geographic distribution of the Barbary macaque. The distribution is based on the IUCN Red List of Threatened Species. Version 2020-2 (Wallis et al. 2020; redrawn with permission from IUCN). (Map template: © Getty Images/iStockphoto)

female not being fertile at the time. There is no sexual dimorphism in the pelage color in the Barbary macaque. In adult and sub-adult Barbary macaques, the pelage coloration on the dorsal surface from the top of the head to the hindquarter is highly variable from light gray, beige, light brown, golden brown, burnt orange, to dark (Fig. 3; Fooden 2007). The ventral fur is generally lighter than the dorsal fur, and it is also not as thick, leaving ventral body parts often visible. To cope with extreme climatic conditions, with hot and dry summer (up to +40 °C) and cold winter (up to −20 °C), Barbary macaques have a thicker fur during autumn and winter, which molts during spring and summer (Fig. 3).

Age determination is difficult in the Barbary macaques, especially at older age, because the different traits used to age a macaque (e.g., dentition, pelage, locomotion, social behavior; Table 2) may significantly vary across individuals of the same age due to factors like dominance rank, diet, and feeding efficiency. In captivity or under provisioned conditions, Barbary macaques may live up to around 20–28 years; the maximum recorded lifespan of a Barbary macaque is 28 years, in captivity (Sue Wiper pers. comm.; Fooden 2007). Macaques are unlikely to live beyond 20 years in the wild in non-provisioned conditions. New noninvasive markers for immune function, urinary neopterin levels, have been recently found to significantly increase with age (Müller et al. 2017), which may provide a useful additional tool to assess age in wild individuals in the future.

Table 1 Body measurements of Barbary macaque by age/sex class

Age/sex class	Weight (kg)		Sitting height (cm)	Head and body length (cm)
Infants <1 year	–	0.9–1.0	–	24.6–28.7
Infants 1 year	1.9–2.1		–	
Juveniles 1–2 years	4.1–4.4	–	–	–
Juveniles 2–3 years	5.7–6.2	6.2–8.6	30–33	50.0–55.3
Juveniles 3–4 years	7.5–8.1		35–38	
Sub-adults 4–5 years	8.6–10.4	–	40–43	–
Sub-adult males 5–6 years	12.2	–	49	–
Sub-adult males 6–7 years	14.3	–	53	–
Adult males	15.3–17.0	10.0–18.0	55–60	60.5–65.3
Adult females	10.2–11.0	8.0–12.0	45	54.0–56.6
Reference	Fa et al. (1984)	Fooden (2007)	Fa et al. (1984)	Fooden (2007)

Fig. 3 Adult pelage of the Barbary macaque during molting and non-molting periods; the pictures depict the same macaque (top/bottom) at the two different periods. Pictures taken from macaques living in Ifrane National park (33°24′N–005°12′W). (© Pictures with permission from Laëtitia Maréchal)

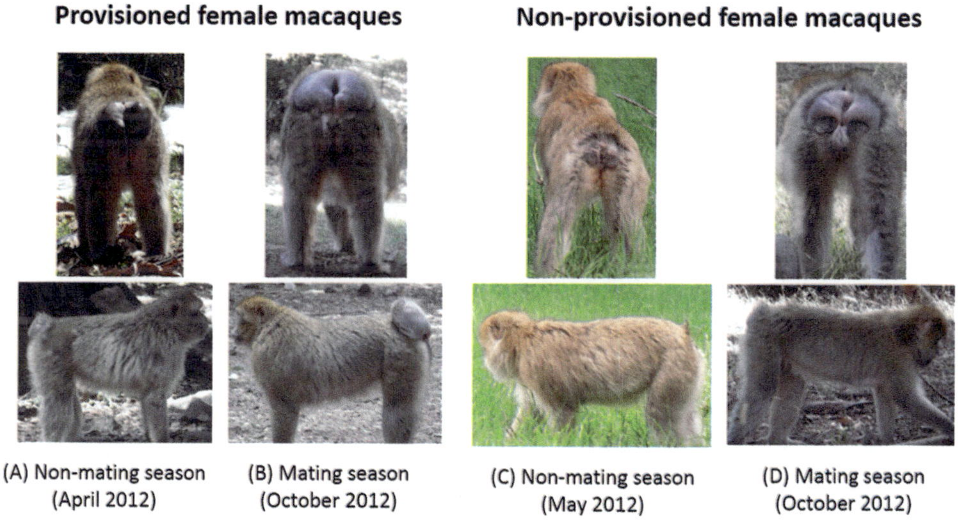

Fig. 4 Sexual swellings in female Barbary macaque. Pictures taken from two females living in Ifrane National park (33°24′N–005°12′W). The female on the left experienced daily provisioning of high caloric food from tourists and local people (A = non-mating season April 2012; B = mating season October 2012); the female on the right fed only on natural resources available in her habitat (C = non-mating season May 2012; D = mating season October 2012). (© Pictures with permission from Laëtitia Maréchal)

Table 2 Age classes of Barbary macaques based on dentition, pelage, body morphology, dexterity, mobility, and behavior (Fa 1984; Fooden 2007; Bissonnette et al. 2009; Maréchal unpublished data)

Stage of development	Age	Dentition	Pelage/body morphology	Dexterity and mobility	Social and other behavior
Infant 1	Birth-145 days	Deciduous teeth only	At birth, the dorsal pelage is dark and gradually becomes brownish between 45–145 days old	Carried ventrally by adults (mother and males) but also dorsally from around 45–75 days old. Stays within short distance from infant carer	Social interactions mostly with mother, gradually interacting with males and other group members under mother's supervision
Infant 2	145 days–1 yr		Light brown infant	Becoming more independent, developing their dexterity and mobility through play	Weaning is usually completed by year 1 but it may last longer if the mother does not have a new infant in the following year. Initial social play with other infants and juveniles
Juvenile	1–3 yrs	Some permanent teeth erupted	Light brown/beige	Mostly independent, developing their dexterity and mobility through play	Frequent play with other juveniles, becoming more independent from their mother
Sub-adult female	3–5 yrs	Molar 3 incompletely erupted	Small face, no beard, slender body	Reaching peak of their dexterity and mobility with great interindividual differences due to previous injuries, diet, rank, and personality	Still engage in play, initial attempts to mate with males
Sub-adult male	3–5 yrs	Canines incompletely erupted	Body grows from adult female size to large body size, but high inter-individual differences emerging (pelage color, body size)		Initial attempts to mate with females during the mating season, often prevented by adult males and/or rejected by high ranking females
Young adult female	5–6 yrs	All permanent teeth completely erupted	Small face, small dark beard under the chin. Sexual swelling appearing.		Stable social bonds with other group members and dominance rank
Young adult male	5–7 yrs		Larger body and full testicular size		Actively participating in sexual behavior, not tolerated by high ranking males
Prime adult female	6–15 yrs	Variation in dentition condition due to diet, rank, competition	Reaching full body size, beard becoming larger and darker, nipples longer, large sexual swelling.	Peak of their dexterity and mobility with great interindividual differences due to previous injuries, diet, rank, and personality	Peak of their sexual activity
Prime adult male	7–12/15 yrs		Reaching full body size with great interindividual		Peak of sociosexual competition and sexual activity

(continued)

Table 2 (continued)

Stage of development	Age	Dentition	Pelage/body morphology	Dexterity and mobility	Social and other behavior
			differences (pelage, body size)		
Old adult female	15 yrs. +	Deterioration of dentition, missing, worn, or broken teeth	Deterioration of the overall body condition with body size reduction, arched back	Degradation in mobility; limping, and arthritis possible	Less socially active
Old adult male	12/15 yrs. +	Deterioration of dentition, missing, worn, or broken teeth	Deterioration of the overall body condition with body size reduction, arched back		Socially more active than females of same age, progressively becoming more peripheral to the core group

Physiology

The Barbary macaque lives in extremely variable ecosystems in terms of food availability, habitat type, and climate. A large proportion of Barbary macaques live in anthropogenic habitats where interactions with humans are frequent (Waterman et al. 2020). Barbary macaques exert flexible metabolic strategies in response to these environmental and social conditions.

Barbary macaques have demonstrated a clear metabolic flexibility to cope with extreme environmental variations. Cristóbal-Azkarate et al. (2016) found that fecal thyroid hormone levels, an index for basal metabolic rate, were negatively related to temperature and food availability, indicating higher energetic expenditure when temperatures were low, and food scattered. In addition, social factors were found to modulate Barbary macaque metabolism responses to environmental stress. For instance, the degree of sociality has been positively associated with an individual's chance of survival in extremely cold winter in Morocco (McFarland and Majolo 2013b). Huddling behavior was also suggested to be a behavioral strategy to modulate thermoregulation and energy in Barbary macaques (Campbell et al. 2018).

Although living in social groups helps individuals to regulate the impacts of environmental conditions, Barbary macaques have shown a distinct metabolic flexibility to deal with social demands. For example, social tension due, for example, to within group aggressive conflicts, is marked by an increase in physiological stress levels in Barbary macaques, but such metabolic responses seem to be buffered by strong social bonds in males (Young et al. 2014b), and by taking part in grooming interactions (Shutt et al. 2007). As many mammals, the gestation period in Barbary macaque is associated with higher physiological stress levels compared to non-gestational states (Maréchal et al. 2016a). In addition, sexual hormones have been shown to influence different behaviors and morphological traits during mating season (see section "Genetics"). The size of female sexual swelling and sexual behavior are positively related to estrogen/progestogen ratio (Möhle et al. 2005; Brauch et al. 2007; Young et al. 2013a). Female fertile phase also appears to affect male sexual behavior and metabolism (Heistermann et al. 2008; Young et al. 2013a). Fecal glucocorticoid metabolite levels increase during the mating season for males but not females (Maréchal et al. 2016a). Fecal thyroid hormone levels and androgen levels also show a significant increase before the onset of the mating season and are generally higher throughout the season compared to outside the mating period (Cristóbal-Azkarate et al. 2016;

Rincon et al. 2017). These hormonal changes appear to coincide with males' body size growth (Maréchal et al. 2016a) and suggest that the mating season requires a major energetic expenditure for male Barbary macaques to cope with. Male-infant grooming during highly social stress periods, such as the mating season and periods of rank instability, was associated with lower androgen levels (Rincon et al. 2017).

Barbary macaques are often encountered in, or in proximity of human-modified landscapes, increasing opportunities for human–macaque interactions. However, such close interactions with humans have been shown to bear often negative impacts on the metabolism and body condition of Barbary macaques. Agonistic interactions with humans increase the physiological stress levels of male Barbary macaques with negative consequences on their health and reproduction (Maréchal et al. 2011). Food provisioning, linked to unregulated tourism, has also shown to have a negative impact on the health of Barbary macaques in Gibraltar and Morocco, increasing their body size, parasite load, stress levels and mortality, changing dietary patterns and nutrition, negatively affecting body condition, and potentially lowering reproductive rates in provisioned macaques in comparison to animals not/less exposed to tourists (Fuentes 2006; Schurr et al. 2012; Borg et al. 2014; Maréchal et al. 2016a).

Genetics

The Barbary macaques live in extremely fragmented sub-populations. The Algerian and Moroccan populations diverged around 1.6 million years ago (Modolo et al. 2005). A study on the phylo-geography of the Barbary macaques found that the Algerian population is paraphyletic whereas the Moroccan population is monophyletic (Modolo et al. 2005). Moreover, genetic analyses found evidence of genetic fragmentation across sub-populations and some sub-populations are genetically isolated from the others (Scheffrahn et al. 1993; Von Segesser et al. 1999). Therefore, such a fragmentation is a major concern for the conservation of this species. However, genetic analyses on the Algerian population estimated the genetic heterozygosity of Algerian Barbary macaques to be comparable to other mammalian species, thus suggesting no restriction to the genetic diversity of the Barbary macaque (Von Segesser et al. 1999).

All of the extant Gibraltarian mtDNA haplotypes have also been found in the Moroccan and Algerian populations (Modolo et al. 2005). Indeed, various introductions of macaques, captured in Morocco or Algeria have taken place in Gibraltar up until at least the eighteenth century; the last reported reintroduction of macaques, probably coming from Morocco, took place during the Second World War (Fa 1984).

Life History

The Barbary macaque is a seasonal breeder (Paul and Kuester 1988). The mating season starts around the beginning of September and lasts approximately until January. The birth season in Gibraltar is between May and August (Fa 1984); in Morocco and Algeria the birth season is between March and June (Ménard and Vallet 1993a; Young et al. 2013b).

Barbary macaques form consortships during the mating season that can last from a few minutes to several hours; during these consortships a male and female spend time in proximity, co-feeding, exchanging grooming and mating. Females are most likely to form consortships when in estrus; however, males are not always capable to detect the peak of a female's fertile period (Young et al. 2013a; see section "Description"). Sexual mounts in the Barbary macaques are dorso-ventral (Fig. 5) and composed of a series of single mount-dismount sequence during which usually the male ejaculates after a few pelvic thrusts (Fooden 2007). Toward the end of the mount-dismount sequence and during ejaculation, the female often produces a copulation call, a series of low frequency grunts. These copulation calls may indicate the female's receptive status to males and/or elicit male–male competition for access to estrous females (Semple 1998).

Fig. 5 A sexual mount in the Barbary macaque in the Middle Atlas Mountains of Morocco. (© Picture with permission from Christopher Young)

Barbary macaques usually give birth to one infant per birth season; twins are rare. Gestation period ranges from 145 to 177 in captive Barbary macaques (Paul and Kuester 1987); in wild macaques in the Moroccan Middle Atlas Mountains the average gestation length was calculated to be 170 ± 3.8 days (Young et al. 2013a). Gestation period appears to be longer for female infants (165.3 ± 4.6) than for male infants (163.2 ± 5.4; Fooden 2007). In Gibraltar, females give birth to their first infant when they are between 4.05 and 6.99 years old (Fa 1984). In non-provisioned Barbary macaques in Algeria, females give birth to their first infant at an average age of 5.4 years (Ménard and Vallet 1993a). Female fecundity reaches a peak at 7–8 years but then declines rapidly as females become older (Paul et al. 1993). The majority of inter-birth intervals fall within a 1-year period, particularly in provisioned macaques (Fa 1984), but it can be longer (2–3 years) in non-provisioned macaques, especially in years of high ecological pressure due, for example, to low food and/or water availability. Inter-birth intervals become longer the older the female (Paul et al. 1993). In the Moroccan Rif Mountains, the adult female birth rate is 0.58 per year (Mehlman 1989). In Morocco and Algeria males become sexually active between the age of 5–7 years old (Deag 1980; Ménard et al. 1985).

Infants start eating solid food when around 1–2 months old. Weaning is completed when infants are between 1 and 2 years old (Fa 1984). Mortality rate is highest in infants, especially in the first 6 months of their life, in both provisioned and non-provisioned macaques, then it decreases until senescence (Fa 1984; Mehlman 1989; Ménard and Vallet 1993a).

The overall sex ratio (male/female) in Gibraltar is 0.99; the sex ratio for macaques above 5 years of age is 0.66 (Shaw and Cortes 2006). In the Moroccan Rif Mountains the sex ratio was 0.73 (Mehlman 1989); in the southernmost Moroccan population of Barbary macaques in the High Ourika valley, High Atlas Mountains, the adult sex ratio is 1 (Namous et al. 2017). At two sites in Algeria (Tigounatine and Akfadou), the adult sex ratio was 1.05 (Ménard et al. 1990).

Habitat and Diet

The Barbary macaque inhabits evergreen and deciduous forests, scrubs, grasslands, and rocky areas with herbaceous plants. The vegetation of the Upper Rock Reserve in Gibraltar is composed of patches of medium/tall plants (mainly *Olea europaea, Rhamnus alaternus*, *Pistacia lentiscus*, *Pistacia terebinthus,* and *Osyris lanceolata*) and areas covered by herbaceous plants (Fa 1984). In Morocco and Algeria the dominant tree species where the Barbary macaque lives are cedar (*Cedrus atlantica*) and oak trees (*Quercus ilex, Quercus faginea,* and *Quercus afares*), and the Spanish fir (*Abies pinsapo*; Fooden 2007). The majority of macaques are found in mountainous forests and, in those habitats, macaques are reluctant to move to open areas away from trees. However, it is not clear whether macaques prefer forested habitats or whether the present-day geographic distribution of the species is due to competition for land with humans and livestock, which forced macaques to move to areas less accessible to humans (Fa 1984).

The elevation distribution of the Barbary macaque lies at an altitude between zero and 2300 meters above sea level. The Barbary macaque mostly lives within the Mediterranean climate. In Gibraltar, and in the Moroccan and Algerian sub-populations living close to the sea, winter is generally mild, but the summer is hot and dry. In the mountainous areas of Morocco and Algeria the Barbary macaque experiences strong seasonal differences, from dry and hot summer months, where temperature can reach up to 40 °C, to wet and snowy winter months with snow and temperatures often going well below 0 °C (Majolo et al. 2013; El Alami et al. 2013).

The Barbary macaque is a diurnal and mostly a terrestrial species, spending between 58% and 100% of the daylight hours on the ground (Fooden 2007). The Barbary macaque is a non-territorial species; the size of its home range varies greatly depending on geographic location, group size, season, and presence of neighbor groups. The "Middle Hill" group in Gibraltar had a home range of 0.25km^2 (Brauch et al. 2007); in Morocco and Algeria the home range varies from 1.8km^2 to a maximum of 12km^2 (Ménard and Vallet 1993b; Ménard et al. 2014a; Fooden 2007). There may be limited to extensive home range overlap between neighboring groups. In the Middle Atlas Mountains of Morocco, the average percentage of home range overlap of four groups was 35% (Majolo et al. 2013a) whereas in the Rif Mountains 85% of the home range of the study groups overlapped with that of other groups (Mehlman 1989). In the Rif Mountains, the average daily traveled distance of the monkeys was 1.43 km (Mehlman 1986).

As a possible consequence of some strong food availability seasonality, the Barbary macaque is a very eclectic forager (Ménard 2002; Ménard et al. 2014a). Macaques are known to eat more than 150 different plant species (Table 3); their primary food source is plant matter (fruit, leaves, seeds, roots, and flowers), which accounts for approximately 70%–90% of their diet (Ménard 2002). Macaques supplement their diet with insects, spiders, small vertebrates, mealworms, and bird eggs (Fooden 2007; Young et al. 2012). Macaques also rely on low energetic food (e.g., bark and lichens) in areas where macaques compete intensively with livestock and during periods of low availability of other food sources (Ménard et al. 2014a).

Behavior

The Barbary macaque lives in multimale-multifemale groups (Ménard 2002), composed of individuals of different age classes (Table 1). Group size ranges from 9 to 88 individuals. The provisioned groups of Gibraltar are composed of around 40–80 individuals. In Djebela North Morocco, average group size was calculated to be 9.9 from nine groups (Waters et al. 2007). In the High Atlas Mountains of Morocco the average group size is 21 macaques (range 7–42; El Alami et al. 2013). The percentage of immature individuals in a group is around 50% (range 41%–62%; Ménard 2002). Groups usually fission when they reach the maximum sustainable group size in relation to the local availability of resources (Ménard and Vallet 1993b). In Algeria, adult females played a key role in group fission; maternally related macaques tended to stay in the same group following fission (Ménard and Vallet 1993b). During daylight hours, the Barbary macaques spend a significant proportion of time feeding, foraging, and traveling, and approximately 40% of the time resting or engaged in social behavior (Majolo et al. 2013). Their activity budget varies across the season (Fig. 6) and it is affected by climatic conditions and snow coverage (Majolo et al. 2013).

In the Barbary macaque, females are the philopatric sex and males are the dispersing sex (Thierry et al. 2004). Females stay in their natal group throughout their lives whereas males tend to emigrate to new groups when they reach sexual maturity and can continue to emigrate to new groups during their lives. There is no record of males forming bachelor groups. Barbary macaques experience both scramble and contest food competition (Thierry et al. 2004): scramble

Table 3 List of plant species eaten by the Barbary macaques (a: Deag 1974; b: Drucker 1984; c: Mehlman 1988; d: Ménard et al. 2014a)

Plant species (scientific name)	Food item						Reference
	Stem & bark	Leaf	Flower & strobilus	Fruit	Seed & acorn	Root, bulb & tuber	
Abies pinsapo	x	x	x		x		c
Acer granatense			x	x			b
Acer monspessulanum		x					d
Acer opalus		x	x				c
Agropyron junceum					x		d
Agropyron marginatum		x		x	x		a, c
Agropyron panormitanum		x		x	x		a, c
Allium paniculatum		x				x	c
Alysum sp.						x	d
Ampelodesmos mauritanicus	x	x					d
Anthemis chia			x				c
Anthoxanthum odoratum		x					c
Arabis alpina caucasica		x	x				d
Armeria plantaginea		x					d
Arrhenatherum elatius						x	d
Asperula cynanchica					x		d
Asphodeline lutea					x		d
Asphodelus microcarpus	x	x			x		d
Asteraceae		x	x		x		d
Astragalus armatus			x		x		c, d
Atractylis sp.					x		d
Avena macrostachya					x		d
Balansea glaberrima		x				x	d
Bellis perennis	x						d
Bellis sylvestris		x	x			x	a, d
Berberis hispanica		x					c, d
Biscutella laevigata					c		c
Bonium alpinum						x	a
Brassica gravinea		x					d
Brisa maxima					x		d
Bromus rigidus					x	x	d
Bromus squarosus					x		d
Bromus sterilis		x		x	x		a, c, d
Bromus tectorum		x		x	x		a, c
Bunium alpinum						x	d
Bupleurum atlanticum		x					d
Calicotome spinosa		x					d
Cardamine hirsuta		x				x	d
Cardunculus pinnatus					x		d
Carduus nutans		x			x		d
Carlina atlantica					x		d

(continued)

Table 3 (continued)

Plant species (scientific name)	Food item						Reference
	Stem & bark	Leaf	Flower & strobilus	Fruit	Seed & acorn	Root, bulb & tuber	
Carlina corymbosa					x		c
Carthemis arborescens					x		c
Carum montanum						x	d
Catananche caerulea	x	x			x	x	a, d
Cedrus atlantica	x	x	x	x	x		a, b, c, d
Centaurea incana		x				x	d
Centaurea nana		x				x	a
Centaurea pullata						x	a
Centaurea tougourensis					x		d
Cephalanthera rubra					x		d
Cerastium gibraltaricum		x					d
Cerastium glomeratum		x					d
Cerastium vulgatum		x					c
Chamaepeuce casabonae		x				x	a
Chenopodium murale		x			x		c
Chrysanthemum sp.	x						d
Cirsium acarna		x			x	x	c
Cirsium syriacum		x			x		d
Cotoneaster racemiflora				x			d
Crataegus laciniata		x	x	x			a, b, c, d
Crataegus monogyna		x					b
Crocus salzmannii						x	c
Crocus nudiflorus						x	a
Crocus nevadensis						x	a
Cruciata pedemontana		x					d
Cynara hystrix				x	x		a
Cynosurus echinatus		x			x		c
Cynosurus elegans				x	x		a, c
Dactylis glomerata		x			x	x	c, d
Daphne laureola				x			c, d
Dasyphyllum breviarstatum					x		d
Dianthus caryophyllus		x			x		d
Diplotaxis catholica			x				d
Erinacea anthyllis		x					d
Erodium sp.						x	c, d
Eryngium campestre		x			c		c, d
Erysimum bocconei		x				x	a, d
Euphorbia characias					x		c
Ferula communis		x					d
Festuca geniculata					x		d
Festuca paniculata				x	x		a
Festuca triflora		x		x	x		a, c

(continued)

Table 3 (continued)

Plant species (scientific name)	Food item						Reference
	Stem & bark	Leaf	Flower & strobilus	Fruit	Seed & acorn	Root, bulb & tuber	
Ficaria verna		x					d
Gagea arvensis						x	c
Gagea foliosa						x	d
Galium perralderii		x					d
Genista tricuspidata		x					d
Geranium atlanticum		x					a
Geranium lucidum		x					a
Geranium malvaeflorum		x					d
Geranium molle	x	x	x				a
Geum sylvaticum						x	a
Geum urbanum						x	a
Hedera helix		x		x			a, c, d
Helictotrichon jahandiezii	x						a
Heracleum spondilium	x						d
Holcus lanatus					x		c
Hyoseris radiata		x	x			x	d
Hypochaeris laevigata		x	x				d
Hypochaeris radicata		x	x				d
Iberis sempervirens		x					d
Ilex aquifolium		x	x	x			a, b, c, d
Juniperus communis			x				d
Juniperus phoenicea		x					b, c
Juniperus oxycedrus		x	x	x			a, b, c, d
Juniperus thurifera		x	x	x			b, d
Jurinea humilis		x					d
Knautia arvensis			x				d
Lamiaceae sp.		x					d
Lactuca intricata						x	d
Lamium album		x	x				c
Linaria heterophila		x			x		d
Linaria sp.							d
Lobularia maritima					x		c
Lonicera biflora		x					d
Lonicera implexa				x			d
Lotus corniculatus		x					d
Luzula nodulosa		x					d
Mantisalca salmantica		x			x		c, d
Marrubium ayardii						x	d
Medicago falcata						x	c
Medicago minima						x	c
Medicago suffruticosa	x	x	x			x	d
Melica cupanii					x		d
Microlonchus salmanticus	x						a

(continued)

Table 3 (continued)

Plant species (scientific name)	Food item						Reference
	Stem & bark	Leaf	Flower & strobilus	Fruit	Seed & acorn	Root, bulb & tuber	
Moehringia trinervia		x					d
Muscari comosum			x			x	a, c
Muscari grandiflorum			x				a
Narcissus bulbocodium		x				x	c
Narcissus romieuxii						x	d
Narcissus tazeta		x					d
Onobrychis sp.					x		d
Ononis aragonensis		x		x	x		d
Onopordum acaulis			x			x	d
Ornithogalum tenuifolium						x	c
Ornithogalum umbellatum						x	d
Phagnalon saxatile		x					d
Phlomis bovei		x					d
Picnomon acarna			x	x	x		a
Pinus clusiana	x	x	x		x		d
Pinus pinaster		x		x	x		c, c
Plantago sp.						s	d
Poa bulbosa		x		x	x	x	a, c, d
Populus nigra		x	x				d
Prunus prostrata				x			d
Ptilostemon casabone		x			x	x	c
Quercus ilex	x	x	x	x	x		a, b, c, d
Quercus faginea	x	x	x	x	x		b, c
Raffenaldia primuloides						x	a
Ranunculus calandrinioides						x	a
Ranunculus muricatus			x				c
Ranunculus rupestris						x	a
Ranunculus paludosus			x				a
Rhamnus alaternus				x			d
Ribes grossularis		x		x			d
Romulea bulbocodium						x	a, c
Rosa canina		x		x		x	c
Rosa pouzinii	x	x	x				a, b
Rubus sp.	x	x		x			b, d
Rumex acetosella		x					d
Rumex bucephalophorus					x		d
Ruscus aculeatus		x		x			d
Salvia argentea			x				d
Salvia verbenaca		x					c
Sanguisorba minor		x					c
Saponaria sicula		x					d
Saxifraga granulata	x		x				a, c

(continued)

Table 3 (continued)

Plant species (scientific name)	Food item						Reference
	Stem & bark	Leaf	Flower & strobilus	Fruit	Seed & acorn	Root, bulb & tuber	
Saxifraga globulifera		x					d
Saxifraga numidica		x					d
Scandix australis		x					d
Scilla sp.					x	x	d
Scolymus grandiflorus					x		d
Sedum album		x	x				a, c
Sedum multiceps	x						d
Senecio gallicus		x					d
Senecio perralderianus		x					d
Silene italica			x		x		d
Silene vulgaris		x			x		c, d
Smyrnium olusatrum						x	d
Smyrnium perfoliatum		x				x	d
Sorbus torminalis		x		x			b, d
Stellaria media	x	x	x				a, c
Sternbergia colchiciflora						x	c
Taraxacum obovatum		x					d
Taraxacum officinale		x	x				c
Taxus baccata			x	x			b, c, d
Teucrium flavum		x					d
Teucrium polium		x					c
Thapsia villosa	x	x				x	a, c
Thlaspi perfoliatum	x						d
Thymelaea virgata	x					x	d
Torilis arvensis						x	a
Torilis elongata		x					d
Trifolium campestre			x				d
Trifolium hirtum		x					d
Trifolium phleoides			x				d
Triticum sp.					x		d
Tulipa sylvestris	x	x				x	a, d
Umbilicus pendulinus		x					d
Umbiculus rupestris		x					a
Valeriana tuberosa						x	a, d
Veronica hederifolia	x	x	x				a, d
Viburnum lantana		x					d
Vicia onobrychoides	x	x					a
Viola sp.						x	d
Vulpia geniculata				x	x		a

competition occurs when macaques, either belonging to the same group or from different groups, feed on the same food source at different time intervals with no direct interaction between the competing animals. Contest competition occurs when macaques aggressively compete over access or monopolization of a food source. Contest competition is often observed over rare

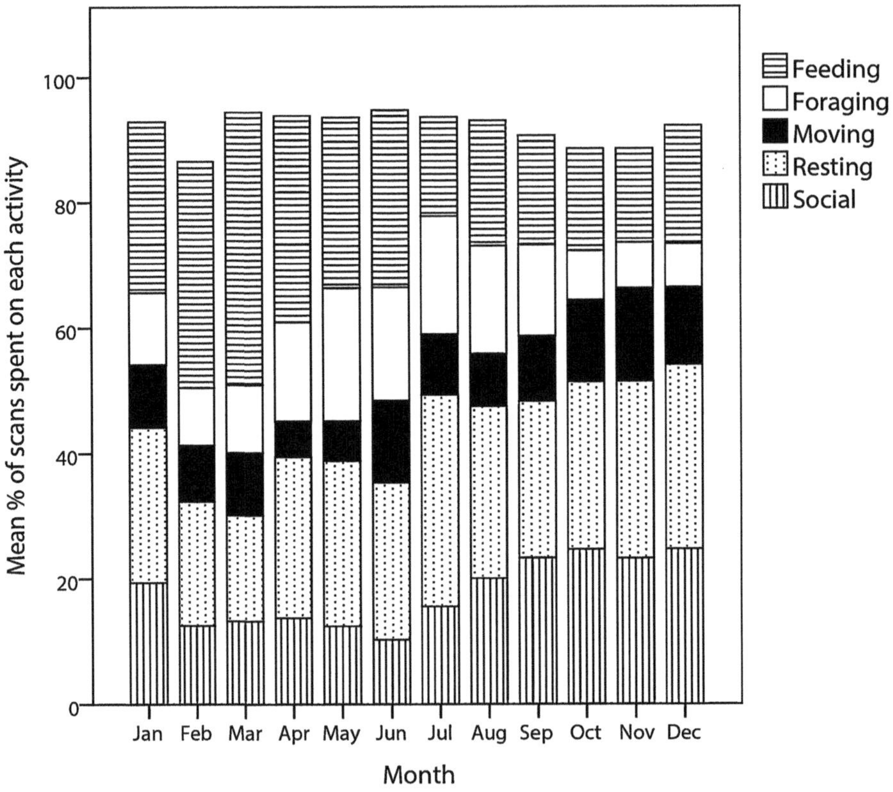

Fig. 6 Activity budget of the Barbary macaque. Mean monthly percentage of times the Barbary macaques spent on each activity during the year in the Middle Atlas Mountains, Morocco, calculated for the period June 2008–January 2011 from four groups of macaques. (From Majolo et al. (2013) with permission from Springer Nature)

and/or energetically rich food sources. Interactions between neighbor groups are often neutral, but macaques from different groups may exchange aggressive display from a distance or physical aggression involving bites in 20–50% of between-group encounters (Deag 1973; Mehlman 1986).

The personality of Barbary macaques has been studied in the Gibraltar population and in the wild in Morocco (Konečná et al. 2012; Adams et al. 2015). Four dimensions appear to describe personality in this species: Friendliness, Activity/Excitability, Confidence, and Opportunism. Barbary macaques have a complex range of facial and body displays, vocal and gestural communication during feeding and social interactions with other group members and/or other social groups (Hesler and Fischer 2007).

During the mating season, intra- and intersexual competition for mating partners is observed in the Barbary macaque (Small 1990). Females mate promiscuously but also show mate choice, preferentially consorting prime males (Bissonnette et al. 2011). Males compete for estrous females and form male–male coalitions to increase their reproductive success. The interplay of male and female reproductive strategies results in a low reproductive skew in the Barbary macaque (Bissonnette et al. 2011). Indeed, in Gibraltar there is no difference in male and female reproductive success between dominant and subordinate macaques (Kümmerli and Martin 2005). However, in Morocco dominant males have greater mating success than subordinates (Young et al. 2013b).

The Barbary macaque is one of the few primate species where there is extensive infant care from adult males. Males appear to selectively choose specific infants to interact with (Ménard et al. 2001; Kubenova et al. 2017). Interactions between adult males and infants involve infant carrying, grooming, and protection of the infant from danger (e.g., predators) and aggressive group members (Taub 1980). Infant care from adult males may be a form of parental investment (Trivers 1972), whereby males preferentially care for infant that is likely to be their own offspring. However, Barbary macaque males do not seem to preferentially care for infants that are their own (Paul et al. 1996; Ménard et al. 2001). Alternatively, males may care for infant to increase their mating opportunities in the future, if females preferentially mate with males who provide infant care (Ménard et al. 2001). Triadic male–infant–male interactions, where two males jointly manipulate an infant, are also frequently observed in the Barbary macaques (Taub 1980). Triadic male–infant–male interactions allow males to form social bonds with other group males while reducing the risk of aggression from those males due to the presence of infant, that is, infants function as "social buffers" (Deag and Crook 1971; Paul et al. 1996; Kubenova et al. 2017).

Barbary macaques establish differentiated social relationships with their group companions, varying from overly aggressive and competitive interactions to strong social bonds. The main affiliative/friendly behaviors that Barbary macaques use to establish and maintain social bonds are grooming exchange (the picking through and/or slow brushing aside the fur of another individual with one or both hands), feeding in proximity, huddling, and agonistic support (e.g., McFarland and Majolo 2011a; Young et al. 2014a). Social bonds give a number of benefits to individuals, including increased chances to survive cold winters, improved resilience to stressors, coalition formation against competitors, reconciliation, and increased thermoregulation (McFarland and Majolo 2011c; McFarland and Majolo 2013b; Young et al. 2014b; Campbell et al. 2018). Aggression and competition between group members is relatively common, especially during the mating season or during periods of food shortage. The majority of aggressive interactions does not involve physical contact but can sometimes result in severe injuries; agonistic support and coalitions are often observed during these aggressive interactions (Bissonnette et al. 2009; Young et al. 2014a). Barbary macaques display a range of post-conflict interactions, from renewed aggression or reconciliation between former opponents (Aureli and de Waal 2000) to third-party affiliation (McFarland and Majolo 2012; McFarland and Majolo 2013a). Grooming, huddling, proximity, lip-smacking, and social play are some of the main behaviors used to analyze reconciliation and third-party affiliation in the Barbary macaque (Patzelt et al. 2009; McFarland and Majolo 2013a).

Species of the genus *Macaca* have been organized into four different categories (grades) depending on their dominance style; such grading system has been proposed to describe differences across macaque species in the type of social interactions females have with other females in their social group (Thierry 2000; Thierry et al. 2004). According to this grading system, species belonging to Grade I (e.g., the Japanese macaque, *Macaca fuscata*) are described as being "despotic": female macaques in this grade form steep dominance hierarchies where the outcome of a dyadic conflict between two females is predicted by their dominance rank; thus dyadic conflicts should always/often be "decided" (have a clear winner/loser). There should be a low frequency of counter-aggression in despotic species. Moreover, there should be a high degree of kin-bias in grooming, coalition, and aggression whereby females should preferentially bond with their kin females (their matriline) and be aggressive toward group females belonging to other matrilines. As a consequence of this, females from the same matriline should all rank close together in the dominance hierarchy. At the opposing end of the system, in Grade IV, "tolerant" species (e.g., Tonkean macaques, *Macaca tonkeana*), females should have a shallow hierarchy where the outcome of dyadic conflicts is context-dependent and does not only depend on the dominance rank of the female opponents, counter-aggression should be frequent and there should be a low degree of kin-bias in grooming, coalitions, and aggression. The Barbary macaque is considered relatively tolerant species (i.e., Grade III; Thierry 2000). However, the Barbary macaques also share some social traits that are typically observed in more despotic species. For example, in wild macaques living in the Middle-Atlas Mountains of Morocco only 5% of conflicts were undecided and less than 4% resulted in counter-aggression (McFarland and Majolo 2011b, 2011c). In the same population, females present a steep and linear dominance hierarchy (Kaburu et al. 2012). Indeed, species belonging to Grade III and IV may not be easily categorized into this four grading system (Balasubramaniam et al. 2012) and it is not clear whether the same dominance gradients can be used to describe male–male social interactions across macaque species.

Parasites and Diseases

Barbary macaques are susceptible to similar zoonotic pathogens as other macaque species, including rabies or Tuberculosis (Honess et al. 2006). However, only some pathogens have been reported in the species (Table 4). Limited information on parasites and pathogens is available for wild populations in Morocco and Algeria. This is particularly problematic as primate tourism and associated close interactions between humans and macaques are growing, increasing the risks for potential pathogen transmission between the two species (Carne et al. 2017).

Population Ecology

According the latest CITES report (September 2016), the overall population of Barbary macaque in Morocco and Algeria is estimated between 6,500–13,500 individuals, 7– <10 individuals per km^2. This estimation varies considerably between geographic regions and surveys. In

Table 4 Summary of recorded parasites and infections in Barbary macaques, in zoos, semi-free ranging conditions, and in the wild

Parasites and infectious diseases			References
Endoparasites	Protozoa	*Entamoeba coli, E. histolytica/dispar, E. hartmanni, E. polecki, lodamoeba butschlii*	Borg et al. (2014)
	Helminths	*Trichostrongylus* spp., *Trichuris* spp., *Physaloptera* sp., *Dicrocoelium* spp., *Strongyloides* spp., *Ascaris* spp., *Capillaria* spp., *Oesophagostomum* spp., *Ancylostoma* spp., *Necator* spp.	Canelli et al. (2010), Borg et al. (2014), Müller et al. (2017)
Ectoparasites	Phthiraptera	Anoplura *Pediculus* sp.	Cohn et al. (2007)
Infectious diseases	Virus	Distemper virus (*Paramyxoviridae*, *Picornaviridae*), hepatitis, simian foamy virus, West Nile virus, Encephalomyocarditis virus, cowpox virus, simian virus 40	Martin (1950), de Turckheim and Merz (1984), Ølberg et al. (2004), Honess et al. (2006), Martina et al. (2006), Engel et al. (2008), Verschoor et al. (2008), Cardeti et al. (2016)
	Bacteria	*Salmonella* spp., *Staphylococcus aureus, Shigella* spp., *Escherichia coli, Klebsiella pneumoniae*	de Turckheim and Merz (1984), Banish et al. (1990) Bachiri et al. (2017)
	Meningitis	Unknown	de Turckheim and Merz (1984)
	Pneumonia	Unknown	
	Peritonitis	Unknown	
	Colitis Cystica superficialis	Unknown	Scott (1978)

Morocco, the population is estimated between 5,000–6,000 individuals, 7 individuals per km^2 (Lowest estimate provided by van Lavieren and Wich 2010), while other studies suggest that the population is between 8,000–9,000 individuals, <10 individuals per km^2. This highest estimate in Morocco is for the sub-population of the Middle Atlas Mountains, estimated at 5,000 individuals, 9 individuals per km^2 (Ménard et al. 2014b), in addition to 1,000 individuals from the High Atlas Mountains, and up to 2,000 individuals from the Northern sub-population of the Rif area (based on recent surveys by Waters et al., unpublished data). In Algeria, the current overall population is unknown (Benrabah 2015). The population in Gibraltar is maintained to around 200 individuals (Shaw, pers. comm.) and it is regularly monitored.

Although the population estimation varies, all experts agree that the wild population of Barbary macaques has dramatically declined since early 1980s from an estimated figure of up to 23,000 individuals (Fa et al. 1984) to fewer than 13,500 individuals in 2016 (CITES 2016). An increase in systematic population surveys across the Barbary macaque range in Morocco and Algeria would give a more accurate estimate of the state of the overall population.

Very little is known about interspecific interactions (e.g., feeding association) with other mammals. Barbary macaques feed on a number of animal species (see section "Life History" above). African wolfs, large raptors, golden jackals, and red foxes are all believed to occasionally prey on Barbary macaques: macaques often give alarm calls when spotting these potential predators but direct observations of attacks have never been reported with the exclusion of dogs. In the Middle Atlas Mountains of Morocco several attacks of dogs to macaques have been observed and at least three attacks have resulted in the disappearance or death of the attacked macaque (Majolo & Maréchal unpublished data).

There are not enough detailed records to link climate change and global warming to changes in the population ecology and feeding behavior of the Barbary macaques. However, global warming is expected to prolong the draught period that macaques experience during the summer and make winter and summer temperatures more extreme. For example, at two sites in Algeria (Tigounatine and Akfadou), the mean annual precipitation decreased from 1205 mm in the period 1914–1938 to 800 mm in the period 1968–1976 (Ménard and Vallet 1993a). In an exceptionally harsh and snowy winter in the Middle Atlas Mountains, in 2008/2009, resulted in the death of approximately half of the adults in two groups of Barbary macaques (McFarland and Majolo 2013b).

Conservation Status

Barbary macaques have been listed in Appendix II of the Convention on International Trade in Endangered Species for the first time in 1975, and listed in Annex B of the European Union Council Regulation in 1997 (CITES, EC 338/97, Annex B). However, the status of the species has recently been upgraded to Appendix I (CITES 2017) due to the steady population decline estimated in recent years. In 2008 the Barbary macaque was classified for the first time as 'Endangered' by the International Union for Conservation of Nature (IUCN) (Wallis et al. 2020). The species is also listed on the Wildlife Trade Regulation of the European Union (EU, L27/60, 1 February 2017). Overall, these international legislations aim to increase the protection of the species within its natural habitat as well as limit trade that seriously affects the species recovery.

In Gibraltar, the macaque population is managed by the Gibraltar Ornithological & Natural History Society (GONHS) and the Gibraltar Veterinary Clinic under agreements with the Government of Gibraltar, and is protected by Gibraltarian and EU laws. In Algeria the Barbary macaque is protected under the Executive Decree No. 83-509 (20 August 1983), and the Executive Decree No. 12-235 (24 May 2012), which list all protected non-domesticated animal species. Such conservation status prohibits poaching, capture, poisoning of the species, as well as define the modality for the protection of the species in its natural habitat. However, despite its classification as endangered species by the IUCN (Wallis et al. 2020), to date the Barbary macaque is not listed as endangered in the Ordinance No. 06-5 (15 July 2006), which would provide the highest protection status of the species in Algeria (Bergin et al. 2018). Therefore, this limits the protection of the species at a national level. An action plan for the conservation of the Barbary macaque in Algeria has recently been published (Alcazar et al. 2019).

In Morocco, the Barbary macaque has been protected under the agricultural minister Decree since 1962, prohibiting capture, hunting, possession, and sale. In addition, the trade of this species is protected under Moroccan Act No. 29-05 (June 2015), which gives legal protection for the species against importation, capture, sale, offer for sale, or killing without a specific license (Van Uhm 2016). Recently the Action Plan for the Conservation of the Barbary in Morocco has been developed by the Haut Commissariat des Eaux et Forêt et à la Lute Contre la Désertification to implement measures to protect the species in Morocco (Moroccan Primate Conservation Foundation 2012).

In Morocco and Algeria habitat destruction due to farming, overgrazing and logging is the main threat for the Barbary macaque. The human–macaque conflict is intense in some parts of Morocco and Algeria and the illegal trade of Barbary macaques is a major concern for the conservation of the species (Van Uhm 2016).

Management

The management of the Barbary macaques is highly related to their relationships with humans (local communities, government, and tourists). These relationships range from very positive to confrontational, and the development of strategies to reduce human–macaque issues and improve positive attitude toward the Barbary macaque are at the core of the management strategy of the species. Throughout their range Barbary macaques often come into contact with humans, particularly with tourists and shepherds, and with shepherds and stray dogs (Waters et al. 2017; Waterman et al. 2020).

The two major threats to the Barbary macaque are habitat loss and degradation, and illegal trade (Wallis et al. 2020). Habitat degradation and overuse by local people lead to an increase in human-primate competition, which are the main factors of the rapid decline of the species. This competition over resources between Barbary macaques and local communities has been reported in Gibraltar, Morocco, and Algeria. Groups of Barbary macaques living at the edge of forests have been reported crop foraging in Morocco (El Alami et al. 2013) or town-foraging in Gibraltar (Perez and Bensusan 2005). Macaques living in oak-cedar forests have been observed bark-stripping creating issues with the forest exploitation industry and forestry authorities in Morocco (Ménard and Qarro 1999; Ciani et al. 2001). In addition, the populations of Barbary macaques live in fragmented forests that are steadily shrinking in size because they are used by shepherds and local communities for their livestock, resulting in overgrazing and fire wood collection (Ciani et al. 2001, Majolo et al. 2013b). The competition over forest between the Barbary macaques and humans explain why this species was considered as pest in some regions in Morocco and Algeria. There, several culls were undertaken to manage the populations of Barbary macaques and reduce their damage before the laws to protect the species were implemented. Recently, translocation plans of macaque groups from farming/living areas to more remote locations or zoos have been considered to reduce conflicts with the local people, but such management solutions have rarely been implemented. One of the few examples of translocation happened in 2015, where a group of macaques where translocated from Gibraltar to a zoo in Scotland (BBC 2014). Translocations to zoos are a debated strategy to save this species in the wild. Moreover, translocations in more remote areas may not be an effective/viable conservation tool for a species those habitats are on the decline.

The illegal trade is another significant factor for the decline of the Barbary macaque in Morocco (van Lavieren 2008; Bergin et al. 2018) and Algeria (Bergin et al. 2018), where young macaques are often sold to national and international tourists in open markets and online.

Sustained efforts are made by governments and NGOs to reduce such trade, but the interest in buying and selling Barbary macaques remain a serious concern for its conservation.

Despite human–macaque coexistence issues reported in all three countries, the Barbary macaque is perceived with pride and a sense of ownership by some local people in Gibraltar (Radford et al. 2018) and with great curiosity and interest by shepherds in Morocco (Waters et al. 2017). In Algeria several places are named after the monkey presence stating the environmental importance of the macaque for the local communities. Barbary macaques can also be perceived by the local community as positive (El Alami and Chait 2015), as they are often related to tourism and the potential benefits in lower-income mountainous regions. For example, with 700,000–1,000,000 visitors per month (GONHS 2018), Gibraltar is the primate tourism hotspot of the endangered and emblematic Barbary macaque. Barbary macaques live freely on the Upper Rock of Gibraltar, which allows visitors to see these animals without any physical barriers. Visitors are commonly observed interacting too closely with the macaques, taking pictures/selfies touching/petting the macaques, or giving food to them, despite these behaviors having negative impacts on macaque welfare, and the latter behavior being prohibited (Perez and Bensusan 2005). To visit the natural reserve in Gibraltar where the macaques are located, a fee is requested (GONHS 2018). After Gibraltar, Morocco is the second most popular destination to encounter wild Barbary macaques, followed by Algeria with a mainly local tourism and occurring at a much lower rate in comparison to Gibraltar and Morocco. Primate tourism in Morocco and Algeria is currently unregulated, and there is no fee to go to see the Barbary macaque in its natural habitat.

Recently, there have been plans to develop regulated primate tourism in Morocco and Algeria as a tool for the conservation of the species. Primate tourism is a growing industry, which has the potential to benefit the local and national economy, and the conservation of endangered species, by increasing positive public awareness and financial support to conservation efforts. However, serious concerns for animal welfare need to be taken into consideration for primate tourism to be beneficial. Indeed, the negative impacts of unregulated tourism on wildlife and Barbary macaques have been well documented (Moroccan Primate Conservation Foundation 2012, Majolo et al. 2013b; Maréchal et al. 2016a). The presence of tourists and their frequent close interactions with the macaques can affect the behavior of the animals, leading to a reduction in social behavior and an increase in aggressive behavior in the macaques (Majolo et al. 2013b, El Alami et al. 2013; Maréchal et al. 2016b). Some serious negative impacts on the metabolism and body condition of Barbary macaques have also been reported (see details section "Physiology").

Future Challenges for Research and Management

In Gibraltar the regulations in place seem to provide an effective means to manage the population of Barbary macaques. Research is challenging on the Gibraltar Rock, due to the steep terrain that makes it difficult to follow the macaques for research purposes. However, with its high competition for space between humans and the Barbary macaque, management strategies and eco-tourism programs, Gibraltar can be at the forefront of research on effective strategies to manage and protect nonhuman primates.

Future challenges for the management of the Barbary macaque populations in Morocco and Algeria will be to fully integrate the species as part of the management and protection of the environment by finding a compromise between habitat usage by humans and the macaques (Waterman et al. 2020). To do so, Morocco and Algeria are each developing a conservation action plan for the conservation of the Barbary macaque. The main actions of these plans will be to increase scientific research on the Barbary macaque behavior and ecology, and conducting systematic population surveys across the Barbary macaque range to have a more accurate estimate of the status of the species. The application of international and national protection laws is currently limited and has yet to mitigate the threats to the species, which continues to dramatically decline. However, NGOs and other governmental organizations have developed some local projects to improve the conservation of the Barbary macaque in Morocco and Algeria.

These projects can help develop effective tourism management that takes place in areas inhabited by the Barbary macaques. Reducing or removing the feeding of macaques by tourists should mitigate the impacts related to stress and reduce the risks of zoonotic and diet-related diseases.

Acknowledgments We would like to thank Professor Klaus Hackländer and Dr. Frank Zachos for inviting us to contribute to this volume, Barbara Wolf and Veronika Mang, at Springer Nature, for their editorial support. We are extremely grateful to Christopher Young for comments on an earlier draft of this chapter.
For our work in Morocco, we are grateful to the Moroccan Haut Commissariat aux Eaux et Forets et a la Lutte Contre le Desertification for permission to conduct research, and Professor Mohamed Qarro for his help and support in the field. We are grateful to Ifrane National Park and the local communities in and around Azrou (Morocco) for their support.
For our work in Algeria, we are grateful to the Algerian officials and authorities, to the Direction Generale de Forets, Gendarmerie, the Algerian national park directors and staff.
We thank the following charities, Institutions, and parks for funding some of our research in Morocco and Algeria: University of Lincoln, University of Roehampton, the Leakey Foundation (grant ID: 28786), the Primate Society of Great Britain, La Montagne des Singes and La Foret des Singes (France), Affenberg Salem (Germany), and Trentham Monkey Forest (Great Britain). Finally, we would like to thank all the colleagues and students who worked with us to study the fascinating lives of Barbary macaques.

References

Adams MJ, Majolo B, Ostner J, Schülke O, De Marco A, Thierry B, Widdig A, Gerald MS, Weiss A (2015) Personality structure and social style in macaques. J Pers Soc Psychol 109:338–353

Alcazar E, Benazouz A, Benkheira A, Benrabah L, Boucekkine O et al (2019) Stratégie et plan d'action pour la conservation du magot (*Macaca sylvanus*) en Algérie 2018–2027. International Union for the Conservation of Nature

Aureli F, de Waal FBM (eds) (2000) Natural conflict resolution. University of California Press, Berkeley

Bachiri T, Bakour S, Ladjouzi R, Thongpan L, Rolain JM, Touati A (2017) High rates of CTX-M-15-producing *Escherichia coli* and *Klebsiella pneumoniae* in wild boars and Barbary macaques in Algeria. J Glob Antimicrob Re 8:35–40

Balasubramaniam KN, Dittmar K, Berman CM, Butovskaya M, Cooper M, Majolo B, Ogawa H, Schino G, Thierry B, de Waal FBM (2012) Hierarchical steepness and phylogenetic models: phylogenetic signals in *Macaca*. Anim Behav 83:1207–1218

Banish LD, Bush M, Montali RJ, Sack D (1990) Shigellosis in a zoological collection of primates. J Zoo Wildlife Med 21:302–309

BBC (2014) Gibraltar banishes troublesome monkeys. http://www.bbc.co.uk/news. Accessed 5 Sept 2018

Benrabah LM (2015) Geographic distribution and population size of Barbary macaque (*Macaca sylvanus*) in Algeria. Unpublished Master thesis. University of Lincoln, Lincoln

Bergin D, Nijman V (2014) Open, unregulated trade in wildlife in Morocco's markets. Traffic Bull 26:65–70

Bergin D, Atoussi S, Waters S (2018) Online trade of Barbary macaques *Macaca sylvanus* in Algeria and Morocco. Biodivers Conserv 27(2):531–534

Bissonnette A, de Vries H, van Schaik CP (2009) Coalitions in male Barbary macaques, *Macaca sylvanus*: strength, success and rules of thumb. Anim Behav 78(2):329–335

Bissonnette A, Bischofberger N, van Schaik CP (2011) Mating skew in Barbary macaque males: the role of female mating synchrony, female behavior, and male–male coalitions. Behav Ecol Sociobiol 65(2):167–182

Borg C, Majolo B, Qarro M, Semple S (2014) A comparison of body size, coat condition and endoparasite diversity of wild Barbary macaques exposed to different levels of tourism. Anthrozoös 27(1):49–63

Brauch K, Pfefferle D, Hodges K, Möhle U, Fischer J, Heistermann M (2007) Female sexual behavior and sexual swelling size as potential cues for males to discern the female fertile phase in free-ranging Barbary macaques (*Macaca sylvanus*) of Gibraltar. Horm Behav 52(3):375–383

Campbell LAD, Tkaczynski PJ, Lehmann J, Mouna M, Majolo B (2018) Social thermoregulation as a potential mechanism linking sociality and fitness: Barbary macaques with more social partners form larger huddles. Sci Rep 8:6074

Canelli E, Luppi CLavazza CLelli D, Sozzi E, Moreno-Martin CGelmetti D, Pascotto E, Sandri C, Magnone W, Cordioli P (2010) Encephalomyocarditis virus infection in an Italian zoo. Virol J 7(64):1–7

Cardeti G, Mariano V, Eleni C, Aloisi M, Grifoni G, Sittinieri S, Dante G, Antognetti V, Foglia EA, Cersini A, Nardi A (2016) Encephalomyocarditis virus infection in *Macaca sylvanus* and *Hystrix cristata* from an Italian rescue centre for wild and exotic animals. Virol J 13(1):193

Carne C, Semple S, MacLarnon A, Majolo B, Maréchal L (2017) Implications of tourist–macaque interactions for disease transmission. EcoHealth 14(4):704–717

Ciani AC, Martinoli L, Capiluppi C, Arahou M, Mouna M (2001) Effects of water availability and habitat quality on bark-stripping behavior in barbary macaques. Conserv Biol 15(1):259–265

CITES (2016) Consideration of proposal for amendment of appendices I and II. Convention on international trade in endangered species of wild fauna and flora Seventeenth meeting, Johannesburg, South Africa CoP 17 Prop 13

Cohn DL, Smith V, Pizarro M, Jones-Engel L, Engel G, Fuentes A, Shaw E, Cortes J (2007) Pediculosis in *Macaca sylvanus* of Gibraltar. Vet Parasitol 145(1–2):116–119

Cristóbal-Azkarate J, Maréchal L, Semple S, Majolo B, MacLarnon A (2016) Metabolic strategies in wild male Barbary macaques: evidence from faecal measurement of thyroid hormone. Biol Lett 12(4):20160168

de Turckheim G, Merz E (1984) Breeding Barbary macaques in outdoor open enclosures. In: Fa JE (ed) The Barbary macaque. Springer, Boston, pp 241–261

Deag JM (1973) Intergroup encounters in the wild Barbary macaque, *Macaca sylvanus* L. In: Michael RP, Crook JH (eds) Comparative ecology and behaviour of primates. Academic, London, pp 315–373

Deag JM (1974) A study of the social behaviour and ecology of the wild Barbary macaque, *Macaca sylvanus* L. Ph.D. Thesis, University of Bristol

Deag JM (1980) Interactions between males and unweaned Barbary macaques: testing the agonistic buffering hypothesis. Behaviour 75(1):54–80

Deag JM, Crook JH (1971) Social behavior and 'agonistic buffering' in the wild Barbary macaque *Macaca sylvanus* L. Folia Primatol 15:183–200

Drucker GR (1984) The feeding ecology of the Barbary macaque and cedar forest conservation in the Moroccan Moyen atlas. In: Fa JE (ed) The Barbary macaque. Springer, Boston, pp 135–164

El Alami A, Chait A (2015) Distribution of the endangered Barbary macaque and human-macaque interaction in the tourist region of Ouzoud, Central high atlas of Morocco. Afr J Ecol 53(3):375–377

El Alami A, van Lavieren E, Aboufatima R, Chait A (2013) A survey of the endangered Barbary macaque *Macaca sylvanus* in the central high Atlas Mountains of Morocco. Oryx 47(3):451–456

Elton S, O'Regan HJ (2014) Macaques at the margins: the biogeography and extinction of *Macaca sylvanus* in Europe. Quat Sci Rev 96:117–130

Engel GA, Pizarro M, Shaw E, Cortes J, Fuentes A, Barry P, Lerche N, Grant R, Cohn D, Jones-Engel L (2008) Unique pattern of enzootic primate viruses in Gibraltar macaques. Emerg Infect Dis 14(7):1112

EU (2017) L27, Official Journal of the European Union, legislation, volume 60, 1 February 2017

Fa JE (ed) (1984) The Barbary macaque: a case study in conservation. Plenum Publishing Corporation, New York

Fa JE (1989) The genus *Macaca*: a review of taxonomy and evolution. Mammal Review 19(2):45–81

Fa JE, Taub DM, Ménard N, Stewart PJ (1984) The distribution and current status of the Barbary macaque in North Africa. In: Fa JE (ed) The Barbary macaque. Springer, Boston, pp 79–111

Fooden J (2007) Systematic review of the Barbary macaque, *Macaca sylvanus* (Linnaeus, 1758). Fieldiana Zool 2007(113):1–60

Fuentes A (2006) Patterns and context of human-macaque interactions in Gibraltar. In J.K. Hodges & J. Cortes (Eds.), The Barbary macaque: biology, management and conservation (pp. 169–184). Nottingham: Nottingham University Press

GONHS. Barbary Macaques. http://www.gonhs.org/macaques.htm. Accessed 5 Sept 2018

Heistermann M, Brauch K, Möhle U, Pfefferle D, Dittami J, Hodges K (2008) Female ovarian cycle phase affects the timing of male sexual activity in free-ranging Barbary macaques (*Macaca sylvanus*) of Gibraltar. Am J Primatol 70(1):44–53

Hesler N, Fischer J (2007) Gestural communication in Barbary macaques (*Macaca sylvanus*): an overview. In: Call J, Tomasello M (eds) The gestural communication of apes and monkeys. Lawrence Erlbaum, Mahwah, pp 159–196

Honess P, Pizarro M, Sene N, Wolfensohn SE (2006) Disease transmission in Barbary and other macaques: risks and implications for management and conservation. In: Hodges JK, Cortes J (eds) The Barbary macaque: biology, management and conservation. Nottingham University Press, Nottingham, pp 149–168

Kaburu SSK, MacLarnon A, Majolo B, Qarro M, Semple S (2012) Dominance rank and self-scratching among wild female Barbary macaques (*Macaca sylvanus*). Afr Zool 47:74–79

Konečná M, Weiss A, Lhota S, Wallner B (2012) Personality in Barbary macaques (*Macaca sylvanus*): temporal stability and social rank. J Res Pers 46(5):581–590

Kubenova B, Konečná M, Majolo B, Smilauer P, Ostner J, Schülke O (2017) Triadic awareness predicts partner choice in male-infant-male interactions in Barbary macaques. Anim Cog 20:221–232

Kümmerli R, Martin RD (2005) Male and female reproductive success in *Macaca sylvanus* in Gibraltar: no evidence for rank dependence. Int J Primatol 26(6):1229–1249

Li J, Han K, Xing J, Kim H, Rogers J, Ryder OA, Disotell T, Yue B, Batzer MA (2009) Phylogeny of the macaques (Cercopithecidae: *Macaca*) based on Alu elements. Gene 448(2):242–249

Majolo B, McFarland R, Young C, Qarro M (2013a) The effect of climatic factors on the activity budgets of Barbary macaques (*Macaca sylvanus*). Int J Primatol 34(3):500–514

Majolo B, van Lavieren E, Maréchal L, MacLarnon A, Marvin G, Qarro M, Semple S (2013b) Out of Asia: the singular case of the Barbary macaque. In: Radhakrishna S, Huffman MA, Sinha A (eds) The macaque connection. Springer, New York, pp 167–183
Maréchal L, Semple S, Majolo B, Qarro M, Heistermann M, MacLarnon A (2011) Impacts of tourism on anxiety and physiological stress levels in wild male Barbary macaques. Biol Conserv 144(9):2188–2193
Maréchal L, Semple S, Majolo B, MacLarnon A (2016a) Assessing the effects of tourist provisioning on the health of wild Barbary macaques in Morocco. PLoS One 11(5):e0155920
Maréchal L, MacLarnon A, Majolo B, Semple S (2016b) Primates' behavioural responses to tourists: evidence for a trade-off between potential risks and benefits. Sci Rep 6:32465
Martin L (1950) Susceptibility of monkeys (*Macaca sylvanus*) to distemper. Bull Acad Vet Fr 23:457–462
Martina BE, Van Doornum G, Dorrestein GM, Niesters HG, Stittelaar KJ, Wolters MC, Van Bolhuis HG, Osterhaus AD (2006) Cowpox virus transmission from rats to monkeys, the Netherlands. Emerg Infect Dis 12(6):1005
McFarland R, Majolo B (2011a) Exploring the components, asymmetry and distribution of relationship quality in wild Barbary macaques (*Macaca sylvanus*). PLoS One 6(12):e28826. https://doi.org/10.1371/journal.pone.0028826
McFarland R, Majolo B (2011b) Grooming coercion and the post-conflict trading of social services in wild Barbary macaques. PLoS One 6(10):e26893. https://doi.org/10.1371/journal.pone.0026893
McFarland R, Majolo B (2011c) Reconciliation and the costs of aggression in wild Barbary macaques (*Macaca sylvanus*): a test of the integrated hypothesis. Ethology 117:928–937
McFarland R, Majolo B (2012) The occurrence and benefits of post-conflict bystander affiliation in wild Barbary macaques (*Macaca sylvanus*). Anim Behav 84:583–591
McFarland R, Majolo B (2013a) The importance of considering the behavioral form of reconciliation in studies of conflict resolution. Int J Primatol 34(1):15–29
McFarland R, Majolo B (2013b) Coping with the cold: predictors of survival in wild Barbary macaques, *Macaca sylvanus*. Biol Lett 9(4):20130428
Mehlman PT (1986) Population ecology of the Barbary macaque (*Macaca sylvanus*) in the fir forests of the Ghomara Moroccan Rif Mountains, Ph.D. dissertation, University of Toronto, Toronto
Mehlman PT (1988) Food resources of the wild Barbary macaque (*Macaca sylvanus*) in high-altitude fir forest, Ghomaran Rif, Morocco. J Zool 214(3):469–490
Mehlman PT (1989) Comparative density, demography, and ranging behavior of Barbary macaques (*Macaca sylvanus*) in marginal and prime conifer habitats. Int J Primatol 10(4):269–292
Ménard N (2002) Ecological plasticity of Barbary macaques (*Macaca sylvanus*). Evol Anthropol 11:95–100
Ménard N, Qarro M (1999) Bark stripping and water availability: a comparative study between Moroccan and Algerian Barbary macaques (*Macaca sylvanus*). Rev Ecol 54:123–132
Ménard N, Vallet D (1993a) Population dynamics of *Macaca sylvanus* in Algeria: an 8-year study. Am J Primatol 30(2):101–118
Ménard N, Vallet D (1993b) Dynamics of fission in a wild Barbary macaque group (*Macaca sylvanus*). Int J Primatol 14(3):479–500
Ménard N, Vallet D, Gautier-Hion A (1985) Démographie et reproduction de *Macaca sylvanus* dans différents habitats en Algérie. Folia Primatol 44(2):65–81
Ménard N, Hecham R, Vallet D, Chikhi H, Gautier-Hion A (1990) Grouping patterns of a mountain population of *Macaca sylvanus* in Algeria–a fission-fusion system? Folia Primatol 55(3–4):166–175
Ménard N, von Segesser F, Scheffrahn W, Pastorini J, Vallet D, Gaci B, Gautier-Hion A (2001) Is male-infant caretaking related to paternity and/or mating activities in wild Barbary macaques (*Macaca sylvanus*)? C R Acad Sci Série III, Sciences de La Vie 324:601–610
Ménard N, Motsch P, Delahaye A, Saintvanne A, Le Flohic G, Dupé S, Vallet D, Qarro M, Tattou MI, Pierre JS (2014a) Effect of habitat quality on diet flexibility in Barbary macaques. Am J Primatol 76(7):679–693
Ménard N, Rantier Y, Foulquier CQarro M, Chillasse L, Vallet D, Pierre JS, Butet A (2014b) Impact of human pressure and forest fragmentation on the endangered Barbary macaque *Macaca sylvanus* in the middle atlas of Morocco. Oryx 48(2):276–284
Modolo L, Salzburger W, Martin RD (2005) Phylogeography of Barbary macaques (*Macaca sylvanus*) and the origin of the Gibraltar colony. Proc Nat Acad Sci 102(20):7392–7397
Möhle U, Heistermann M, Dittami J, Reinberg V, Hodges JK (2005) Patterns of anogenital swelling size and their endocrine correlates during ovulatory cycles and early pregnancy in free-ranging Barbary macaques (*Macaca sylvanus*) of Gibraltar. Am J Primatol 66(4):351–368
Moroccan Primate Conservation Foundation (2012) Conservation action plan for the Barbary macaque (*Macaca sylvanus*) in Morocco. Royaume du Maroc, Haut Commissariat aux Eaux et Forêts et à la Lutte Contre la Désertification
Müller N, Heistermann M, Strube C, Schülke O, Ostner J (2017) Age, but not anthelmintic treatment, is associated with urinary neopterin levels in semi-free ranging Barbary macaques. Sci Rep 7:41973
Namous S, Znari M, Baamrane MAA, Naimi M, Aourir M, Siess J, Mokhtari S (2017) Size and structure of the southernmost population of the endangered Barbary macaque *Macaca sylvanus* in the western Moroccan high atlas. Oryx 51(4):694–700
Ølberg RA, Barker IK, Crawshaw GJ, Bertelsen MF, Drebot MC, Andonova M (2004) West Nile virus

encephalitis in a Barbary macaque (*Macaca sylvanus*). Emerg Infect Dis 10(4):712

Patzelt A, Pirow R, Fischer J (2009) Post-conflict affiliation in Barbary macaques is influenced by conflict characteristics and relationship quality, but does not diminish short-term renewed aggression. Ethology 115(7):658–670

Paul A, Kuester J (1987) Sex ratio adjustment in a seasonally breeding primate species: evidence from the Barbary macaque population at Affenberg Salem. Ethology 74(2):117–132

Paul A, Kuester J (1988) Life history patterns of Barbary macaques (*Macaca sylvanus*) at Affenberg Salem. In: Fa JE, Southwick CH (eds) Ecology and behavior of food-enhanced primate groups, Monographs in primatology, vol 11. Alan R. Liss, New York, pp 199–228

Paul A, Kuester J, Podzuweit D (1993) Reproductive senescence and terminal investment in female Barbary macaques (*Macaca sylvanus*) at Salem. Int J Primatol 14(1):105–124

Paul A, Kuester J, Arnemann J (1996) The sociobiology of male-infant interactions in Barbary macaques, *Macaca sylvanus*. Anim Behav 51:155–170

Perez CE, Bensusan KJ (2005) Upper rock nature reserve: a management action plan. Gibraltar Ornithological and Natural History Society. p 297

Radford L, Alexander S, Waters S (2018) On the rocks: using discourse analysis to examine relationships between Barbary macaques (*Macaca sylvanus*) and people on Gibraltar. Folia Primatol 89(1):30–44

Rincon AV, Maréchal L, Semple S, Majolo B, MacLarnon A (2017) Correlates of androgens in wild male Barbary macaques: testing the challenge hypothesis. Am J Primatol 79(10):e22689

Scheffrahn W, Ménard N, Vallet D, Gaci B (1993) Ecology, demography, and population genetics of Barbary macaques in Algeria. Primates 34(3):381–394

Schurr MR, Fuentes A, Luecke E, Cortes J, Shaw E (2012) Intergroup variation in stable isotope ratios reflects anthropogenic impact on the Barbary macaques (*Macaca sylvanus*) of Gibraltar. Primates 53(1):31–40

Scott GBD (1978) Colitis cystica superficialis: a report of a case in a Barbary ape (*Macaca sylvanus*). J Pathol 124(1):23–25

Semple S (1998) The function of Barbary macaque copulation calls. Proc R Soc Lond B 265:287–291

Shaw E, Cortes J (2006) The Gibraltar macaques: origin, history and demography. In J.K. Hodges & J. Cortes (Eds.), The Barbary macaque: biology, management and conservation (pp. 185–198). Nottingham: Nottingham University Press

Shutt K, MacLarnon A, Heistermann M, Semple S (2007) Grooming in Barbary macaques: better to give than to receive? Biol Lett 3(3):231–233

Small MF (1990) Promiscuity in Barbary macaques (*Macaca sylvanus*). Am J Primatol 20:267–282

Taub DM (1980) Testing the 'agonistic buffering' hypothesis. Behav Ecol Sociobiol 6:187–197

Taub DM (1984) A brief historical account of the recent decline in geographic distribution of the Barbary macaque in North Africa. In: Fa JE (ed) The Barbary macaque. Springer, Boston, pp 71–78

Thierry B (2000) Covariation of conflict management patterns across macaque species. In: Aureli F, de Waal FBM (eds) Natural conflict resolution. University of California Press, Berkeley, pp 106–128

Thierry B, Singh M, Kaumanns W (eds) (2004) Macaque societies: a model for the study of social organization. Cambridge University Press, Cambridge

Trivers RL (1972) Parental investment and sexual selection. In: Campbell B (ed) Sexual selection and the descent of man. Aldine-Atherton, Chicago, pp 136–179

van Lavieren E (2008) The illegal trade in Barbary macaques from Morocco. Traffic Bull 21(3):81–87

van Lavieren E, Wich SA (2010) Decline of the endangered Barbary macaque *Macaca sylvanus* in the cedar forest of the Middle Atlas Mountains, Morocco. Oryx 44(1):133–138

van Uhm DP (2016) Monkey business: the illegal trade in Barbary macaques. JTOCS 2(1):36–49

Verschoor EJ, Niphuis H, Fagrouch Z, Christian P, Sasnauskas K, Pizarro MC, Heeney JL (2008) Seroprevalence of SV40-like polyomavirus infections in captive and free-ranging macaque species. J Med Primatol 37(4):196–201

Von Segesser F, Ménard N, Gaci B, Martin RD (1999) Genetic differentiation within and between isolated Algerian subpopulations of Barbary macaques (*Macaca sylvanus*): evidence from microsatellites. Mol Ecol 8(3):433–442

Wallis J, Benrabah ME, Pilot M, Majolo B, Waters S (2020) *Macaca sylvanus*. The IUCN red list of threatened species 2020: e.T12561A50043570. https://doi.org/10.2305/IUCN.UK.2020-2.RLTS.T12561A50043570.en. Downloaded on 2 Sept 2020

Waterman JO, Campbell LA, Maréchal L, Pilot M, Majolo B (2020) Effect of human activity on habitat selection in the endangered Barbary macaque. Anim Conserv 23:373–385

Waters SS, Aksissou M, El Harrad A, Hobbelink ME, Fa JE (2007) Holding on in the Djebela: Barbary macaque *Macaca sylvanus* in northern Morocco. Oryx 41(1):106–108

Waters S, El Harrad A, Chetuan M, Bell S, Setchell JM (2017) Dogs disrupting wildlife: domestic dogs harass and kill Barbary macaques in Bouhachem forest, North Morocco. Afr Primates 12:55–58

Young C, Schülke O, Ostner J, Majolo B (2012) Consumption of unusual prey items in the Barbary macaque (*Macaca sylvanus*). Afr Primates 7(2):224–229

Young C, Majolo B, Heistermann M, Schülke O, Ostner J (2013a) Male mating behaviour in relation to female sexual swellings, socio-sexual behaviour and hormonal changes in wild Barbary macaques. Horm Behav 63(1):32–39

Young C, Hähndel S, Majolo B, Schülke O, Ostner J (2013b) Male coalitions and female behaviour affect male mating success independent of dominance rank and female receptive synchrony in wild Barbary macaques. Behav Ecol Sociobiol 67(10):1665–1677

Young C, Majolo B, Schülke O, Ostner J (2014a) Male social bonds and rank predict supporter selection in cooperative aggression in wild Barbary macaques. Anim Behav 95:23–32

Young C, Majolo B, Heistermann M, Schülke O, Ostner J (2014b) Responses to social and environmental stress are attenuated by strong male bonds in wild macaques. Proc Nat Acad Sci 111:18195–18200

Further Reading

Hodges JK, Cortes J (eds) (2006) The Barbary macaque: biology, management and conservation. Nottingham University Press, Nottingham

European Rabbit *Oryctolagus cuniculus* (Linnaeus, 1758)

2

Miguel Delibes-Mateos, Heiko G. Rödel, Carlos Rouco, Paulo C. Alves, Miguel Carneiro, and Rafael Villafuerte

Contents

M. Delibes-Mateos (✉) · R. Villafuerte
Instituto de Estudios Sociales Avanzados (IESA-CSIC), Córdoba, Spain
e-mail: mdelibes@iesa.csic.es; rafael.villafuerte@csic.es; rafael.villafuerte@uclm.es

H. G. Rödel
Laboratoire d'Ethologie Expérimentale et Comparée (LEEC), UR4443, Université Sorbonne Paris Nord, Villetaneuse, France
e-mail: heiko.rodel@univ-paris13.fr

C. Rouco
Área de Ecología, Departamento de Botánica, Ecología y Fisiología Vegetal, Universidad de Córdoba, Córdoba, Spain

Sociedad, Ecología y Gestión del Medio Ambiente, UCO-IESA, Unidad Asociada al CSIC, Córdoba, Spain
e-mail: crouco@uco.es

P. C. Alves
CIBIO, Centro de Investigação em Biodiversidade e Recursos Genéticos, InBIO, Universidade do Porto, Vairão, Portugal

BIOPOLIS Program in Genomics, Biodiversity and Land Planning, CIBIO, Vairão, Portugal

Departamento de Biologia, Faculdade de Ciências, Universidade do Porto, Porto, Portugal

Estação Biológica de Mértola (EBM), Mértola, Portugal
e-mail: pcalves@fc.up.pt

M. Carneiro
CIBIO, Centro de Investigação em Biodiversidade e Recursos Genéticos, InBIO, Universidade do Porto, Vairão, Portugal

BIOPOLIS Program in Genomics, Biodiversity and Land Planning, CIBIO, Vairão, Portugal
e-mail: miguel.carneiro@cibio.up.pt

K. Hackländer, P. C. Alves (eds.), *Primates and Lagomorpha*, Handbook of the Mammals of Europe,
https://doi.org/10.1007/978-3-030-34043-8_13

Common Names

English	European rabbit, Coney
German	Wildkaninchen
French	Lapin de garenne, Lapin commun
Spanish	Conejo europeo, Conejo de monte, Conejo común
Italian	Coniglio selvatico europeo
Russian	Дикий кролик

Taxonomy, Systematics and Paleontology

Oryctolagus cuniculus (Linnaeus, 1758) is one of the 63 species in the family Leporidae and the only one in the genus *Oryctolagus* (Smith et al. 2018). This genus appeared in the fossil record before any other living leporid genus (Alves et al. 2008); a single tooth found in the region of Granada, southern Spain, marks its appearance 6 Ma. The first recorded *Oryctolagus* species is *O. laynensis*. This ancient rabbit has been recorded from 3.5 to 2.5 Ma in Spain, where it was associated with an arid, warm, subtropical savannah-type fauna (Alves et al. 2008). This species appears closer to the modern *O. cuniculus* than to the extinct *O. lacosti*, a larger-sized rabbit species recorded between 2.5 and 0.6 Ma in south-western Europe (Alves et al. 2008). The potential transitional species between *O. laynensis* and *O. cuniculus* could be *O. giberti*, a species that was identified in the Lower Pleistocene levels (around 1.4 Ma) of Cueva Victoria in Murcia, eastern Spain (De Marfà 2008). Remains attributable to *O. giberti* have been recently found in Bois-de-Riquet (south-eastern France), which suggests that it occurred beyond the Iberian Peninsula (Pelletier et al. 2015). This species was smaller than extant rabbits but does share several modern characters (Pelletier et al. 2015). The oldest known fossil of the current forms of *O. cuniculus* was found in southern Spain at Cúllar Baza, and it is placed in the Middle Pleistocene (about 0.6 Ma). Nevertheless, the discovery of remains of *O. cuniculus* in a Middle Paleolithic site in northern Morocco suggests that a North African origin of the species cannot be completely discarded (Wrinn 1998 cited in Callou 2003), but see Alves et al. (2008) for a detailed discussion. In the Late Pleistocene, only *O. cuniculus* survived, spreading to Mediterranean area and northern Europe (Alves et al. 2008). A review of the mammals encountered in excavation sites between 0.3 and 0.01 Ma indicates that *O. cuniculus* was abundant in Spain, Portugal, and southern France; rabbit densities were two order of magnitude higher than similar-sized rodents and larger than for any other mammal species (Fa et al. 2013). During the maximum glacial period and Early Holocene, the species seems to be again confined to the Iberian Peninsula and southern France (Alves et al. 2008). The fossil record shows that different species of *Oryctolagus*, including *O. cuniculus*, occurred in southern France before this glacial confinement. However, the lower levels of genetic diversity observed in southern France than in the other side of the Pyrenees suggest that rabbits colonized that region after the last glacial maximum; throughout history, episodes of extinction and recolonization like this seem to have been frequent in southern France due to climatic variation (Alves et al. 2008).

Currently, two rabbit subspecies are well-recognized *O. c. cuniculus* Linnaeus, 1758 and *O. c. algirus* Loche, 1858. Multiple genetic evidence indicates that they evolved independently, approximately 2 Ma during the Quaternary glaciations (Alves et al. 2008), and have different geographical ranges (Fig. 1). In addition to these evidence, noticeable differences between rabbit subspecies in parasitology, behavior, reproduction, and morphology have been reported (see below in section "Genetics"). These differences, together with recent evidence of genomic incompatibilities and partial reproductive isolation (see more details in section "Genetics"), have raised the question of whether *O. c. algirus* and *O. c. cuniculus* are nascent or already well-separated species (Delibes-Mateos et al. 2018a). Another relevant issue is rabbit domestication, which occurred around 1000 years ago from the subspecies *O. c. cuniculus*, and the only domestication process occurring in Europe (Alves et al. 2008). Domestic rabbits are distributed all over the world; it is an important protein source for humans, but also relevant pet, thus having and enormous economic and social relevance. After domestication, feral rabbits have been introduced in all continents, except Antarctica, being an exotic species, and in the majority of the situations considered a pest (Thompson and King 1994).

Current Distribution

The fossil record suggests that the European rabbit was likely originated in the Iberian Peninsula (Alves et al. 2008), from where it would have expanded naturally to southern France and northern Africa. Natural populations of this species are currently widespread in Portugal, Spain, southern France, and possibly in northern Africa. The subspecies *O. c. algirus* only occurs in south-western Iberia (Portugal and southern Spain) and as introductions to the Azores, Madeira, and Canary Islands, whereas *O. c. cuniculus* is naturally present in northeastern Iberia and in southern France (Fig. 1). This subspecies has been introduced (as feral type) in several other central European countries and in the United Kingdom (Fig. 1). Moreover, the high adaptability of the feral forms of *O. c. cuniculus* promoted the successful establishment of the rabbit in all continents, except Antarctica, as well as in 800 islands worldwide (Thompson and King 1994). Rabbit spread in Europe was reviewed by Thompson and King (1994). First introductions referred to the introduction in Menorca by Neolithic settlers as early as 1400–1300 BCE. Later, the Phoenicians and the Romans spread the species around Mediterranean Europe. In the Middle Ages, rabbit gardens and warrens were built in France, Britain, and Germany, because their meat was highly

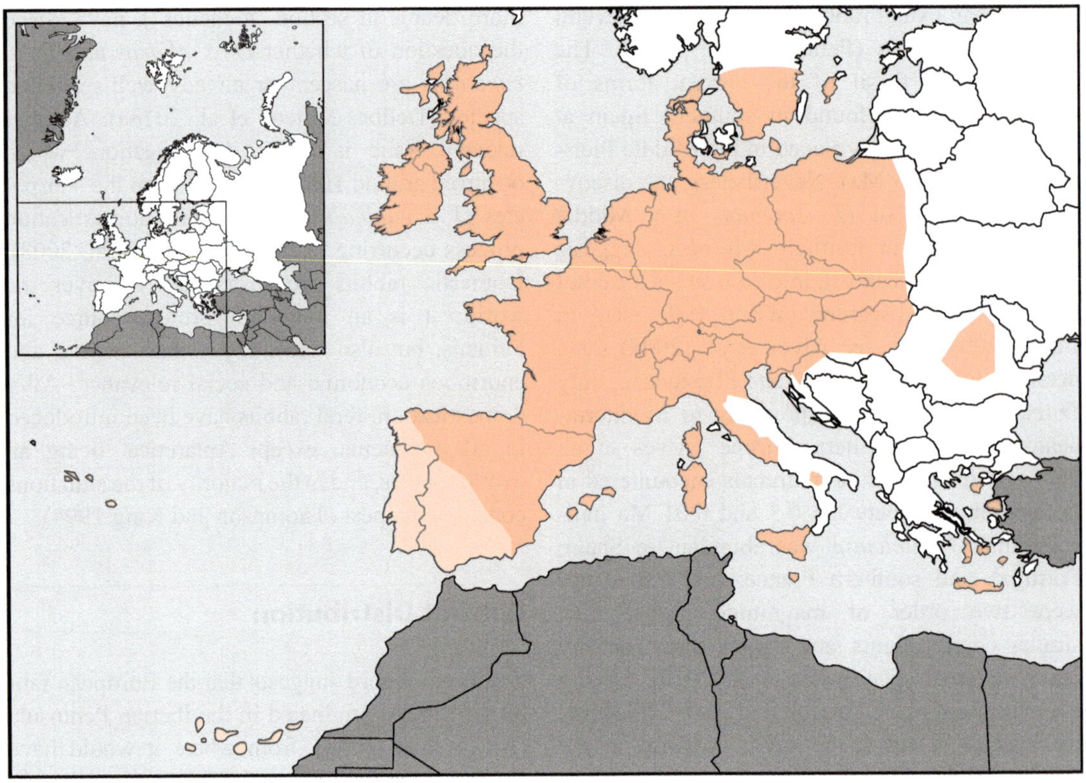

Map template: © Getty Images/iStockphoto

Fig. 1 Distribution of the European rabbit in Europe according to the information recorded in the Atlas of European Mammals (https://www.european-mammals.org/osm/earthmap.php) and the IUCN webpage (https://www.iucnredlist.org/species/41291/10415170). The distribution of both rabbit subspecies is shown; apricot: *O. c. cuniculus*, and light apricot: *O. c. algirus*. (Modified from Geraldes et al. 2008). (Map template: © Getty Images/iStockphoto)

appreciated. Multiple human introductions resulted in the slowly spread of feral rabbits over much of Europe. As a consequence, the species currently occurs in many European countries as an exotic feral species, including Albania, Austria, Belgium, Bulgaria, Croatia, Czech Republic, Denmark, Germany, Greece, Hungary, Republic of Ireland, Italy, Luxembourg, The Netherlands, Norway, Poland, Romania, Russian Federation, Slovakia, Slovenia, Sweden, Switzerland, and the United Kingdom (Fig. 1). It is also present in many European islands like Canary Islands, Mallorca and Menorca, Corsica, Sardinia, Sicily or Crete. In the archipelagos of Azores and Madeira, the introduced subspecies was the *O. c. algirus*, being the only places where this subspecies was successfully introduced in Europe.

Historically, European rabbits reached very high numbers both in their native range and in areas where the species was introduced. For example, the geographer Strabo (63/64 BCE–24 CE) wrote that settlers in the Balearic Islands were so unhappy about the very abundant rabbits that they asked the Emperor August either to send a Roman legion to clear the land of rabbits or to give the settlers other land elsewhere. Similarly, Pliny the Elder (23/24–79 CE) told in his Natural History that rabbit warrens undermined the ramparts of the city of Tarragone, northern Spain. It is not well known how rabbit populations fluctuated since those ancient times to the twentieth century, when the first rabbit monitoring efforts were done. Nevertheless, most evidence suggests that for centuries the species was abundant in many European regions. For example, rabbits were introduced in Canary Islands in the fifteenth century and they became so abundant that special gangs had to combat the lagomorph to palliate the huge damage they caused to crops. In the seventeenth century, the French prime minister during the kingdom of Louis XIV ordered the destruction of rabbits in all royal forests or alternatively the compensation for rabbit damage when their eradication was not possible (Thompson and King 1994). Records of the number of game animals harvested at the beginning of the twentieth century indicate that rabbits were still abundant in several European regions. For example, hunting bags of over 10 rabbits/ha were common in northern France from 1920 to 1950 (Thompson and King 1994). In Britain, the number of rabbits killed per year at the beginning of the twentieth century was more than double of that recorded during the first decade of the twenty-first century (Aebischer et al. 2011). However, during the second half of the twentieth century, the advent of two viral diseases (myxomatosis and rabbit hemorrhagic disease (RHD)) together with changes in land use and subsequent habitat loss devastated most rabbit populations across Europe (see below in section "Parasites and Diseases" as well as in section "Conservation Status"). In the Iberian Peninsula, recent negative population trends have been reported in the distribution area of *O. c. algirus*, while more stable or even positive trends have been reported in the distribution area of *O. c. cuniculus* (Vaquerizas et al. 2020).

Description

The European rabbit is intermediate in size among leporids. The main size measures are: head and body 340–450 mm, tail 40–60 mm, hind foot 70–85 mm, and ear 65–80 mm (Delibes-Mateos et al. 2018a). In the Iberian Peninsula, adult rabbit weight ranges from 900 g to 1400 g. Body mass increases with latitude both in the native range (Ferreira et al. 2015) and across Europe (Sharples et al. 1996), supporting Bergmann's rule. In fact, European rabbits in northern Europe commonly exceed 1500 g. Domesticated forms bear little resemblance to the original wild stock; some are small and those bred for meat production can be very large and heavy (Rosell and de la Fuente 2012). In addition, Ferreira et al. (2015) showed that there are unequivocal biometric differences between rabbit subspecies even when controlling for the effect of body mass: ear length is shorter for adult *O. c. algirus* than for adult *O. c. cuniculus* (average ± SD *algirus*: 72.76 ± 5.75 mm; *cuniculus*: 78.72 ± 4.82 mm), and hint foot length follows the same pattern (*algirus*: 56.38 ± 3.42 mm; *cuniculus*: 59.45 ± 3.26 mm). *O. c. algirus* individuals are on average lighter than those of *O. c. cuniculus* (*algirus*: 1043 ± 137 g; *cuniculus*: 1234 ± 169 g; Ferreira et al. 2015; Fig. 2). Furthermore, there are slight differences in cranial measurements between the subspecies (Sharples et al. 1996). Most traditional studies suggest that rabbits are not generally sexually dimorphic (Sharples et al. 1996). However, according to a recent study, females are slightly heavier than males in rabbit's native range (Ferreira et al. 2015). Such reversed sexual dimorphism has been observed also in domestic rabbits (Rosell and de la Fuente 2012), as well as in other lagomorphs, such as *Sylvilagus* rabbits and hares (Alves et al. 2008; Davis and Roth 2008).

European rabbits, and other lagomorph species like North American cottontails (*Sylvilagus floridanus*), are superficially alike. In fact, non-specialists can find difficulties to differentiate European rabbits from other rabbits or hares. For example, 60% of the hunters interviewed by Cerri et al. (2016) in Tuscany (Italy) considered the introduced cottontail rabbits as a subspecies of European rabbits. European rabbits have strong hind limbs, adapted to running (Fig. 3). The feet have large claws, and hind feet have four toes and forefeet have five. The ears are relatively long, although shorter than those of most hare species. The coat is normally greyish-brown, with rufous fur on the neck and light grey belly fur, the short tail is white below and brownish-black above, and the ears are the same color as the coat, but they lack the black tip typical of the hares that occur in Europe (Fig. 3; Delibes-Mateos et al. 2018a). The pelage is very variable in the domestic forms of the rabbit.

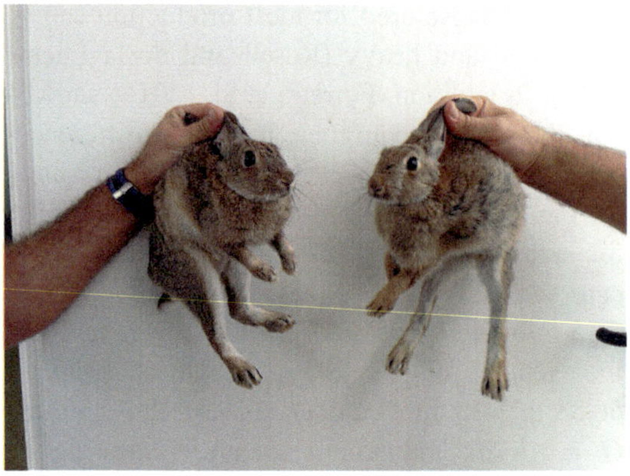

Fig. 2 European rabbits of the subspecies *Oryctolagus cuniculus algirus* (right) vs. *Oryctolagus cuniculus cuniculus* (left). Wild animals kept in captivity in the Research center of wild lagomorphs facility (Córdoba, Spain). (© Rafael Villafuerte with kind permission)

Fig. 3 European rabbit in northern Spain. (© Marco A. Escudero with kind permission)

It is not infrequent that people erroneously categorize rabbits as rodents, ignoring that lagomorphs have an extra pair of incisors. As pikas and hares, rabbits have six incisors teeth: two are on the lower jaw and two on the upper jaw, as with the rodents, but there are also two small teeth behind the upper incisors (Figs. 4 and 5); these are usually known as auxiliary incisors or "peg teeth." In addition to these six incisors, rabbits have other 22 teeth: three pairs of premolars on the maxilla, two pairs of premolars on the mandible, and three pairs of molars on the maxilla and also on the mandible (Figs. 4 and 5). Premolars and molars are often known as "cheek teeth" and rabbits use them to grind their food. Therefore, the normal dental formula is $I^2{}_1\ C^0{}_0\ P^3{}_2\ M^3{}_3$ for 28 teeth.

Several methods are used for the determination of rabbit age. One of this is checking the ossification of rabbit distal epiphysis of the ulna, which only occurs in adults (Villafuerte 1994). Also, a threshold of rabbit size is often used to differentiate between adult and juvenile rabbit; in the Iberian Peninsula, rabbits are considered as adults when their weight is higher than 800–900 g (e.g., Villafuerte 1994; Gonçalves et al. 2002). A combination of both methods has also been employed in the literature (e.g., Rouco et al. 2018). However, most body parameters are dependent on a number of factors, including the environment, physical condition, nutritional history, gender, or the subspecies to which the individual belongs (Ferreira et al. 2015). From this perspective, the weight of the dried eye-lens seems to be the most reliable indicator of the age of the rabbits (Augusteyn 2007), as in many other mammal species. This approach was first proposed for rabbits by Dudzinski and Mykytowycz (1961), who suggested that the relationship between rabbit age and the dry weight of eye-lens could be described using a logistic equation. Since then, this method has been widely used with differences in the constants necessary for the equation and in the drying process: temperatures of 80 or 85 °C and drying times from 24 h to 14 days (e.g., Wheeler and King 1980). Augusteyn (2007) suggested that eyes should be fixed for at least 4 weeks and fixed lenses should be dried for at least 2 weeks at 85 °C or preferably for 3 days at 100 °C. Recent research suggests that the logistic curves used in previous studies do not work well for *O. c. algirus*, and therefore that a specific equation is needed for this subspecies (Vaquerizas et al. 2021).

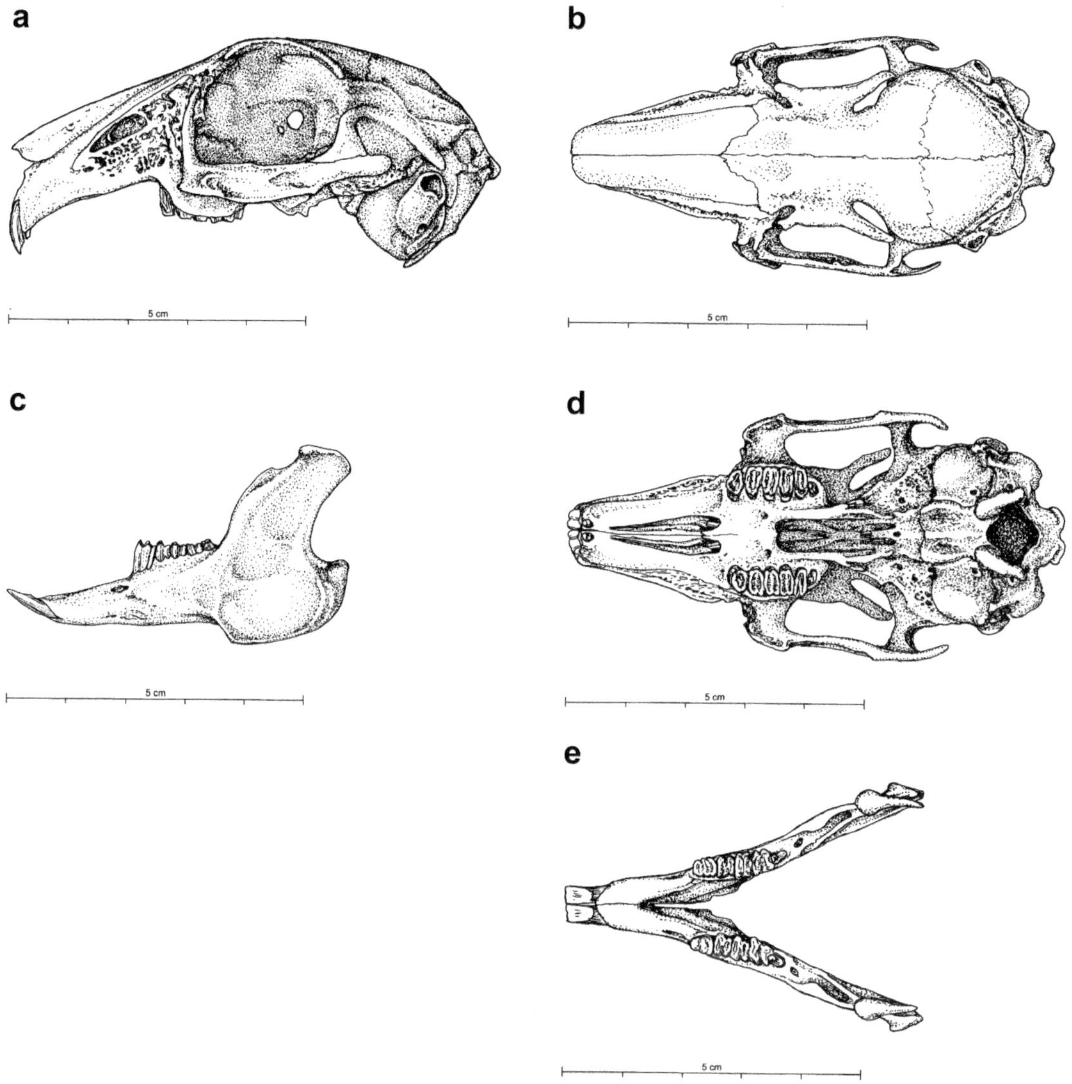

Fig. 4 Skull of *Oryctolagus c. cuniculus*; (**a**, **b**, **d**) lateral, dorsal, and ventral view of skull, (**c**) lateral view of left mandible (view from outside), (**e**) dorsal view of mandible. (© Franz Müller with kind permission)

Physiology

Metabolism, Growth, and Huddling Behavior of Pups

Newborn rabbit pups lack the capacity to thermoregulate efficiently (Hull and Hull 1982). As rabbit mothers only visit their young for a brief nursing event once per day and do not brood them (see details below in the sections "Life History" and "Behavior"), rabbits have evolved other adaptations. On the one hand, mothers build well-insulated nests out of plant materials, lined with mother's abdominal fur (Ross et al. 1963, more details in section "Behavior"). On the other hand, pups huddle together with their littermates. In doing so, pups can efficiently reduce the energetic costs of thermoregulation, thus enabling them to allocate more energy into growth (Gilbert et al. 2007). However, having a larger number of siblings does not only bring advantages. Each pup

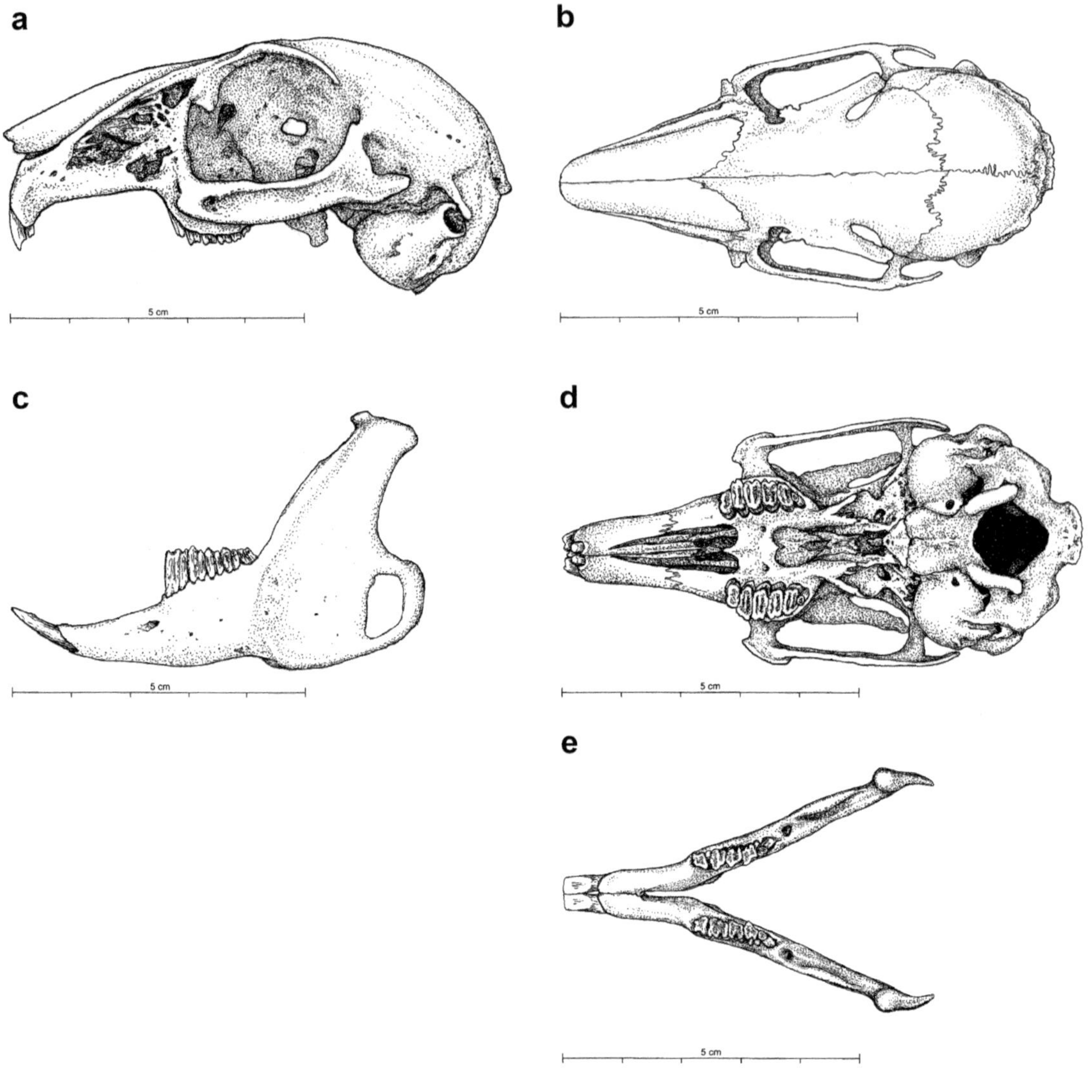

Fig. 5 Skull of *O. c. algirus* (specimen reference EBD 06043M). (**a**, **b**, **d**) Lateral, dorsal, and ventral view of the skull, (**c**) lateral view of the left mandible (view from outside), (**e**) dorsal view of the mandible. (© Franz Müller with kind permission)

receives a lower share of milk and thus shows a lower preweaning growth when the number of littermates (i.e., litter size) is higher (details in section "Life History"). A study on rabbit pups in natural burrows showed that the balance between costs and benefits of having larger numbers of litter siblings depends on the ambient temperature conditions. Individual growth was highest in litter sizes of 3 when the soil temperature around the nest chambers was higher than 10 °C. However, in early spring when the soil temperature was lower (even as low as 5–6 °C), the optimal litter size for an efficient preweaning growth was four pups, as under such conditions the energetic benefits of huddling with a larger number of siblings partially outweigh the costs of competing with them for their share of mother's milk (Rödel et al. 2008a).

Already at birth, there is notable variation in pup body mass within a litter, which can be attributed to a large extent to the fetal implantation site in the uterus horn. Fetuses implanted in more peripheral or distal positions in the horn show a better prenatal growth than fetuses in more central positions (Bautista et al. 2015a). This variation has important consequences, as pups with a lower birth mass have a lower chance to reach the warmer and thus energetically more favorable center of the litter huddle. Thus, these individuals need to allocate more energy into thermoregulation as exemplified by their higher depletion of brown adipose fat tissue (García-Torres et al. 2015). These factors all contribute to a lower growth until weaning in lighter pups compared to their heavier litter mates, as it has been show in domestic as well as in wild European rabbit pups (Bautista et al. 2015b; Rödel et al. 2017).

Differences in Body Condition

There can be tremendous variation in body mass and body condition among different populations, years, or even between birth cohorts within a given year due to adverse environmental conditions. In Mediterranean-climate habitats, periods of droughts can cause decreases in growth rate and even considerable losses in body mass (Kontsiotis et al. 2014). Rainfall, by promoting the growth of green pasture, has positive effects on growth and body condition (Dudzinski and Mykytowycz 1960). Winters with longer periods of cold, snow cover, and low food availability are also a major challenge and can cause significant body mass losses including decreases in body fat content, in particular in young individuals (Wallage-Drees 1986; Rödel 2005).

Juvenile rabbits are particularly sensitive to adverse climatic conditions and the combination of wetness and low temperatures significantly increases their resting metabolic rates, as exemplified by the increased oxygen consumption rate of around 2.13 O_2/g h under wet conditions at 5 °C compared to the lower rate of around 1.25 O_2/g h under dry conditions at 23 °C (Seltmann et al. 2009). The increased metabolic costs under rainy and colder weather during the following weeks after weaning might be one of the mechanisms why European rabbit pups exposed to such weather conditions show lower age-specific body masses in the end of the vegetation period, as it has been reported by a study in a temperate zone habitat (Rödel et al. 2004a). A further mechanism contributing to the association between rainy weather experienced early in life and growth might be that such weather conditions promote infestations of juvenile rabbits with endoparasites, such as intestinal nematodes (Rödel and Starkloff 2014).

There are also notable changes in body condition along the reproductive season related to the rabbit's social behavior. During the early breeding season, social ranks are acquired and maintained, sometimes by fierce fights. This leads to an increased occurrence of wounds and increased levels of stress hormones (mainly corticosterone) during this time of the year (von Holst et al. 1999). In this sense, Cabezas et al. (2007) showed that following exposure to long-term stress, elevated serum corticosterone levels in *O. c. algirus* were negatively associated with body condition.

Energetic Costs of Reproduction

Due to the rabbit's postpartum estrus, the pregnancy frequently overlaps with the period of lactation (Brambell 1944, details below in section "Life History"). This imposes high energetic costs to females in such conditions (Martínez-Gómez et al. 2004). In accordance, a study in a field enclosure showed significantly longer feeding times (increased by 10–13%) of females with concurrent pregnancy and lactation compared to pregnant and nonlactating females, and compared to nonreproducing females (Rödel et al. 2016).

Genetics

Chromosomes: 2n = 44: including 13 pairs of meta and submetacentric, 4 pairs of subtelocentric, 4 pairs of acrocentric, X-chromosome is submetacentric and Y-chromosome subtelocentric (Hsu and Benirschke 1967). Rabbit chromosomes 12, 19, and X were found to be homologous to human chromosomes 6, 17, and X, respectively (Korstanje et al. 1999).

The European rabbit is probably one of the nonmodel species with the most comprehensive genetic information available and definitely the lagomorph species with the highest number of publications in genetics and evolution. The studies in genetic diversity started with the foundations of molecular genetics, by using traditional molecular markers as the alloenzymes, serum proteins, and polymerase chain reaction-restriction fragment length polymorphisms (PCR-RFLPs) (e.g., Ferrand and Rocha 1992; Hardy et al. 1995; Branco et al. 2002). The number of studies was highly accentuated due to the existence of a domestic form, the domestic rabbit, which is also used as a laboratory animal (see Manning et al. 1994). All these studies have demonstrated that the European rabbit is one of the mammal species with highest levels of genetic variability (Carneiro et al. 2009, 2014a). This high genetic variability is often associated with the great capacity of adaptation to new environments, making this species an amazing successful colonizer, and thus occurring in artic, boreal, temperate, and desert ecosystems, and over 800 islands (see section "Current Distribution"). In addition, several immunogenetic studies have also revealed high genetic variability, namely in the immunoglobulins (e.g., Pinheiro et al. 2019), which may be associated with a tremendous resilience to severe viral outbreaks (over 90%), namely to the myxomatosis and the RHD. Alves et al. (2019) reveal how the standing genetic variation observed in genes associated with the immune system allowed populations to rapidly evolve resistance to novel and highly virulent pathogen, as the myxoma virus, which caused high rates of mortalities (details in section "Parasites and Diseases"), and describe the molecular and genetic basis of an iconic example of natural selection.

The European rabbit was the first lagomorph species where the whole genome was sequenced. The Broad Institute completed a deep coverage draft of the rabbit genome in 2005 and in 2009 released the OryCun 2.0, with an assembly size of 2.6 Gb (Carneiro et al. 2014a). The rabbit genome project was considered a significantly contribution for the study of immunology, and medical research in general, but also for the annotation of the human genome. Several studies have used a variety of genetic markers and genomic information to infer the evolutionary history of the European rabbit, namely the time of divergence between the two rabbit subspecies (*O. c. cuniculus* and *O. c. algirus*), which was estimated to occurred around 1.8 million/year BP (Branco et al. 2000; Carneiro et al. 2009, 2014b). Several other studies have focused on the history and genetic basis of domestication (e.g., Carneiro et al. 2014a). These studies estimated that the domestication from the subspecies *O. c. cuniculus* likely occurred around 1000 years ago, and found that several genes affecting brain and neuronal development were targeted during rabbit domestication, which is consistent with the view that critical phenotypic changes during the initial steps of domestication probably involved behavioral traits that allowed animals to tolerate humans.

More recently, a comprehensive genomic rabbit database was established, containing both genome and transcriptome data from many studies, and the variations from several samples, including wild European rabbits (both subspecies) and different domestic rabbit breeds (Zhou et al. 2018). As reported by Carneiro et al. (2014a) and Zhou et al. (2018), a clear pattern of decreasing genetic diversity is observed from the European rabbits in the Iberian Peninsula towards populations in southern France, and a second reduction is also observed during domestication, which created other European domestic breeds.

The European rabbit provides an excellent and model example on evolutionary and speciation processes in mammals. Genomic information reveals levels of differentiation between the two described subspecies (*O. c. algirus* and *O. c. cuniculus*) in the Iberian Peninsula, yet also showing extensive introgression. Nevertheless, some genomic regions show a substantial differentiation between the subspecies, suggesting that genes in these regions are strongly involved in the maintenance reproductive isolation between the forms (Carneiro et al. 2009, 2014b; Rafati et al. 2018). This pattern of mosaic genome differentiation puts the rabbit subspecies in the range of species divergence, but complicates simple assessments of their taxonomic status. Nevertheless, Geraldes et al. (2008) reported high level of differentiation in the Y chromosome between the two subspecies on either side of a hybrid zone, which suggests a role for the Y chromosome in reproductive isolation. Other unpublished studies suggest reduced fertilization and higher mortality in hybrids between the two subspecies, stressing the validity of *O. c. cuniculus* and *O. c. algirus* being different species. The apparent narrow and stable hybrid zone between the two subspecies, with a diagonal from NW to SE Iberian Peninsula, is also an indirect confirmation of the validly of reproductive isolation between the two taxa (Geraldes et al. 2008; Carneiro et al. 2013; Rafati et al. 2018).

Life History

Characteristics of Social Groups

European rabbits are polygynous and are usually organized in social groups formed by around 1–5 adult males and 1–8 adult females (Cowan 1987a; von Holst et al. 2002). Groups are characterized by linear social rank hierarchies, separately in adult males and females (Mykytowycz and Fullagar 1973; von Holst et al. 2002). Related to the acquisition and maintenance of high social rank positions, the onset of the breeding season is typically characterized by a peak in intraspecific aggression in both sexes, including fights in particular between adult males within and among groups (von Holst et al. 1999). Rabbits show sex-specific differences in natal dispersal. While usually the majority of young males disperse to other groups before reaching maturity, a high proportion of females remains in the group of birth (Parer 1992; Webb et al. 1995; Künkele and von Holst 1996). As a consequence, the adult females of a social group typically show a high degree of genetic relatedness (e.g., Webb et al. 1995).

Reproduction

The European rabbit is a seasonal breeder, although the start and the length of the breeding season show considerable variability among populations in different parts of the world (reviewed in Thompson and King 1994; Tablado et al. 2009). In warmer, Mediterranean climates with dry summers, the breeding season of *O. c. algirus* usually lasts from autumn/winter until late spring/early summer, with a reproductive pause between July and September (Gonçalves et al. 2002). In temperate zones characterized by colder winters, breeding usually starts around early spring and lasts until summer/autumn.

Such differences among populations in different habitats are also apparent with respect to the age at first reproduction. Generally, European rabbits (males and females) may reach sexual maturity at the age of 3–4 months (reviewed in Tablado et al. 2009). In southern Spain, females already start reproducing at this age, often within their breeding season of birth (Soriguer 1981). Also, in more temperate zones, such as some areas in France, rabbits born early in the season might start reproducing already during their season of birth when reaching an age of around 6 months (Thompson and King 1994). In a field enclosure study in Bavaria/Germany, females always started to reproduce during the following year at the age of around 7–12 months (von Holst et al. 2002).

Female rabbits show postpartum estrus, that is, they can be fertilized shortly after parturition (Brambell 1944; von Holst et al. 2002). Thus, females frequently carry a pregnancy during the period of lactation, which occurred in 80–100% of cases during the peak breeding season in a study in Great Britain (Brambell 1944). Ovulation is induced by the stimulation caused by mating (Friedman 1929). The gestation period lasts around 28–31 days (Thompson and King 1994; von Holst et al. 2002). Superfetation as found in hares, that is, the development of a second pregnancy while the previous litter is not delivered, has not been found in wild European rabbit populations. Field data on the timing of weaning are only anecdotal; studies on domestic rabbits indicate that the young are usually weaned by the mother at around postnatal day 28, before she gives birth to a consecutive litter (Hudson et al. 1996).

The average litter size varies considerably between different populations, from around 3.2 to up to 7.3 pups per litter (reviews in Thompson and King 1994; Tablado et al. 2009). The variation at the individual level is even more substantial, and females produce litters of 1 up to 9 pups

(Brambell 1944; von Holst et al. 2002). Studies from different climate zones agree that the average litter size changes along the annual breeding season, starting with smaller litters then reaching a maximum during mid-season and frequently declining again in the end of the breeding season, with average differences between peak and low reproduction of 1–2 pups per litter (Parer et al. 1987; Eccard and Rödel 2011).

Also, the average number of litters per female and season shows a high variation between populations, from an average of 2.3 to 7.0, and mainly depends on the length of the breeding season (reviewed in Tablado et al. 2009). Furthermore, there is strong evidence for reproductive differences between the two European rabbit subspecies with a smaller average litter size (3.90 ± 0.50) for the *O. c. algirus* subspecies (Gonçalves et al. 2002). There are indications that the fecundity of females is density-dependent, that is, females tend to produce a lower number of litters per season when the population density is high (Rödel et al. 2004b). The annual reproductive output of females differs strongly among individual females within populations. For example, in a field enclosure in Germany, where females produced on average 2.5 litters per season, individual values ranged from 0 to up to 6 or 7 litters per season (von Holst et al. 2002; Rödel et al. 2009a). The offspring sex ratio at birth does not significantly deviate from a 1:1 distribution (Brambell 1944; von Holst et al. 2002).

The breeding season of the rabbit has been reported to depend on day length and photoperiod change (Tablado et al. 2009). A meta-analysis of studies on European rabbit reproduction also highlights the importance of temperature and food availability (Tablado et al. 2009). In Mediterranean-climate environments and in arid areas, the onset of the breeding season is determined by the green-up of vegetation, and the length of the breeding season is determined by the availability of green vegetation related to soil moisture (Villafuerte et al. 1997; Wood 1980; Gonçalves et al. 2002). In temperate zones, the annual onset of breeding in spring can be delayed by several weeks when the preceding winter was particularly harsh or long and thus the growth of green pasture is delayed (Bell and Webb 1991; Rödel et al. 2005).

Due to the high offspring mortality in the rabbit (see below in this section), the lifetime reproductive success of females in terms of the number of offspring surviving to maturity is typically low and highly right-skewed, that is, in a high proportion of females (>50% in a field enclosure study in Germany) none of the offspring produced over lifetime managed to survive to reproductive age (von Holst et al. 2002; Rödel et al. 2009a). At the individual level, female reproductive performance is strongly affected by age (see below for reproductive senescence), body mass, and social rank. Young females (first-season breeders) produce a lower number of litters per season than older ones, start reproducing later within the season, have smaller litter sizes, and show higher prenatal mortality rates (Fraser 1988; Rödel et al. 2004b). Females with lower body mass produce litters with lower mass, and pups of lighter mothers show lower growth rates than the offspring of heavier mothers (Rödel et al. 2008b, 2016). Low ranking females show a later onset of seasonal reproductive activities, smaller litter sizes, lower offspring masses, lower numbers of litters per season and have lower offspring survival than dominant females (Mykytowycz and Fullagar 1973; von Holst et al. 2002; Rödel et al. 2009a, b). Furthermore, it has been shown that females that managed to obtain a high social rank position early in life had a longer reproductive lifespan and an about 60% higher lifetime reproductive success than lower-ranking ones (von Holst et al. 2002). Female age, body mass, and social rank might not necessarily act independently, as younger females are typically not fully grown and thus lighter when starting to reproduce, and usually occupy low social ranks within their groups (Rödel et al. 2004b).

A further characteristic of the rabbit's reproductive biology is its system of "absentee mothering" as mothers only return to nurse the young once a day for around 3–4 min (details in section "Behavior"). That is, the pups need to quickly find the nipples of the mother to efficiently suckle during these brief nursing events. This mother-offspring interaction is supported by the existence of a pheromone excreted by glands at the mother's teats, which helps to guide the pups to the nipples (Hudson and Distel 1983; Schaal et al. 2003).

Early Growth and Development

Rabbits are altricial; that is, the young are born with closed eyes, without any fur, and with a limited capacity to thermoregulate independently during the first postnatal days (see details in section "Physiology"). Studies in Germany report body masses on postnatal day 1 of on average 44 g, with individual values ranging from 32 g to 60 g (Rödel et al. 2008b). The birth mass is negatively correlated with litter size, that is, pups from smaller litters are usually heavier than pups from larger litters, and there are no notable differences in birth masses between males and females (von Holst et al. 2002).

During nursing, young rabbits can ingest high amounts of milk, up to more than 25% of their body mass. Overall, the growth of European rabbit pups during the first 2–3 postnatal weeks follows a linear shape, and pups show average growth rates of around 7.2–7.8 g per day (Rödel et al. 2017).

As mentioned above in section "Physiology," litter size is the main driver of variation in postnatal growth among litters, as pups in larger litters typically receive a lower share of milk that pups in smaller litters. The increase in body mass from postnatal days 1 to 12 was on average 62 g in litters of 7 pups and 95 g in litters of 2 pups in a field enclosure study in Germany (Rödel et al. 2008a). Mainly as a consequence of such litter size effects, and as litters are usually smaller during the early breeding season (see above), pups born early in the season are heavier (Dudzinski and Mykytowycz 1960).

At around postnatal day 10, the pups open their eyes, already show a well-developed fur cover, and start to actively explore the environment inside the nursery burrow (Rödel et al. 2015). Around this time, they also start to feed on the nest material and on feces deposited by the mother inside the nest (Mykytowycz and Ward 1971).

When the young leave the nursery burrow at around postnatal day 18–21 (von Holst et al. 2002), their average body mass has been reported to range between 180 and 250 g for the subspecies *O. c. cuniculus* (Tyndale-Biscoe and Williams 1955), but is smaller for the *O. c. algirus* subspecies ranging from 120 to 200 g (Villafuerte unpublished data). During the first few months after emergence above ground, the growth of juveniles is approximately linear, at least until around day 70 (Ferreira and Ferreira 2014). For the subspecies *O. c. cuniculus,* Myers (1958) reported growth rates of around 16.5 g/day from emergence above ground until day 70, and Wood (1980) reported growth rates of weaned rabbits of around 10 g/day until they reach a body mass of around 750 g. The growth rates of the subspecies *O. c. algirus* are lower, with around 7.7 g/day until the animals have reached a body mass of around 600 g (Ferreira and Ferreira 2014). Until maturity, there are no notable differences in growth rate or body mass between males and females (Webb 1993; Rödel et al. 2004a). Adult females exhibit larger variation in body mass than adult males and are typically heavier during the breeding season due to pregnancies (more details above in section "Description").

Mortality

Intrauterine mortality can be high in European rabbit populations, and a study in England showed that 42–45% of litters are resorbed before the last third of the gestation period (Brambell 1942). Postnatal mortality during the nesting period occurs in around 43% of litters, affecting either single pups (10% of cases) but mostly the whole litter in 32% of cases. Peak nest mortality occurs during the first few postnatal days, including cases of stillbirth (von Holst et al. 2002; Rödel et al. 2009b; Seltmann et al. 2017). Mortality factors during the nest period are manifold and can be, for example, due to flooding after heavy rains, predation, low ambient temperatures, insufficient maternal behavior, and infanticide (Wood 1980; Palomares 2003; Rödel et al. 2009b). In the south of the Iberian Peninsula, where the subspecies *O. c. algirus* occurs, it has been shown that predation by different carnivores, such as red fox (*Vulpes vulpes)*, Egyptian mongoose (*Herpestes ichneumon*), and European badger (*Meles meles*), strongly affects the nest mortality of rabbits (>20% mortality; Villafuerte 1994; Viñuela and Villafuerte 2003).

Mortality of juveniles during the following weeks or months after leaving the burrow is also high and can mainly be attributed to predation, diseases, and to starvation during periods of drought (more details in section "Population Ecology"). Webb (1993) reported mortalities of 80–90% during 8 months after emergence above ground in a field study in England. In a field enclosure study in Germany, around 24% of juveniles died or disappeared during the first 30 days after emergence above ground (Rödel et al. 2015). Monthly mortality in southern Spain (Doñana National Park) was estimated to be 72% and was mainly attributed to predators by raptors and carnivores (Villafuerte 1994; Viñuela and Villafuerte 2003). Studies in large enclosures free from terrestrial predators in southern Spain further highlight the impact of predation by raptors on *O. c. algirus* rabbits, as juvenile mortality was considerably decreased by experimentally improving the cover around the warrens (Ferreira et al. 2014a).

The body mass the animals reach around emergence above ground predicts their survival during the following months; heavier juveniles with body masses between 180 and 210 g had a survival probability of around 40%, while the survival of lighter juveniles with body masses between around 60 and 100 g was lower than 5% (Rödel et al. 2015). Furthermore, juveniles born early in the season survive better than the ones born later (Wood 1980; Webb 1993). Studies in field enclosures showed mortality rates of around 15–20% in early born and of 90–95% in late born individuals (Mykytowycz and Fullagar 1973), and mortality rates of on average 86% in individuals born during the first 2 weeks of the breeding season compared to almost 100% mortality in animals born during the second half of the season (Rödel et al. 2015). In general, juveniles suffer from higher mortality than adults, especially during periods of starvation or low temperature conditions (Wallage-Drees 1986). Furthermore, juvenile mortality is usually higher in males than in females (von Holst et al. 2002; Rödel et al. 2015), by around 10% (Webb 1993). This difference finally contributes to a female-biased sex-ratio as it is frequently observed in adult populations (Cowan 1987b; von Holst et al. 2002). Adult mortality rates can also strongly vary between years and populations; different studies in England report adult mortalities of 29% and 57% during 2 consecutive years (Webb 1993) and sex-specific mortalities of 32–65% in females and 57–70% in males (Cowan 1987b). In southern Spain, adult mortality *for O. c. algirus* has been reported to be up to 86% (Moreno et al. 2004), and predation is one of the main causes. For example, most radio-tracked rabbits found dead by Lombardi et al. (2003) in Doñana National Park, southern Spain, had been predated by carnivores and raptors (48% and 20%, respectively), while approximately another third died as a consequence of diseases. Similarly, predation was the main cause of juvenile mortality in eastern England (Webb 1993) and was negatively related to seasonal population growth rate in the Lemnos Island, Greece (Kontsiotis et al. 2013). Indeed, predation pressure on rabbits can be high, especially in Mediterranean areas within southern Europe where there are many rabbit predator species (see for more details section "Management"). In this sense, Kufner (1986) estimated that in the 1980s, before the initial outbreak of RHD, the entire predator community consumed about 51,000 rabbits (*O. c. algirus*) annually in the 6000 ha Doñana Biological Reserve, southern Spain. Overall, predation by carnivores is considered to be more influential than by raptors (71% vs. 26%; Villafuerte 1994), and opportunistic predator species such as red foxes and kites (*Milvus milvus*, *M. migrans*) are the ones that cause the highest losses (Viñuela and Villafuerte 2003). Rabbit males are more frequently predated by raptors and females by carnivores excluding the Iberian lynx (*Lynx pardinus*) (Villafuerte 1994).

Longevity and Aging

Once a rabbit reaches adulthood, the mortality risk decreases. A field enclosure study in Germany showed that the average life expectancy of females reaching adulthood is around 2.6 years; 40% died during or shortly after their first breeding season and 16% managed to live longer than 5 years (Rödel et al. 2009a). The maximum life span of single individuals can be up to 8 years, as it has been shown by a field enclosure study (1 female of 7.8 years; Rödel et al. 2009a) and by a study in the wild (Peacock and Sinclair 2009). Older females (older than 4–5 years) can show signs of reproductive senescence by a lowered number of offspring per season and by a lower preweaning offspring growth (Rödel et al. 2004b, 2008b). Furthermore, maternal age has been shown to exert intergenerational effects, as female offspring from 1-year-old mothers as well as from mothers older than 5 years showed a lower lifetime reproductive success than offspring from middle-aged mothers (Rödel et al. 2009a). Accurate information in longevity and aging is not available for *O. c. algirus*.

Habitat and Diet

Habitat

The worldwide distribution of rabbits reveals their ecological plasticity and their capacity to subsist in many different habitats (Thompson and King 1994). In their native range, rabbits typically prefer areas with a mosaic of shrub and grasses and/or crops: shrub provides protection against predators but low forage abundance, whereas grassland and crops offer little protective vegetation but high food availability (Calvete et al. 2004). In fact, in Iberian Mediterranean-scrubland ecosystems, rabbits reach the highest abundance in the ecotone, which provides intermediate levels of both resources (Lombardi et al. 2003; Calvete et al. 2004). The species, therefore, benefits from traditional agriculture, which provides small patches of feeding habitats (i.e., crops or pastures) interspersed with areas dominated by shrub. The transformation of agriculture that took place in many European regions during the second half of the twentieth century posed an impoverishment of rabbit-friendly habitats, and therefore, it did not contribute to the recovery of rabbit populations after the crash caused by myxomatosis (Calvete et al. 2004). A good example is found in Andalusia, southern Spain, where habitat favorable for *O. c. algirus* rabbits became impoverished between the 1960s and the 1990s as a consequence of land abandonment in some areas and agriculture intensification in others (Delibes-Mateos et al. 2010). Similar findings were reported in other European regions. For example, in Eastern Scotland changes in land use between 1977 and 1985 reduced the amount of cover available for rabbits to live in, resulting in a decrease in rabbit numbers (Boag 1987), and the destruction of hedges of natural vegetation in farmland areas in France was unfavorable for rabbits (Thompson and King 1994). Currently, some Iberian healthy rabbit populations are still linked to Mediterranean scrubland ecosystems. However, many of the healthiest populations occur in farmland areas, where rabbit numbers were not historically high. In these areas, food availability for rabbits (i.e., crops) is high, and the verges of new human infrastructures like roads and railway lines are ideal to build rabbit warrens (Delibes-Mateos et al. 2018b).

In addition to landscape structure and composition, the distribution of rabbits is also determined by soil hardness, climatological conditions, and topography (Trout et al. 2000; Virgós et al. 2003). Rabbits thrive overall in areas dominated by soft soils (Calvete et al. 2004; Delibes-Mateos et al. 2009), because these make warren building easier, and warrens are essential for rabbit reproduction and as refuge (see below). In the Iberian Peninsula, rabbits are mostly found in Mediterranean climate areas, characterized by hot dry summers and mild winters. Rainfall between 400 and 600 mm annually favors the growth of the pasture species that provide rabbit main food (Delibes-Mateos et al. 2018a). Rabbits rarely occur above an altitude of 1500 m a.s.l. (Delibes-Mateos et al. 2018a).

Home range is a basic but fundamental descriptor of habitat selection and use by all animals. Different studies have shown a great variability in rabbits' home range size across populations living in contrasting environments (Hulbert et al. 1996; Lombardi et al. 2003). For example, free-ranging rabbits occupying chalk grassland tend to be territorial and center their social life around the warrens (Cowan 1987a), whereas rabbits that occupy areas dominated by shrub often do not use warrens and are not territorial (Kolb 1991; Villafuerte and Moreno 1997). This wide variety in social organization makes difficult to generalize about the size of rabbit home ranges (Hulbert et al. 1996). Furthermore, rabbit home range size and shape are determined by the availability and spatial distribution of both an adequate source of food for maintenance and reproduction, as well as shelters and covers for breeding and protection against adverse climatic conditions and predators. Overall, home ranges are highly variable and may range from 0.7 ha to 9.7 ha depending on sex, density, habitats, and season (Hulbert et al. 1996; Devillard et al. 2008).

Diet

As other lagomorphs, European rabbits are herbivorous and therefore only eat vegetation, mainly grasses, and other herbaceous plants (Macdonald 2007). Rabbits are selective grazers that use low-fiber herbaceous plants (Mátrai et al. 1998), being essentially graminivorous (revised in Delibes-Mateos et al. 2008a); they feed mainly on grasses, but a significant proportion of dicotyledons may also occur in their diet (Chapuis 1990). Rabbits tolerate relatively well food like alfalfa or thyme (Saunders 2005). In seasons with poorer trophic supply, rabbits adapt their behavior by reducing their food choices according to local conditions of plant availability (e.g., Chapuis 1990; Marques and Mathias 2001). At the same time, they exploit all kind of resources available in the feeding grounds, such as woody parts of herbs, tree roots, and seeds (Mátrai et al. 1998; Marques and Mathias 2001; Barrio et al. 2010a). Studies on the diet and food habits of rabbits in agricultural landscapes have shown as well that they feed on cultivated grasses, such as wheat and barley, especially during the peak-growing season, which coincides with the rabbit breeding season. Rabbits' diet may also be influenced by the nutritive quality of available food, such as protein, fiber, and water contents, and by the competition with other herbivorous species (Thompson and King 1994). Overall, food choices vary along the seasons and depend mostly on the availability of plants in rabbits' feeding grounds (Marques and Mathias 2001; Barrio et al. 2010a). However, some exceptions can be found under particular conditions such as in insular environments and absence of competitors, where rabbits may select some plants regardless of their availability (Cubas et al. 2019).

Behavior

Antipredatory Behavior

Rabbit responses to predation risk include spatial avoidance and changes in foraging behavior. For example, in a study conducted in southern Spain rabbits tended to abandon the warrens when the perceived predation risk was increased experimentally using red fox odor (Rouco et al. 2011a). In agreement with this, a recent experiment conducted in central Spain showed that rabbit latrine activity and use were reduced when latrines were treated with fox feces (Navarro-Castilla et al. 2019). In addition, rabbits usually restrict their foraging activity to areas beneath or near scrubland when predator pressure is strong, while they feed more often in open spaces when predators are scarce or absent (Jaksic and Soriguer 1981). Also, solitary animals remain closer to scrubland, whereas rabbits in groups (three or more rabbits) forage further from cover, which suggests that grouping is a strategy to defend against predation (Villafuerte and Moreno 1997). Predation risk also influences rabbit temporal activity pattern. In this sense, Bakker et al. (2005) observed that rabbits were predominantly nocturnal in the North-East of The Netherlands, but became more active during the day when mink pellets were added. Similarly, rabbits are mostly diurnal in central Spain where nocturnal red foxes are abundant, whereas they are active mainly during the night where their main predators are diurnal raptors and human hunters (Moreno et al. 1996). Potential differences in activity patterns between both subspecies have been observed in preliminary studies (R. Villafuerte unpublished data). Furthermore, the moon phase has a strong influence on rabbit antipredatory behavior: during the darkest nights, rabbits take the risk to move further from their refuge, while they reduce their activity during periods of full moon (Penteriani et al. 2013).

Resource Competition

Resource competition strongly influences rabbit population dynamics (see also section "Population Ecology"). For example, the interaction between rabbit density and pasture biomass had a strong positive influence on recruitment rates in Turretfield, Australia (Fordham et al. 2012), suggesting that intraspecific density dependence limits rabbit reproduction through food shortage. Also, an increase in rabbit density may reduce the availability of suitable places for breeding (Guerrero-Casado et al. 2013), thus limiting reproduction. Density-dependent reproduction has been indeed demonstrated in rabbit enclosures in Germany and Spain (Rödel et al. 2004b; Ruiz-Aizpurua et al. 2014; more details in section "Life History"), and some evidences have been collected also in free-living populations in central Spain (Fernández de Simón et al. 2015). Interestingly, a reduction in spring rabbit density as a consequence of disease outbreak, for example, may make available vital resources for late breeding-season recruitment (Fordham et al. 2012). Furthermore, rabbit population density can negatively impact survival through increased intraspecific competition for food or through agonistic encounters (Rödel et al. 2004a; Fordham et al. 2012); resident adults often eject violently subadult individuals from their social groups (Lockley 1961), and the occurrence of this behavior is more frequent when density is high.

Interspecific competition for limited resources may be an additional regulatory force of rabbit density. Although European lagomorphs exhibit strong competitive interactions (Leach et al. 2015), most evidences suggest that European hares (*Lepus europaeus*) and rabbits co-occur across Europe without direct competition (Alves et al. 2008). Indeed, their moderate diet overlap throughout the year does not suggest a food competition between both species (Katona et al. 2004). Moreover, differences in habitat use probably enable the co-occurrence of both species (Santilli et al. 2015). On the contrary, a recent study based on species distribution models showed that Iberian hares (*Lepus granatensis*) co-occur less than expected with rabbits, suggesting the existence of potential competitive interactions (Leach et al. 2017). In any case, further field research is needed to confirm this potential competition between rabbits and Iberian hares. Competition with livestock might also affect negatively rabbit populations in some areas, although this is something that deserves further investigation. The biggest threat for rabbits in terms of competitors seems to be ungulates, at least in the rabbit's native range. Red deer (*Cervus elaphus*) and wild boar (*Sus scrofa*) have tremendously increased over the past years in many European regions (Apollonio et al. 2010), reaching extremely high abundances on certain occasions. In the Iberian Peninsula, for example, deer and wild boar numbers can exceed 50 and 90 individuals/km^2, respectively, in some intensively managed hunting estates (e.g., Acevedo et al. 2008), and such extremely high densities may be incompatible with the occurrence of healthy populations of other herbivores like the rabbit as a result of direct competition and/or habitat degradation. In this sense, a negative correlation between rabbit and wild boar abundances has been found in different parts of Spain (Lozano et al. 2007; Cabezas-Díaz et al. 2011; Carpio et al. 2014). High wild boar numbers are associated with a decline in herbaceous coverage and in the occurrence of legumes in the pasture (Carpio et al. 2014), and these plants are indispensable components in the rabbit's diet (Ferreira and Alves 2009; see also section "Habitat and Diet"). In addition, wild boar rooting causes soil disturbance (Carpio et al. 2014), likely compromising the construction of warrens by rabbits. Similarly, red deer abundance was negatively correlated with rabbit numbers in Monfragüe Natural Park, Spain (Lozano et al. 2007). Interestingly, it has been suggested that the high abundance of ungulates could compromise the recovery of small rabbit populations in the Iberian Peninsula through a "competitor pit" (Cabezas-Díaz et al. 2011).

Burrowing Behavior

European rabbits are active burrowers, and their warrens are of great importance for their social behavior, reproduction and survival (Cowan 1987a, b; Parer et al. 1987; Kolb 1991). The rabbit's exceptional ability among leporids to build their own warrens is likely related to the fact that European rabbit young are altricial, that is, born at the least advanced stage of development among lagomorph, and present a limited capacity to thermoregulate independently during the first postnatal days (Parer et al. 1987; Kolb 1991; more details in section "Physiology"). Although rabbits have no obvious morphological adaptations for burrowing, they can dig 2 m of burrow in few hours in suitable soils (Villafuerte 1994). Commonly warrens are described as the place where the European rabbit gives birth to its young. Warrens provide protection against predators and adverse weather and may range from single tubes to complex and large structures (Parer and Libke 1985; Villafuerte 1994). The preferred place to have a litter is in a chamber at the end of a blind tunnel either in nests inside the main warren (see below) or breeding burrows separated from it (Thompson and King 1994). Although on heavy cracking clay soils, in which digging a burrow is not possible, rabbits may have litters in hollow logs or natural holes in the soil. Before birth, the female rabbit excavates a short blind tunnel (length range: 0.30–2.5 m, mean: 0.85; Villafuerte 1994) with a breeding chamber at the end that will be used for the birth and nursing of its pups during the first 3 weeks of life. During this time, the litter will stay embedded in a comfortable special bed built with fur that the mother plucked from her belly few days before the birth adding in many cases also some dried grass (i.e., Villafuerte 1994; Seltmann et al. 2017; Fig. 6).

After the construction, the female always carefully plugs the entrance with soil and smooth over the surface to make it inconspicuous to predators, but also because the important mortality due to infanticide usually committed by other females (Rödel et al. 2008c). This is the reason why these chambers, especially the ones created outside the warren, are called "stops." In some few cases, the nest may be created in the middle of a tunnel of a big warren by blocking off it on either side of a chamber, but obviously, these nests are

Fig. 6 Schema of a typical breeding tunnel (stop) outside a warren. Extracted from Villafuerte (1994). (© Joaquín López-Rojas with kind permission)

more prone to disturbance from other rabbits than those at the ends of blind tunnels.

During the preweaning period, the entrance of the breeding chamber is only opened by the female for nursing the young; the whole activity lasts only few minutes (approx. 10–15) and occurs only once or twice a day (Lloyd and MacCowan 1968). Rarely the same breeding chamber is reused, but a new breeding chamber may be built in it, making longer and more complex the original tunnel, which may become in a complex and large warren in dense populations, given the high philopatry among females (Künkele and von Holst 1996).

Warren density is very variable and depends on a number of factors like soil hardness, habitat type, or presence of predators. For example, Kolb (1991) reported 496 warren entrances per hectare in a sandy area in UK, and Gea-Izquierdo et al. (2005) reported 64.4 warren entrances per hectare in Mediterranean habitats in central Spain. The number of entrances in a warren varies from 1 to more than 100, being likely the highest the 275 entrances reported by Viñuela et al. (1994) in Southern Spain, for *O. c. algirus*. In general, warrens in extremely soft soils are smaller than warrens in heavy soils, especially for those with high carbonate content and low clay content, where the length and number of interconnection between tunnels underground are greater (Parer and Libke 1985; Gea-Izquierdo et al. 2005). Larger warrens (longer tunnels and greater number of entrances) are more usually found in open habitats, where rabbits require this structure to avoid predators and weather inclemency, while smaller warrens occur more frequently in scrubby habitats (e.g., Lombardi et al. 2003). According to Lombardi et al. (2003), these smaller warrens in the scrublands are distributed randomly and are basically for breeding purposes. Therefore, open areas with abundant and uniformly distributed food may favor the formation of groups of rabbits whose lives are centered in large warrens, while scarcity of food in habitats with a dense cover may force a more solitary life where warrens are basically isolated litter holes (Villafuerte 1994; Fig. 7).

In most of the cases, rabbits need to build and maintain warrens to subsist, and the space available in the warrens limits the density of the species in a particular area. However, the relationship between rabbit group size and warren size is not linear because one social group may utilize several small warrens, and several social groups may use a single large warren (approx. 5–6 adults, 3 females; Rouco et al. 2011b). In the latter case, social groups maintain their territories by defending a small number of chambers and tunnels (Parer et al. 1987). Therefore, although the number of warren entrances does not necessarily reflect rabbit abundance in an area, it has been often used as a reliable estimator (Viñuela et al. 1994).

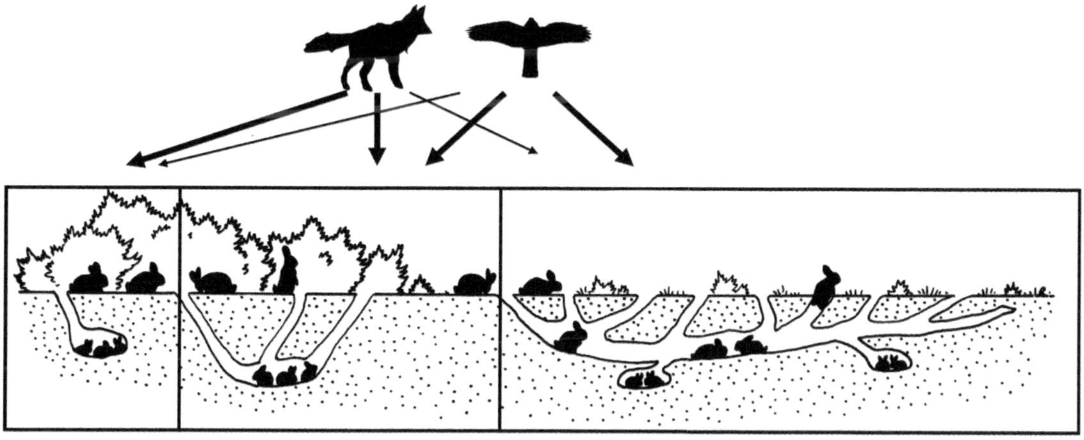

Fig. 7 Carnivore and raptor predation risk (thickness of arrow lines indicate intensity), rabbit warren size and use depending on the structure of the habitat. (Rosell 2000; © Rafael Villafuerte and J. López Rojas with kind permission)

Parasites and Diseases

Parasites

The parasite and pathogen fauna of European rabbits, either as intermediate or as final hosts, has been well documented in several countries of Europe during the last 50 years; the most common species identified in European free living rabbit populations are shown in Table 1 (but see also Foronda et al. 2003; Baker 2008). The prevalence of different parasite species depends ultimately on the presence of the host and the capacity of each parasite species to infect its host. Moreover, the range of parasites prevalence shown in Table 1 varies depending on the geographical region where each study took place. Furthermore, parasite intensity and prevalence can vary with season, year, host age, sex, immune status, the presence of other parasitic organisms, and intraspecific competition (Boag et al. 2001). Thus far, the helminthofauna, represented by nematodes (9 species), cestodes (14 species), and trematodes (2 species), was the most representative rabbit parasites in European countries. In particular, the five most common helminths of the European rabbit are the nematodes *Trichostrongylus retortaeformis*, *Graphidium strigosum,* and *Passalurus ambiguus* and the cestodes *Mosgovoyia pectinata* and *Cittotaenia denticulata*. Blasco et al. (1996) found both qualitative and quantitative differences between the helminth communities of a northern population of *O. c. cuniculus* and a southern population of *O. c. algirus*. Also, four species of Coccidia, four species of Acarina, *Haemodipsus ventricosus* (Diptera), *Spilopsyllus cuniculi* (flea), which is a vector of myxomatosis, *Linguatula serrata* (Pentastomida), and *Trypanosoma nabiasi* (Kinetoplastea) were found to parasitize European rabbits in Europe (Table 1). Among the protozoan rabbit parasites, there are several human pathogenic species like the cryptosporidian *Cryptosporidium cuniculus* (Chalmers et al. 2009), which is not common in free living rabbit populations, and *Toxoplasma gondii* (Coccidia) (Sroka et al. 2003), which can be fatal during human pregnancy. In Spain, the prevalence of *T. gondii* was more than fourfold lower in European rabbits living in dry habitats than in animals inhabiting forestry and humid habitats (Almería et al. 2004).

Overall, parasites itself rarely cause the death of their host, although this depends on several factors, such as the age or the body condition of the host. For example, severe coccidiosis caused by *Eimeria* spp. can kill young rabbits (6–11 weeks old) when population density is high (Baker 2008), but it does not affect adult rabbits and it is inconsequential at low rabbit density. The interaction among several parasites in European rabbits with the two main rabbit viral diseases (i.e., myxomatosis and RHD; see below) seems to play a key role in the species' condition (Cattadori et al. 2007; Bertó-Morán et al. 2013; Boag et al. 2013). For example, Boag (1988) showed that myxomatosis-infected European rabbits had significantly greater populations of the intestines nematodes *Trichostrongylus retortaeformis* and *Passalurus ambiguous* and the cestode *Mosgovoyia pectinate* in Scotland. Further observations suggested that the elevated burden of *T. retortaeformis* was caused by the impact of the myxoma virus, which negatively affected the rabbit immune response to the nematode (Cattadori et al. 2007). So, rabbits infected with myxoma virus are more susceptible to nematode infection, whereas rabbits with existing nematode infestations suffer longer myxoma virus infections (Cattadori et al. 2007). In contrast, Bertó-Morán et al. (2013) found that nematode seemed to play a minor role in the prevalence of antibodies against myxomatosis, which was inversely proportional to coccidian load. Nevertheless, the role that the immune system plays in the regulation of disease susceptibility and the potential differential impact viruses may have on both protozoan and helminth parasites is a complex subject and it will require further laboratory and field investigations to unravel the mechanisms involved (Boag et al. 2013).

Diseases

Nonetheless, without doubt, the most devastating events that dramatically declined the populations

Table 1 Maximum and minimum prevalence (P %) reported of most common parasites that affects free living European rabbit populations in Europe. * indicates presence when prevalence is not given

Species	P (%)	References
Protozoa		
Kinetoplastea		
Trypanosoma nabiasi	58–70	Reglero et al. (2007)
Coccidia		
Cryptosporidium cuniculus	*	Chalmers et al. (2009)
Cryptosporidium parvum	7	Sturdee et al. (1999)
Eimeria spp.	0–100	Voza et al. (2003)
Toxoplasma gondii	6.1–14.6	Almería et al. (2004)
Nematodes		
Dermatoxys hispaniensis	2–79.2	Blasco et al. (1996), Foronda et al. (2003)
Gongylonema neoplasticum	11.8	Eira et al. (2007)
Graphidium strigosum	33–100	Foronda et al. (2003)
Nematodiroides zembrae	32.6–82.3	Blasco et al. (1996)
Nematodirus battus	<1	Boag (1988)
Passalurus ambiguus	11.8–100	Blasco et al. (1996), Foronda et al. (2003)
Prostostrongylus cuniculorum	1.5	Blasco et al. (1996)
Trichostrongylus retortaeformis	17–100	Allan et al. (1999), Foronda et al. (2003)
Trichuris leporis	4.2–17.8	Foronda et al. (2003), Eira et al. (2007)
Cestodes		
Andrya cuniculi	0.6–50	Foronda et al. (2003), Eira et al. (2007)
Andrya rhopalocephala	*	Foronda et al. (2003)
Cittotaenia ctenoides	11–24	Mead-Briggs and Vaughan (1973), Mead-Briggs and Page (1975)
Cittotaenia denticulata	1.3–88	Blasco et al. (1996), Frank et al. (2013)
Cittotaenia pectinata	19–40	Mead-Briggs and Vaughan (1973)
Leporidotaenia pseudowimerosa	3.3–11.3	Blasco et al. (1996)
Leporidotaenia wimerosa	7–24	Blasco et al. (1996)
Monoecocestus americana	*	Hofing and Kraus (1994)
Mosgovoyia perplexa	*	Hofing and Kraus (1994)
Taenia pisiformis	16.1–33.3	Blasco et al. (1996)
Taenia serialis	<1	Boag (1988), Foronda et al. (2003)
Trematodes		
Dicrocoelium dendriticum	17.8	Blasco et al. (1996)
Fasciola hepatica	<1	Boag (1988)
Arthropods		
Acarina		
Cheyletiella parasitivorax	40	Baker (2008), Frank et al. (2013)
Ixodes ricinus	16	Baker (2008), Frank et al. (2013)
Leporacarus gibbus	98	Frank et al. (2013)
Sarcoptes scabiei	13–36	Millán et al. (2012)
Diptera		
Haemodipsus ventricosus	32	Frank et al. (2013)
Siphonaptera (Fleas)		
Spilopsyllus cuniculi	3–86	Frank et al. (2013)

of the European rabbit were the outbreaks of the myxomatosis and the RHD during the twentieth century, in addition to the recent new variant outbreak of the latter in the 2010s (see below in this section). These diseases have reduced rabbit populations over most of the historical range of the species, especially in ecologically less-favorable areas (Delibes-Mateos et al. 2008b).

Myxomatosis is a viral disease caused by a *Leporipoxvirus* of the *Poxviridae* family, known as "myxoma virus" or "Sanarelli virus," which induces a benign disease in its natural host, American *Sylvilagus* rabbits, while it causes "myxomatosis," a fatal and systemic disease in domestic and wild European rabbits. This virus was first detected in 1896 by Sanarelli in Montevideo, Uruguay, in a batch of imported European rabbits accidentally infected with myxoma virus (Fenner and Ratcliffe 1965). In 1952, it was introduced illegally in France to control rabbit pests. As consequence more than 95% of the French rabbit population was killed between 1953 and 1955 (Arthur and Louzis 1988). Fleas and mosquitoes are important vectors for the dissemination of myxoma virus in European rabbits (Fenner and Ratcliffe 1965). From France, it spread rapidly throughout Europe and caused the crash of the wild populations reaching mortality rates of around 90% (Fenner and Ratcliffe 1965). Due to the circulation of attenuated virus strains and the development of genetic resistance and acquisition of immunity in rabbits (Thompson and King 1994; Alves et al. 2019), the mortality decreased significantly the following years after the first outbreaks (Fenner and Ratcliffe 1965). Myxomatosis reached Spain in 1953, just 1 year after its first outbreak in Europe, and caused a significant reduction in European rabbit numbers, which in some extreme cases resulted in local extinctions (Villafuerte et al. 1994; Calvete 2006). Indeed, the virus caused the disappearance of nearly all rabbits in several Spanish provinces (Villafuerte et al. 2017). The effects of the first outbreaks were also catastrophic in Great Britain, where a reduction of 99% of the population numbers was estimated (Aebischer et al. 2011). Rabbit recovery after myxomatosis was very variable (e.g., Kolb 1994), but most European populations never reached the abundance levels previous to the disease outbreak (Thompson and King 1994; Aebischer et al. 2011). Nowadays, the disease is present in several countries within Europe. In Spain, for example, a nation-wide scale study carried out between 2003 and 2009 revealed that 65% of the adult rabbits were positive to antibodies against myxoma virus more than 60 years after its first outbreak (Villafuerte et al. 2017). According to these authors, seroprevalence against myxomatosis tends to be high in high-abundance populations. No differences in seroprevalence were observed between the two different rabbit subspecies present in Spain (i.e., *O. c. algirus* and *O. c. cuniculus*). In sum, myxomatosis can be considered currently as an endemic disease in the Iberian Peninsula and likely in several other countries within Europe (Villafuerte et al. 2017). Interestingly, there are no reported cases of myxomatosis in Azores archipelago, where *O. c. algirus* was introduced by Portuguese colonizers in the fifteenth century (Ferreira et al. 2012).

RHD is an acute, and usually fatal, necrotizing form of viral hepatitis that affects wild and domestic European rabbits. It is caused by the *Lagovirus europaeus*/GI.1/virus, family *Caliciviridae* (Le Pendu et al. 2017; henceforth GI.1). The virus was firstly detected in 1984 in the Jiangsu Province in China within a group of commercially bred Angora rabbits imported from Germany (Liu et al. 1984). The spread of RHD in Europe beginning in 1984 constituted a new landmark in the history of long-term rabbit population trends. It was first reported in Italy (Cancellotti and Renzi 1991) and from there it spread to the rest of Europe, becoming endemic in several countries (Abrantes et al. 2012). The first outbreaks decimated most populations across the continent, and particularly in the species' native range, causing the local extinction of some populations (Abrantes et al. 2012). In the Iberian Peninsula, the disease was first detected in 1988 (Spain) and 1989 (Portugal); the first outbreaks killed between 55% and 75% of the European rabbit populations in Spain (e.g. Villafuerte et al. 1994). By 1993 the disease was present in most of the Spanish populations (Villafuerte et al. 1995). In Doñana National Park (southern Spain), the maximum monthly rabbit density declined from 8.6 rabbits/ha in June 1989 to less than 1.5 a few years after the initial RHD outbreak (Moreno et al. 2007). Similar population crashes were also recorded in other European regions like western Germany (Knauer et al. 2010) and different regions in France (Thompson and King 1994). Rabbit population recovery after the initial impact of GI.1 was uneven and depended on several factors, such as habitat quality, climate conditions, and population management (Delibes-Mateos et al. 2009).

In the early 2010s, another population crash was caused by a novel lagovirus named GI.2/RHDV2/b (henceforth GI.2). It was firstly identified in France in 2010, and it shows a distinct antigenic profile and a reduced case fatality rate when compared to GI.1 (Le Gall-Reculé et al. 2013). Contrary to GI.1, this new virus has been detected in other leporid species (e.g., Velarde et al. 2017; Rouco et al. 2020), causing a fatal disease in these species. One of the main important differences between GI.1 and GI.2 is the differential susceptibility of the different age classes. GI.2 seems to be less virulent for adult rabbits (0–75% mortality) than GI.1 (above 90% mortality) (Abrantes et al. 2012); however, it affects pups as young as 11 days old (Dalton et al. 2012), which does not occur with GI.1. Additionally, GI.2 has shown a great diversity by recombining with pathogenic and nonpathogenic forms of the virus (Silvério et al. 2018), and it seems to have replaced the former circulating viruses, mainly the GI.1 variants, in wild populations (Dalton et al. 2014; Rouco et al. 2018). GI.2 spread in only few years throughout most Europe and other continents and islands worldwide; the most plausible explanation for this rapid spread is human intervention (Rouco et al. 2019a). During the first outbreaks, it caused 60–80% rabbit declines in Portugal (Monterroso et al. 2016) and Spain (Delibes-Mateos et al. 2014a). Similar declines were recorded in several British regions (Harris 2017). Globally, the impact of both variants of RHD has been devastating for many European rabbit populations. In the UK, for example, survey results from The British Trust of Ornithology show that the population of rabbits has declined 60% between 1996 and 2016 (Harris 2017), and it is thought that such decline has been caused, at least to a large extent, by RHD viruses' impact.

Population Ecology

Population Dynamics

As in any other species, rabbit population dynamics depend on recruitment and mortality in addition to immigration and emigration processes (Fig. 8). Rabbit reproduction is largely covered previously in this chapter (see section "Life History"). Nevertheless, it is important to remark that rabbit population recruitment can be high (see below), and this is mainly due to the following reasons: (1) the length of the breeding season is usually large, although it varies depending on photoperiod and food availability and quality, which ultimately depend on climate conditions and habitat characteristics (Tablado et al. 2009; Fig. 8); (2) rabbit's litter size is usually large, namely in *O. c. cuniculus*, reaching values higher than 7; (3) female rabbits become available for mating immediately after giving birth (postpartum estrus); and (4) rabbits are able to reproduce in the year of their birth, although the contribution made by first-year females to recruitment is usually lower.

The rate of population growth is dependent upon the mean number of births per adult rabbit (Fa et al. 1994), that is, the number of new rabbits that are recruited into the population. Rabbit pup recruitment varies both between areas and years. In the UK, for example, Bell and Webb (1991) reported values of 8.3–51.4% for recruitment of pups into the adult population. In general, the highest recruitment of pups into the adult population takes place along the spring, when more vital resources are available, and therefore, the annual population maximum is usually reached in summer (Beltrán 1991; Kontsiotis et al. 2013). In the Iberian Peninsula, it has been estimated that there is often a three- to four-fold increase in rabbit numbers from the minimum to the annual maximum after the breeding season (Thompson and King 1994). The high number of rabbit pups that are usually recruited into the population is often compensated by a low survival rate, particularly among pups and juveniles (see more details in section "Life History"). However, survival rates can be highly variable depending on a number of

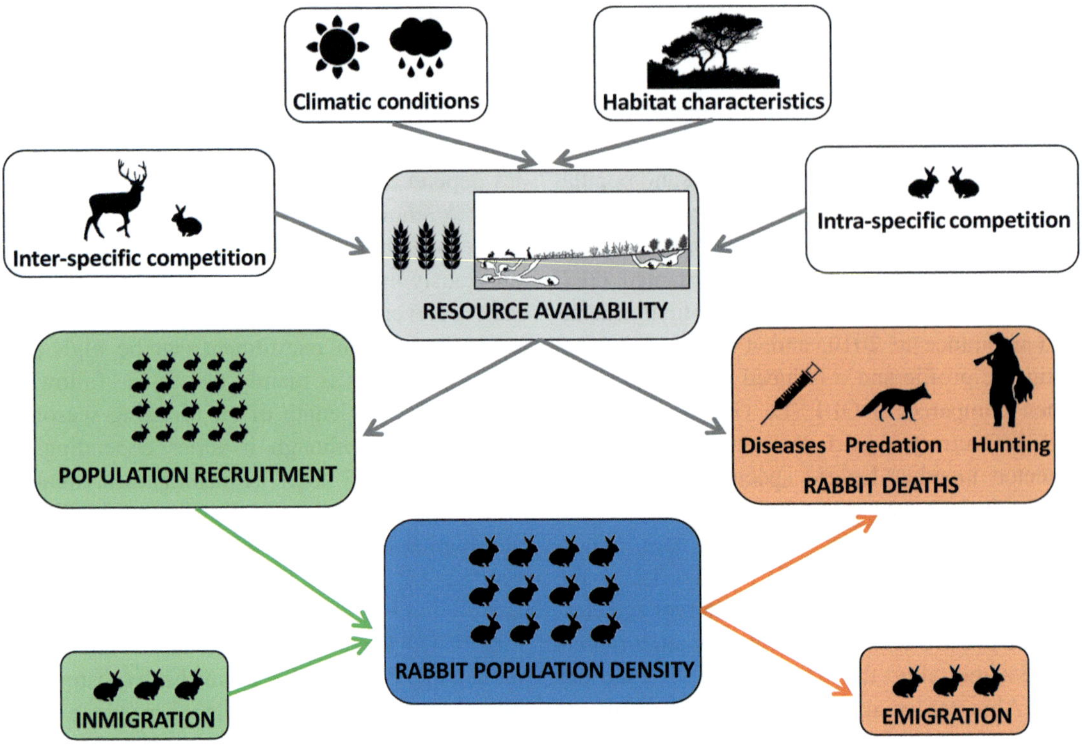

Fig. 8 Diagram showing main factors involving in rabbit population dynamics

factors, such as habitat conditions. In Doñana National Park (southern Spain, where *O. c. algirus* occurs), for example, Lombardi et al. (2003) recorded significantly higher rabbit survival rates in scrubland (0.53) than in grassland (0.14). It is also remarkable that rabbit survival rates usually vary seasonally, for example, in the early 2000s survival rates varied in Doñana National Park between 0.67 in spring to 0.82 in summer, 0.83 in autumn and 0.11 in winter, when RHD killed many rabbits (Moreno et al. 2004).

There are many limiting factors of rabbit population density (Fig. 8), which act synergistically (Tablado et al. 2009, 2012). Many studies have demonstrated that the availability and abundance of high quality food is a critical factor influencing rabbit reproduction (see for more details section "Life History") and thus rabbit population density (Villafuerte et al. 1997; Kontsiotis et al. 2013). Food availability is largely influenced by climatic conditions, which therefore play an essential role for rabbit population recruitment; in general, rabbit fecundity is favored by mild temperatures (Tablado et al. 2009). In addition, feeding resources may be reduced by intra- and interspecific competition (Fig. 8), which is previously discussed in this chapter (see section "Behavior").

In most rabbit populations, the main mortality causes are predation and diseases (Fig. 8). Both factors are largely covered previously in this chapter (see sections "Life History" and "Parasites and Diseases," respectively). With any doubt, viral diseases play a critical role in rabbit population dynamics, namely a disease-mediated bottom-up regulation (Monterroso et al. 2016). However, while some studies suggest that predation has a strong impact on rabbit population dynamics, others point at the opposite direction. The former is supported, for example, by the positive correlation found in England and Wales between predator removal and rabbit numbers (Trout et al. 2000), and the higher recovery (after initial RHD outbreak) of rabbit populations observed in Spanish hunting estates where predator (i.e., fox) control was conducted (Delibes-Mateos et al. 2008b). Also, fox predation was associated with the decline in rabbit abundance noted in a 3-year study carried out in a hunting estate in central Spain (Fernández de Simón et al. 2015). In contrast, rabbit numbers were neither correlated with the abundance of red foxes and common buzzards (*Buteo buteo*) in Mount Etna Park, Italy (Caruso and Siracusa 2001), nor with the abundance of foxes in northern Spain (Williams et al. 2007). The debate about regulation of rabbit population densities by predation is therefore still unsolved, and previous studies suggest that predation effects may be context-dependent. Relative to this, predation at lower rabbit abundance may produce a higher proportional loss in the rabbit population and potentially regulate it, in contrast to what occurs at higher densities. It has been suggested that, in some areas like Doñana National Park, rabbit numbers have been kept at low numbers by predation regulation after the population crash caused by RHD in the late 1980s (Moreno et al. 2007), that is, the so-called "predator pit" theory.

Human-induced mortality may also regulate rabbit population dynamics (Fig. 8), as rabbit hunting is a recreational activity in most of the species' range within Europe (see below in section "Management"). The effects of hunting on rabbit populations have been only evaluated through correlational studies, and hunting pressure has been usually estimated roughly, as the number of shotgun cartridges counted along transects. For example, Kontsiotis et al. (2013) found a negative correlation between rabbit population growth rate and hunting pressure in the Lemnos Island, Greece. Similarly, in northeastern Spain, rabbits recovered better several years after the first RHD outbreak in areas with low hunting pressure (Williams et al. 2007). However, in Spain, highest rabbit abundances are found in hunting estates (Delibes-Mateos et al. 2009), which suggests that hunting may not be of critical importance in rabbit population dynamics, at least in high-abundance populations.

Rabbit survival may be compromised by adverse climatic conditions (Fig. 8). For example, extremely high temperatures can negatively influence rabbit survival (Cooke 1977). Also, low winter temperatures may limit the over-winter survival of subadults (Rödel et al. 2004a). This may explain the increased mortality reported in The Netherlands during a winter with heavy snow fall (Wallage-Drees 1986), and the population decline observed in southern Sweden after severe winter conditions (Erlinge et al. 1984). In addition, there are many evidences for the detrimental impact of heavy rain on rabbit populations, especially through warren flooding and/or collapse (Palomares 2003). For example, the number of rabbits shot in 28 counties of The Netherlands and Germany was lower in years with high precipitation in spring of the respective year (Rödel and Dekker 2012). Similarly, in the 1990s rabbit density significantly decreased in southern Spain during rainy years (Palomares 2003). The negative effect of precipitation on survival decreases in importance in older rabbits (Tablado et al. 2012). Rainy periods usually coincide with the peak of rabbit reproduction, and therefore, many pups may die inside the warrens when these become flooded or collapse. In this sense, floods usually kill almost all younger litters of the population (with an age of less than 10 days; Rödel et al. 2009b). Therefore, the repeated occurrence of heavy rains may have the potential of impacting rabbit population dynamics (Palomares 2003).

Climate Change Biology

There is a growing concern about the impact of climate change on ecosystems and in particular about how it will influence the future distribution of wildlife species. Regarding rabbits, Tablado and Revilla (2012) predicted that their future reproductive periods will tend to decrease in magnitude and increase in variability in most of the current European distribution of the species. According to this study, climate change effect on rabbit reproduction will be especially worrying in the species native range in southern Europe, where the models predict population declines and increased probabilities of extinction. In contrast, climate change will be positive in terms of breeding season length and stability towards the eastern and northern borders of the European rabbit range, where populations are expected to grow and expand to adjacent areas that are currently not occupied by the species (Tablado and Revilla 2012).

Conservation Status

The IUCN Red List of Threatened Species: Global: Endangered (Villafuerte and Delibes-Mateos 2019), Europe: Near Threatened (Smith and Boyer 2007). The decline of Iberian rabbit populations caused by the new variant of RHD in the 2010s (see section "Parasites and Diseases") led to the recent change in the conservation status of the European rabbit by IUCN to Endangered (Villafuerte and Delibes-Mateos 2019). This assessment was mostly based on the declining situation of *O. c. algirus* since information about rabbit population trends in the native range of *O. c. cuniculus* is scarce (Vaquerizas et al. 2020). In areas where the European rabbit has been introduced, it is often considered as a harmful invasive species due to its characteristic of successful colonizer (see sections "Current Distribution," "Management" or "Future Challenges"). This contradiction explains its IUCN classification as Near Threatened in Europe (Smith and Boyer 2007) and highlights the character of the rabbit as a conservation paradox for the twenty-first century (Lees and Bell 2008). In addition, this may suggest the need of distinctive IUCN assessment for both rabbit subspecies.

Management

The European rabbit plays several different roles in Europe; it is an important game species in many countries, it acts as a keystone species in Mediterranean ecosystems, and it is a pest species in some farmland areas (Delibes-Mateos et al. 2014b). In this context, there are two opposite rabbit management strategies: one aimed at boosting rabbit numbers for shooting or to feed predators of conservation concern and the other one directed to decimating rabbit numbers in order to reduce economic loss caused by the species in agricultural areas. Paradoxically, some Iberian rabbit populations are managed simultaneously to increase and reduce their numbers (e.g., through predator control and intensive shooting and trapping, respectively) for hunting and to protect crops, respectively. These contrasting interests lead to frequent tensions between the involved stakeholders: farmers, hunters, and conservationists (Delibes-Mateos et al. 2014b).

Rabbits as Game Species

The European rabbit is a game species not only in its native range in southern Europe, but also in countries where the species has been introduced like in the United Kingdom, Italy, Greece, Poland, Hungary, Slovakia, etc. In the Iberian Peninsula, the rabbit is one of the most sought-after small-game species, and many of the more than 800,000 and 150,000 Spanish and Portuguese hunters, respectively, preferentially hunt rabbits; for example, rabbits are hunted in >95% of the more than 5000 game estates in Castilla-La Mancha region, central Spain (authors unpublished data). The importance of this lagomorph as game in its native range is also well illustrated by the fact that between 6 and 9 million rabbits are killed by hunters annually in Spain. Rabbit hunting season is variable among countries. Outside the native range, hunting seasons range from restricted (e.g., Hungary) to open hunting seasons (e.g., UK) (see factsheets on http://www.face.eu). In the native range (Iberian Peninsula), hunting is restricted to autumn/winter (Angulo and Villafuerte 2003), although special permits are often granted to kill rabbits out of the general hunting period in some agricultural areas where the lagomorph causes damage to crops (see below). The method most frequently employed to hunt rabbits in its native range is walked-up shooting, where hunters, with or without dogs, shoot the rabbits as they encounter them; for example, in central Spain, this method is used in nearly all of the game estates in which rabbits are shot. In addition to shooting, some forms of trapping, including ferreting, are also used throughout Europe (e.g., Ríos-Saldaña et al. 2013).

Rabbits as Keystone Species

In the Iberian Peninsula, mostly in the distribution area of *O. c. algirus*, rabbits are consumed by more than 40 predator species (Delibes-Mateos et al. 2008a), including many of conservation concern. For example, the highly endangered Iberian lynx (*Lynx pardinus*) and Spanish Imperial eagle (*Aquila adalberti*), both endemic to the Iberian Mediterranean ecosystem, are considered as rabbit-specialist predators, because both preferentially consume this prey (Ferrer and Negro 2004); that is, European rabbits constitute more than 85% of the lynx diet (Delibes-Mateos et al. 2008a). In southern Spain, conservation projects of the Iberian lynx and the Spanish Imperial eagle have attracted funding of more than €100 million over the past years, and a part of this funding has been allocated exclusively to the enhancement of European rabbit populations as the primary prey of these top predators (Simón et al. 2012). It is remarkable that European rabbits are also a key prey for predators of conservation concern in areas where this lagomorph has been introduced. For example, introduced rabbits have become a significant prey item for a number of native predators in the United Kingdom, including Western polecats (*Mustela putorius*), European wildcats (*Felis silvestris*), red kites or common buzzards (Lees and Bell 2008). Similarly, several raptors like the Bonelli's eagle (*Aquila fasciata*) and the common buzzard mainly prey upon rabbits in some Greek islands (Kontsiotis et al. 2013). In addition, European rabbits conspicuously alter plant composition and vegetation structure through grazing and seed dispersal, creating open areas, and contributing to plant species diversity, and rabbit burrows provide shelter for several vertebrate and invertebrate species. Moreover, rabbit latrines have an important effect on soil fertility and plant growth and provide feeding resources for different invertebrate species. Through these multiple roles, reviewed in Delibes-Mateos et al. (2008a), rabbits affect ecosystem structure and function in the Mediterranean area of the Iberian Peninsula, which includes the majority of the range of *O. c. algirus*, and has been termed "the rabbit's ecosystem" (Delibes-Mateos et al. 2008a).

Rabbits as Pest Species

The European rabbit (in the great majority of the cases the subspecies *O. c. cuniculus*) is considered one of the most harmful vertebrate pest species in the world, because of the huge ecological and economic impacts it causes where it has been introduced. Although these detrimental effects of rabbit introduction are legendary outside Europe (e.g., Cooke 2012; Cubas et al. 2019), negative ecological and economic impact by European rabbits has been described in Europe, too. For example, the Centre for Agriculture and Bioscience International (CABI) estimated that 40 million rabbits cost to the economy of England, Scotland, and Wales more than £260 million a year including damage to crops, business, and infrastructure, which makes it the most costly invasive species in Great Britain (Williams et al. 2010). Rabbits are also considered a major agricultural pest in other European regions. For example, in the island of Lemnos/Greece, cereal crop damage caused by rabbits was estimated at €312,202 in 2008 (Kontsiotis et al. 2018). Interestingly, farmers also complain often about rabbit damage to crops in agricultural areas within the species' native range (Delibes-Mateos et al. 2014b).

Restocking and Translocations

In southern Europe, restocking and translocation operations are one of the main strategies used by hunters and conservationists to increase rabbit numbers (e.g., Alves et al. 2008). These consist of the release of rabbits either from healthy wild populations (Rouco et al. 2008) and/or from facilities where domestic or wild-domestic rabbit hybrids are raised for such purposes (Piorno et al. 2015). Although precise estimates are not available, millions of rabbits have been possibly released in Spain, Portugal, and France since the first outbreak of RHDV1. A good example of these striking figures is in the Doñana Biosphere Reserve, southern Spain, where more than 52,000 rabbits were released between 2005 and 2015 within predator conservation projects (Carro et al. 2019). The extremely low rate of success of traditional translocation programs has increasingly led hunters and conservationists to use in situ extensive European rabbit captive breeding plots. Generally, one of these plots is an area of 0.5–4 ha, usually fenced, with variable number of artificial warrens, that serves as soft release site and as potential donor population for the neighboring areas (e.g., Rouco et al. 2008; Guerrero-Casado et al. 2013). Restocking and translocations of rabbits with unknown origin are not recommended, and even not allowed in Portugal and Southern Spain, in order to preserve the genetic identity of *O. c. algirus* subspecies, native from these countries.

Habitat Management

In addition to releasing rabbits, hunters and conservationists intend to promoting rabbit reproduction and reducing rabbit mortality through a battery of management actions. For example, in the Iberian Peninsula, the habitat is managed to reproduce traditional landscapes – that is, mosaics of agricultural areas and natural vegetation – that provide suitable food and cover to rabbits (Moreno and Villafuerte 1995). Generally, habitat management consists of creating feeding habitats in dense scrubland and providing shelter to rabbits where refuge is scarce. Creating pastures and planting crops are the most frequently employed habitat management techniques in Spain (Ferreira et al. 2014b). Hunters also use often supplementary feeders for game species, including rabbits (Delibes-Mateos et al. 2008b). The availability of high food quality has the potential to trigger an early onset of the rabbit's annual reproduction (Villafuerte et al. 1997), in addition to improving individual physical condition, which is critical to face predation risk and epizootic events. Also, the construction of artificial warrens is usually employed to enhance rabbit population densities (Rouco et al. 2011b; Guerrero-Casado et al. 2013), since these confer protection against predators and provide refuge again extreme climatic conditions and play a key role in rabbit reproduction and the establishment of social ties (Parer and Libke 1985).

Disease Treatments

Disease prevention treatments have become frequent among Iberian hunters and conservationists. Since periodic outbreaks of RHD and myxomatosis occur and cause high rabbit mortality, vaccination against these viral diseases is one of the most popular management tools to recover rabbit numbers (Cabezas et al. 2006). In the field, vaccines are typically administrated to rabbits, either captured in their warrens (employing ferrets) or brought from a donor area, when associated with restocking operations (Ferreira et al. 2009). Due to economic and logistic constraints, rabbits are usually vaccinated only once without previously assessing their immunological status and without considering their age, sex, or body condition. This is usually known as "blind vaccination" and its effectiveness has been called into question (Ferreira et al. 2009; Rouco et al. 2016).

Predator Control

Hunters often see predators as competitors for game species, and therefore, they usually make efforts to reduce predator numbers. In the rabbit's native range, predator control is conducted in many hunting estates; for example, it was employed in >90% of the small-game states studied by Delibes-Mateos et al. (2013) in central Spain. Rabbit predators that are more frequently killed by hunters legally in Spain and Portugal include red foxes, Egyptian mongooses, feral cats (*Felis catus*), and dogs (*Canis lupus familiaris*). Nevertheless, some hunters also kill illegally protected predator species to increase numbers of game species, and particularly rabbits (Villafuerte et al. 1998).

Self-imposed Hunting Restrictions

Another strategy commonly used by Iberian hunters to reduce rabbit mortality is implementing self-imposed hunting restrictions (Angulo and Villafuerte 2003). Regulatory mechanisms of hunting pressure often include limiting the number of hunters, the number of hunting days, the number of hours per hunting day, and/or the number of animals to be shot per hunter and day. In Andalusia, southern Spain, >70% of the hunters interviewed by Angulo and Villafuerte (2003) applied voluntarily at least one of these hunting restrictions.

Rabbit Control

When rabbits threaten human livelihood, land managers control their populations in order to reduce rabbit numbers to an economically acceptable level. Numerous control options including trapping, gassing, shooting, and ferreting are used to kill invasive pest rabbits (Smith et al. 2006). In addition, land managers employ different tools like fencing or using repellents to reduce rabbit damage to crops (e.g., Mckillop et al. 1992). In Great Britain, the use of shooting and ferreting is more widespread to control rabbits than other options like gassing and fencing because the latter are more expensive (Smith et al. 2006). European rabbits are also controlled in some farmland areas in their native range (Ríos-Saldaña et al. 2013). According to the Spanish and Portuguese legislations (as well as in some other European countries), holders of hunting rights are responsible for damage to crops caused by game species like rabbits. This means that, when land ownership and hunting rights are unbound, hunters have to collaborate with farmers to reduce rabbit numbers or to compensate them financially (Delibes-Mateos et al. 2014b). Shooting is the most widespread control option in Spain and Portugal, although ferreting and trapping are also used (Ríos-Saldaña et al. 2013). Rabbit control is usually very costly both in the species' native range and in areas invaded by the lagomorph. For example, in Great Britain, rabbit control was estimated to be at approximately £5 millions per year in 2006 (Smith et al. 2006).

Future Challenges for Research and Management

The European rabbit is very likely one of the most studied mammal species worldwide. A search in Scopus web engine (http://www.scopus.com, December 2020) reported more than 50,000 articles containing the term "*Oryctolagus cuniculus*" either in the title, abstract, or keywords. This figure is striking comparing to results yielded by similar searches for most other mammal species; for example, *Mus musculus* (~36,000), *Apodemus sylvaticus* (~1300), *Canis lupus* (~4300 articles), *Lynx lynx* (~3200 articles), or *Lepus europaeus* (~1000 articles). The rabbit sparks the interest of multiple and varied researchers, not only because of its wide distribution, but also because of the species' multiple roles. For instance, domestic rabbits are used very often in lab experiments to investigate questions about fundamental biology and health issues. Also, many investigations aim to generate knowledge that helps improve rabbit production in farms because of the species' value as meat source. Furthermore, intensive research is devoted to the study of medical aspects and cares of pet rabbits. Last but not least, European rabbits have been in focus of many studies both in the species' native range and in areas where the rabbit was introduced and causes harmful ecological and economic damage. In this section, we will identify future challenges for research and management on *O. cuniculus* in Europe, paying particular attention to southern Europe where the species is native.

Apart from a few good pioneer studies conducted in Doñana National Park in the first decades of the second half of the twentieth century (e.g., Soriguer 1981), rabbit research had a late start in the Iberian Peninsula in comparison with other regions where the species was introduced (i.e., UK or Australia). Iberian rabbit studies have been increasingly conduced since the late 1990s, likely as a repercussion of the declining population trends experienced after the initial outbreak of

RHD (Ferreira 2012). Most of these investigations have been conducted at a local scale, which implies that many of their management recommendations have limited applicability to other regions (Ferreira 2012). Global approaches are usually instrumental to set management priorities in biodiversity conservation, and therefore, future rabbit research at a wider scale is needed to design management plans that can be applied broadly across the Iberian Peninsula. In this sense, a long-term program for large-scale monitoring of European rabbit abundance and trends should be established urgently in the species' native range. Such program would provide the information required to understand the extend and causes of Iberian rabbit population fluctuations and should be the basis to set large-scale management interventions aimed at either increasing rabbit abundance for hunting or conservation purposes or at controlling rabbits where they cause crop damage. In addition to being carried out locally or regionally, most current rabbit monitoring programs developed in Spain and Portugal use different methodologies (see some examples in Delibes-Mateos et al. 2009), which make their findings hardly comparable. It is problematic that there is currently no consensus among researchers and managers about the most effective methodology to survey wild populations at such large scale, and surprisingly only a few attempts have been done to compare methods that are used more often to estimate rabbit abundance in Iberia, and such studies have been performed only in some specific habitats (Barrio et al. 2010b; Fernández de Simón et al. 2011). Therefore, more efforts are necessary to search the most cost-effective method to survey rabbit population trends at large scale.

This large-scale assessment of rabbit numbers is essential to understand the relationship between rabbit abundance and the species' various roles in the ecosystem. We know how many rabbits are needed to sustain a breeding population of the Iberian lynx (Ferrer and Negro 2004), but we are still far from comprehending, for example, how many rabbits are needed for the successful dispersion of a particular plant species or for the establishment of a population of Mediterranean dug beetles that feed on rabbit pellets. In this sense, Iberian researchers and practitioners have traditionally viewed rabbits as a prey species, and thus much less efforts have been done to investigate other rabbit ecological roles as well as to manage their populations for the benefit of other species rather than predators. It is therefore essential to conduct more experiments to address the multi-functional roles of rabbits in Iberian Mediterranean ecosystems, and this should include the assessment of how rabbits participate in the maintenance and organization of the complete ecological communities they belong to (Ferreira 2012).

In addition, we need to know more about rabbits in Iberian farmland areas where they are controlled as pests and population reductions might have negative consequences for biodiversity conservation. The scarce information on the rabbits' ecology and management in such areas has been collected recently in some local or regional studies (e.g., Barrio et al. 2010a; Ríos-Saldaña et al. 2013), but many questions still have to be addressed. For example, it is essential to identify those environmental and anthropogenic factors that determine rabbit "hotspots" in some Iberian farmland areas (but see Delibes-Mateos et al. 2018b; Rouco et al. 2019b). This knowledge will help design management interventions to control rabbits that damage crops, and it could be also potentially helpful for the management of declining rabbit populations in natural areas. Furthermore, the relationship between rabbit abundance and crop damage needs to be investigated more in depth. Currently, farmers' complaints about rabbit damage are widespread in central-southern Iberia (Rios-Saldaña et al. 2013), even in areas where rabbit numbers are moderate (Delibes-Mateos et al. 2014b). However, it is not well known how the magnitude of rabbit damage to crops varies with rabbit abundance. It is also necessary to investigate how other factors may modulate the damage-abundance relationship. In this sense, Barrio et al. (2013) suggested that in some areas in which rabbits are not very abundant, their damage to crops is largely determined by the low availability of alternative food sources as a consequence of agriculture intensification. Many other factors like crop types, climatic conditions, soil characteristics, or wildlife community composition (including rabbit competitors and/or rabbit predators) may be also involved in damage caused by rabbits, but their roles have not been assessed yet. Information is lacking about which tools are most cost-effective to manage pest rabbits in Iberian farmland areas (but see for example Barrio et al. 2010a), and in particular the effectiveness of hunting to reduce rabbit numbers and mitigate their damage has not been evaluated scientifically, although this is the main method currently employed in Spain and Portugal (Delibes-Mateos et al. 2014b). Economic studies are encouraged to evaluate, for example, if the investment carried out by hunters and farmers to reduce rabbit damage compensates the monetary loss that this lagomorph causes to farmers. More research is also needed to better understand the social context around rabbit damage. This should include the assessment of the attitudes of involved stakeholders (i.e., farmers, hunters, conservationists, and policy makers) towards rabbit damage and towards the potential use of alternative methodologies to mitigate such damage. In conclusion, a better understanding of all these issues is essential for the effective management of rabbit pests in farmland areas within the species' native range. Importantly, this knowledge could be also useful for the management of invasive rabbits in other European regions.

Ferreira (2012) revised the rabbit scientific literature in the Iberian Peninsula and found that near one third of the publications dealt with the rabbit's ecology. According to her review, most research on natural rabbit populations focused on only some few aspects (e.g., feeding and reproductive strategies), while others have been left more unattended (Ferreira 2012). For example, few information exists on seasonal home range sizes and dispersion parameters in natural populations (but see Lombardi et al. 2007). In general, the interaction between rabbit population dynamics and environmental and anthropogenic factors is not well known yet. In fact, on most occasions, we are not able to explain why rabbits are abundant in one particular location while they are absent few kilometers apart. As mentioned previously in this chapter, there is still considerable debate about the effects of predation on rabbit population dynamics (see section "Population Ecology"). Experimental studies are needed to test the predation regulation hypothesis and ideally these should be conducted at a large scale. Similarly, more efforts have to be done to investigate the impact of human hunting on rabbit populations, particularly in areas where this lagomorph has declined and it currently reaches moderate to low numbers. Despite the increasing number of publications of brilliant research on rabbit diseases, there are many aspects that still require further investigation. For example, we do not know well how the viruses evolve and increase their pathogenicity. In addition, it would be interesting to assess how these viruses spread, paying particular attention on the role played by other species including humans (Rouco et al. 2019a). Furthermore, we encourage the investigation of interactions between rabbit viruses and how such synergies impact the survival of rabbits in Europe. A good example in this regard comes from a recent study conducted in Australia which shows that previous exposure to myxoma virus reduces rabbit survival during RHD outbreaks (Barnett et al. 2018). Large-scale assessment of the synergic impact of different factors like predation, hunting, or diseases on rabbit population dynamics would be also very welcome. Overall, all this information is critical for the implementation of successful management interventions. In relation to this, strong empirical evidence is still lacking to support the usefulness and impact of some of the most widespread management actions, such as predator control or restocking operations (but see Carro et al. 2019). In fact, considering the rabbit is a highly managed species, management studies are still relatively scarce (Ferreira 2012). For example, to the best of our knowledge, there is only one study that investigates the cost-effectiveness of tools frequently used to enhance rabbit numbers in Iberia (Ferreira et al. 2014b). It is essential to know which measures are more cost-effective, particularly at large scale, to design management plans devoting human and economic resources more efficiently.

Throughout this chapter, we emphasized the existence of several differences, namely in genetics, reproduction, behavior, parasitology morphology, and population trends, between the two European rabbit subspecies. In addition, unpublished data suggest that there may be other ecological aspects that would confirm such differentiation between rabbit subspecies. These include, for example, different activity patterns, use of warrens, and spatial and antipredatory behavior, among others. In our opinion, all these differences support the separation of rabbit subspecies as different species, as already proposed (Delibes-Mateos et al. 2018a). This species split is critically important to design and implement urgently a management program for the conservation of the *O. c. algirus* subspecies, since its geographic range is restricted to Portugal and Southern Spain, and most populations of this subspecies are currently declining (e.g., Delibes-Mateos et al. 2014a; Guerrero-Casado et al. 2016; Monterroso et al. 2016; Vaquerizas et al. 2020).

Finally, the potential impact of climate change on rabbit populations in Europe has received very little attention in the scientific literature. As discussed previously in this chapter, Tablado and Revilla (2012) suggested that climate change will have negative effects on rabbit populations in south-western Europe, while conditions for rabbits will be improved in north-eastern Europe. These scenarios are worrisome not only because native rabbit populations will be highly compromised, but also because the predicted colonization by rabbits of new areas will likely bring catastrophic effects both for ecosystems and human interests like agriculture. From this perspective, further research is urgently needed to explore more in depth the effects of climate change on rabbit population dynamics.

References

Abrantes J, van der Loo W, Le Pendu J et al (2012) Rabbit haemorrhagic disease (RHD) and rabbit haemorrhagic disease virus (RHDV): a review. Vet Res 43:12

Acevedo P, Ruiz-Fons F, Vicente J et al (2008) Estimating red deer abundance at a wide range of management situations in Mediterranean habitats. J Zool 276:37–47

Aebischer NJ, Davey PD, Kingdon NG (2011) National Gamebag census: mammal trends to 2009. Game & Wildlife Conservation Trust, Fordingbridge. Available via http://www.gwct.org.uk/ngcmammals. Accessed 15 June 2019

Allan JC, Craig PS, Sherington J et al (1999) Helminth parasites of the wild rabbit *Oryctolagus cuniculus* near Malham Tarn, Yorkshire, UK. J Helminth 73:289–229

Almería S, Calvete C, Pagés A et al (2004) Factors affecting the seroprevalence of *Toxoplasma gondii* infection in wild rabbits (*Oryctolagus cuniculus*) from Spain. Vet Parasit 123:265–270

Alves PC, Ferrand N, Hackländer K (eds) (2008) Lagomorph biology: evolution, ecology and conservation. Springer, Berlin

Alves JM, Carneiro M, Cheng JY et al (2019) Parallel adaptation of rabbit populations to myxoma virus. Science 363:1319–1326

Angulo E, Villafuerte R (2003) Modelling hunting strategies for the conservation of wild rabbit populations. Biol Conserv 115:291–301

Apollonio M, Andersen R, Putman R (eds) (2010) European ungulates and their management in the 21st century. Cambridge University Press, Cambridge, UK

Arthur CP, Louzis C (1988) A review of myxomatosis among rabbits in France. Rev Sci Tech 7:959–976

Augusteyn RC (2007) On the relationship between rabbit age and lens dry weight: improved determination of the age of rabbits in the wild. Mol Vis 13:2030–2034

Baker DG (ed) (2008) Flynn's parasites of laboratory animals, 2nd edn. Wiley-Blackwell, Hoboken

Bakker ES, Reiffers RC, Olff H et al (2005) Experimental manipulation of predation risk and food quality: effect on grazing behaviour in a central-place foraging herbivore. Oecologia 146:157–167

Barnett LK, Prowse TAA, Peacock DE et al (2018) Previous exposure to myxoma virus reduces survival of European rabbits during outbreaks of rabbit haemorrhagic disease. J Appl Ecol 55:2954–2962

Barrio IC, Bueno GC, Tortosa FS (2010a) Alternative food and rabbit damage in vineyards of southern Spain. Agric Ecosys Environ 138:51–54

Barrio IC, Acevedo P, Tortosa FS (2010b) Assessment of methods for estimating wild rabbit population abundance in agricultural landscapes. Eur J Wildlife Res 56:335–340

Barrio IC, Bueno CG, Villafuerte R et al (2013) Rabbits, weeds and crops: can agricultural intensification promote wildlife conflicts in semiarid agro-ecosystems? J Arid Environ 90:1–4

Bautista A, Rödel HG, Monclús R et al (2015a) Intrauterine position as a predictor of postnatal growth and survival in the rabbit. Physiol Behav 138:101–106

Bautista A, Zepeda JA, Reyes-Meza V et al (2015b) Contribution of within-litter interactions to individual differences in early postnatal growth in the domestic rabbit. Anim Behav 108:145–153

Bell DJ, Webb NJ (1991) Effects of climate on the reproduction in the European wild rabbit (*Oryctolagus cuniculus*). J Zool 224:639–648

Beltrán JF (1991) Temporal abundance pattern of the wild rabbit in Doñana, SW Spain. Mammalia 55:591–599

Bertó-Moran A, Pacios I, Serrano E et al (2013) Coccidian and helminth infections influence the sero-conversion against myxoma and rabbit haemorrhagic disease viruses in European wild rabbit. J Wildl Dis 49:10–17

Blasco S, Torres J, Feliu C et al (1996) The helminthfauna of *Oryctolagus cuniculus* (Linnaeus, 1758) in the Iberian Peninsula faunistic and ecological considerations. Parasite 4:327–333

Boag B (1987) Reductions in numbers of the wild rabbit (*Oryctolagus cuniculus*) due to changes in agricultural practices. Crop Prot 6:347–351

Boag B (1988) Observations on the seasonal incidence of myxomatosis and its interactions with helminth parasites in the European rabbit (*Oryctolagus cuniculus*). J Wildl Dis 24:450–455

Boag B, Lello J, Fenton A et al (2001) Patterns of parasite aggregation in the wild European rabbit (*Oryctolagus cuniculus*). Int J Parasitol 31:1421–1428

Boag B, Hernandez AD, Cattadori IM (2013) Observations on the epidemiology and interactions between myxomatosis, coccidiosis and helminth parasites in a wild rabbit population in Scotland. Eur J Wildlife Res 59: 557–562

Brambell FWR (1942) Intrauterine mortality of the wild rabbit, *Oryctolagus cuniculus*, (L.). Proc Zool Soc Lond 130:462–479

Brambell FWR (1944) The reproduction of the wild rabbit, *Oryctolagus cuniculus*, (L.). Proc Zool Soc Lond 114: 1–45

Branco M, Ferrand N, Monnerot M (2000) Phylogeography of the European rabbit (*Oryctolagus cuniculus*) in the Iberian Peninsula inferred from RFLP analysis of the cytochrome b gene. Heredity 85:307–317

Branco M, Monnerot M, Ferrand N et al (2002) Postglacial dispersal of the European rabbit (*Oryctolagus cuniculus*) on the Iberian Peninsula reconstructed from nested clade and mismatch analyses of mitochondrial DNA genetic variation. Evolution 56:792–803

Cabezas S, Calvete C, Moreno S (2006) Vaccination success and body condition in the European wild rabbit: application for conservation strategies. J Wildl Manag 70:1125–1131

Cabezas S, Blas J, Marchant TA et al (2007) Physiological stress levels predict survival probabilities in wild rabbits. Horm Behav 51:313–320

Cabezas-Díaz S, Virgós E, Mangas JG et al (2011) The presence of a "competitor pit effect" compromises wild rabbit (*Oryctolagus cuniculus*) conservation. Anim Biol 61:319–334

Callou C (2003) De la garenne au clapier: étude archéozoologique du lapin en Europe occidentale. Muséum National d'Histoire naturelle, Paris

Calvete C (2006) The use of immunization programs in wild populations: modelling effectiveness of vaccination campaigns against rabbit haemorrhagic disease. Biol Conserv 130:290–300

Calvete C, Estrada R, Angulo E et al (2004) Habitat factors related to wild rabbit conservation in an agricultural landscape. Landsc Ecol 19:531–542

Cancellotti FM, Renzi M (1991) Epidemiology and current situation of viral haemorrhagic disease of rabbits and the European brown hare syndrome in Italy. Rev Sci Tech OIE 10:409–422

Carneiro M, Ferrand N, Nachman MW (2009) Recombination and speciation: loci near centromeres are more differentiated than loci near telomeres between subspecies of the European rabbit (*Oryctolagus cuniculus*). Genetics 181:593–606

Carneiro M, Baird SJ, Afonso S et al (2013) Steep clines within a highly permeable genome across a hybrid zone between two subspecies of the European rabbit. Mol Ecol 22:2511–2525

Carneiro M, Rubin CJ, Di Palma F et al (2014a) Rabbit genome analysis reveals a polygenic basis for phenotypic change during domestication. Science 345:1074–1079

Carneiro M, Albert FW, Afonso S et al (2014b) The genomic architecture of population divergence between subspecies of the European rabbit. PLoS Genet 10: e1003519

Carpio AJ, Guerrero-Casado J, Ruiz-Aizpurua L et al (2014) The high abundance of wild ungulates in a Mediterranean region: is this compatible with the European rabbit? Wildl Biol 20:161–166

Carro F, Ortega M, Soriguer RC (2019) Is restocking a useful tool for increasing rabbit densities? Global Ecol Conserv 17:e00560

Caruso S, Siracusa AM (2001) Factors affecting the abundance of wild rabbits (*Oryctolagus cuniculus*) in agroecosystems of the Mount Etna Park. Hystrix 12:45–49

Cattadori IM, Albert R, Boag B (2007) Variation in host susceptibility and infectiousness generated by co-infection: the myxoma-*Trichostrongylus retortaeformis* case in wild rabbits. J R Soc Interface 4:831–840

Cerri J, Ferretti M, Tricarico E (2016) Are you aware of what you are doing? Asking Italian hunters about an invasive species they are introducing. J Nat Conserv 34:15–23

Chalmers RM, Robinson G, Elwin K et al (2009) Cryptosporidium rabbit genotype, a newly identified human pathogen. Emerg Infect Dis 15:829–830

Chapuis JL (1990) Comparison of the diets of two sympatric lagomorphs, *Lepus europaeus* (Pallas) and *Oryctolagus cuniculus* (L.) in an agroecosystem of the Ile-de-France. Z Säugetierkd 55:176–185

Cooke BD (1977) Factors limiting the distribution of the wild rabbit in Australia. Proc Ecol Soc Aust 10:113–120

Cooke BD (2012) Rabbits: manageable environmental pests or participants in new Australian ecosystems? Wildl Res 39:279–289

Cowan DP (1987a) Aspects of the social organization of the European wild rabbit (*Oryctolagus cuniculus*). Ethology 75:197–210

Cowan DP (1987b) Group living in the European rabbit (*Oryctolagus cuniculus*) – mutual benefit or resource localization. J Anim Ecol 56:779–795

Cubas J, Irl SD, Villafuerte R et al (2019) Endemic plant species are more palatable to introduced herbivores than non-endemics. Proc R Soc B 286:20190136

Dalton KP, Nicieza I, Balseiro A et al (2012) Variant rabbit hemorrhagic disease virus in young rabbits, Spain. Emerg Infect Dis 18:2009–2012

Dalton KP, Nicieza I, Abrantes J et al (2014) Spread of new variant RHDV in domestic rabbits on the Iberian Peninsula. Vet Microbiol 169:67–73

Davis CM, Roth VL (2008) The evolution of sexual size dimorphism in cottontail rabbits (*Sylvilagus*, Leporidae). Biol J Linn Soc 95:141–156

De Marfà R (2008) *Oryctolagus giberti* n. sp. (Lagomorpha, Mammalia) du Pléistocène inférieur de Cueva Victoria (Murcie, Espagne). CR Palevol 7: 305–313

Delibes-Mateos M, Delibes M, Ferreras P et al (2008a) Key role of European rabbits in the conservation of the Western Mediterranean Basin Hotspot. Conserv Biol 22:1106–1117

Delibes-Mateos M, Ferreras P, Villafuerte R (2008b) Rabbit populations and game management: the situation after 15 years of rabbit haemorrhagic disease in Central-Southern Spain. Biodivers Conserv 17:559–574

Delibes-Mateos M, Ferreras P, Villafuerte R (2009) European rabbit population trends and associated factors: a review of the situation in the Iberian Peninsula. Mammal Rev 39:124–140

Delibes-Mateos M, Farfán MA, Olivero J et al (2010) Land-use changes as a critical factor for long-term wild rabbit conservation in the Iberian Peninsula. Environ Conserv 37:169–176

Delibes-Mateos M, Díaz-Fernández S, Ferreras P et al (2013) The role of economic and social factors driving predator control in small game estates in Central Spain. Ecol Soc 38:18

Delibes-Mateos M, Ferreira C, Carro F et al (2014a) Ecosystem effects of variant rabbit hemorrhagic disease virus emergence, Iberian Peninsula. Emerg Infect Dis 20:2166–2168

Delibes-Mateos M, Ferreira C, Rouco C et al (2014b) Conservationists, hunters and farmers: the European rabbit *Oryctolagus cuniculus* management conflict in the Iberian Peninsula. Mammal Rev 40:190–203

Delibes-Mateos M, Villafuerte R, Cooke B et al (2018a) *Oryctolagus cuniculus* (Linnaeus, 1758). In: Smith AT, Johnston CH, Alves PC et al (eds) Lagomorphs: pikas, rabbits and hares of the world. John Hopkins University Press, Baltimore, pp 99–104

Delibes-Mateos M, Farfán MA, Rouco C et al (2018b) A large-scale assessment of European rabbit damage to agriculture in Spain. Pest Manag Sci 74:111–118

Devillard S, Aubineau J, Berger F et al (2008) Home range of the European rabbit (*Oryctolagus cuniculus*) in three contrasting French populations. Mammal Biol 73: 128–137

Dudzinski ML, Mykytowycz R (1960) Analysis of weights and growth rates of an experimental colony of wild rabbits, *Oryctolagus cuniculus* (L.). Wildl Res 5:102–115

Dudzinski ML, Mykytowycz R (1961) The eye lens as an indicator of age in the wild rabbit in Australia. Wildl Res 6:156–159

Eccard JA, Rödel HG (2011) Optimising temperament through litter size in iteroparous mammals in seasonal environments. Dev Psychobiol 53:585–591

Eira C, Torres J, Miquel J et al (2007) The helminth parasites of the wild rabbit *Oryctolagus cuniculus* and their effect on host condition in Dunas de Mira, Portugal. J Helminth 81:239–246

Erlinge S, Göransson G, Högstedt G et al (1984) Can vertebrate predators regulate their prey? Am Nat 123: 125–133

Fa JE, Sharples CM, Bell DJ et al (1994) An individual-based model of rabbit viral haemorrhagic disease in European wild rabbits (*Oryctolagus cuniculus*). Ecol Model 144:121–128

Fa JE, Stewart JR, Lloveras L et al (2013) Rabbits and hominin survival in Iberia. J Hum Evol 64:233–241

Fenner F, Ratcliffe F (1965) Myxomatosis. Cambridge University Press, Cambridge, UK

Fernández de Simón J, Díaz-Ruíz F, Cirilli F et al (2011) Towards a standardized index of European rabbit abundance in Iberian Mediterranean habitats. Eur J Wildl Res 57:1091–1100

Fernández de Simón J, Díaz-Ruíz F, Rodríguez de la Cruz M et al (2015) Can widespread generalist predators affect keystone prey? A case study with red foxes and European rabbits in their native range. Popul Ecol 57: 591–599

Ferrand N, Rocha J (1992) Demonstration of serum albumin (ALB) polymorphism in wild rabbit (*Oryctolagus cuniculus*) by means of isoelectric focusing. Anim Genet 23:275–278

Ferreira C (2012) European rabbit research in the Iberian Peninsula: state of the art and future perspectives. Eur J Wildl Res 58:885–895

Ferreira C, Alves PC (2009) Influence of habitat management on the abundance and diet of wild rabbit (*Oryctolagus cuniculus algirus*) populations in Mediterranean ecosystems. Eur J Wildl Res 55:487–496

Ferreira A, Ferreira AJ (2014) Post-weaning growth of endemic Iberian wild rabbit subspecies, *Oryctolagus cuniculus algirus*, kept in a semi-extensive enclosure: implications for management and conservation. World Rabbit Sci 22:129–136

Ferreira C, Ramírez E, Castro F et al (2009) Field experiment vaccination campaigns against myxomatosis and their effectiveness in the wild. Vaccine 27:6998–7002

Ferreira C, Rodrigues T, Leitão M et al (2012) Gestão de Recursos Cinegéticos no Arquipélago dos Açores – O Coelho-bravo. CIBIO-UP e Direcção Regional dos Recursos Florestais, Porto

Ferreira C, Villafuerte R, Villar N et al (2014a) Experimental study on the effect of cover and vaccination on the survival of juvenile European rabbits. Popul Ecol 56: 195–202

Ferreira C, Touza J, Rouco C et al (2014b) Habitat management as a generalized tool to boost European rabbit (*Oryctolagus cuniculus*) populations in the Iberian Peninsula: a cost-effectiveness analysis. Mammal Rev 44: 30–43

Ferreira CC, Castro F, Piorno V et al (2015) Biometric analysis reveals major differences between the two subspecies of the European rabbit. Biol J Linn Soc 116:106–116

Ferrer M, Negro JJ (2004) The near extinction of two large European predators: super specialists pay a price. Conserv Biol 18:344–349

Fordham DA, Sinclair RG, Peacock DE et al (2012) European rabbit survival and recruitment are linked to epidemiological and environmental conditions in their exotic range. Austral Ecol 37:945–957

Foronda P, Valladares B, Lorenzo-Morales J et al (2003) Helminths of the wild rabbit (*Oryctolagus cuniculus*) in Macaronesia. J Parasitol 89:952–957

Frank R, Kuhn T, Mehlhorn H et al (2013) Parasites of wild rabbits (*Oryctolagus cuniculus*) from an urban area in Germany, in relation to worldwide results. Parasitol Res 112:4255

Fraser KW (1988) Reproductive biology of rabbits, *Oryctolagus cuniculus* (L.), in Central Otago, New Zealand. New Zeal J Ecol 11:79–88

Friedman MH (1929) The mechanism of ovulation in the rabbit. I. the demonstration of a humoral mechanism. Am J Phys 89:438–442

García-Torres E, Hudson R, Castelán F et al (2015) Differential metabolism of brown adipose tissue in newborn rabbits in relation to position in the litter huddle. J Therm Biol 51:33–41

Gea-Izquierdo G, Muñoz-Igualada J, San Miguel-Ayanz A (2005) Rabbit warren distribution in relation to pasture communities in Mediterranean habitats: consequences for management of rabbit populations. Wildl Res 32: 723–731

Geraldes A, Carneiro M, Delibes-Mateos M et al (2008) Reduced introgression of the Y chromosome between subspecies of the European rabbit (*Oryctolagus cuniculus*) in the Iberian Peninsula. Mol Ecol 17: 4489–4499

Gilbert C, Blanc S, Giroud S et al (2007) Role of huddling on the energetic of growth in a newborn altricial mammal. Am J Physiol Reg I 293:R867–R876

Gonçalves H, Alves PC, Rocha A (2002) Seasonal variation in the reproductive activity of the wild rabbit (*Oryctolagus cuniculus algirus*) in a Mediterranean ecosystem. Wildl Res 29:165–173

Guerrero-Casado JM, Ruiz-Aizpurua L, Carpio AJ et al (2013) Factors affecting wild rabbit production within extensive breeding enclosures: how can we optimize efforts? World Rabbit Sci 21:193–199

Guerrero-Casado J, Carpio AJ, Ruiz-Aizpurua L et al (2016) Recent negative trends of wild rabbit populations in southern Spain after the arrival of the new variant of the rabbit hemorrhagic disease virus RHDV2. Mammal Biol 81:361–364

Hardy C, Callou C, Vigne JD et al (1995) Rabbit mitochondrial DNA diversity from prehistoric to modern times. J Mol Evol 40:227–237

Harris S (2017) The breeding bird survey 2017. Technical report. BTO, RSPB and JNCC. Available via http://www.bto.org/bbs-report. Accessed 15 June 2019

Hofing GL, Kraus AL (1994) Arthropod and helminth parasites. In: Manning PJ, Ringler DH, Newcomer CE (eds) The biology of the laboratory rabbit. Academic, London, pp 231–258

Hsu TC, Benirschke K (1967) An atlas of mammalian chromosomes. Springer, New York

Hudson R, Distel H (1983) Nipple location by newborn rabbits: evidence for pheromonal guidance. Behaviour 82:260–275

Hudson R, Bilkó Á, Altbäcker V (1996) Nursing, weaning and the development of independent feeding in the rabbit (*Oryctolagus cuniculus*). Z Säugetierkd 61:39–48

Hulbert IAR, Iason GR, Elston DA et al (1996) Home-range sizes in a stratified upland landscape of two lagomorphs with different feeding strategies. J Appl Ecol 33:1479–1488

Hull J, Hull D (1982) Behavioral thermoregulation in newborn rabbits. J Comp Psychol 96:143–147

Jaksic FB, Soriguer RC (1981) Predation upon the European rabbit (*Oryctolagus cuniculus*) in Mediterranean habitats of Chile and Spain: a comparative analysis. J Anim Ecol 50:269–281

Katona K, Biro Z, Hahn I et al (2004) Competition between European hare and European rabbit in a lowland area, Hungary: a long-term ecological study in the period of rabbit extinction. Folia Zool 53:255–268

Knauer F, Küchenhoff H, Pilz S (2010) A statistical analysis of the relationship between red fox *Vulpes vulpes* and its prey species (grey partridge *Perdix perdix*, Brown hare *Lepus europaeus* and rabbit *Oryctolagus cuniculus*) in western Germany from 1958 to 1998. Wildl Biol 16:56–65

Kolb HH (1991) Use of burrows and movements by wild rabbits (*Oryctolagusc cuniculus*) in an area of hill grazing and forestry. J Appl Ecol 28:892–905

Kolb HH (1994) Rabbit *Oryctolagus cuniculus* populations in Scotland since the introduction of myxomatosis. Mammal Rev 24:41–48

Kontsiotis VJ, Bakalaudis DE, Tsiompanoudis AC (2013) Key factors determining the seasonal population growth rate of European wild rabbits and their implications for management. Eur J Wildl Res 59:495–503

Kontsiotis VJ, Bakaloudis DE, Tsiompanoudis AC et al (2014) Body condition variation of wild rabbit population in the north-East Mediterranean island of Lemnos – Greece. Folia Zool 63:87–94

Kontsiotis VJ, Bakalaudis DE, Liordos V (2018) Impact of European wild rabbits foraging in different habitat and vegetation types in an insular environment. Mammalia 82:193–196

Korstanje R, O'Brien PC, Yang F et al (1999) Complete homology maps of the rabbit (*Oryctolagus cuniculus*) and human by reciprocal chromosome painting. Cytogenet Cell Genet 86:317–322

Kufner MB (1986) Tamaño, actividad, densidad relativa y preferencia de hábitat de los pequeños y medianos mamíferos de Doñana, como factores condicionantes de su tasa de depredación. PhD thesis, Universidad Autónoma de Madrid, Madrid

Künkele J, von Holst D (1996) Natal dispersal in the European wild rabbit. Anim Behav 51:1047–1059

Le Gall-Reculé G, Lavazza A, Marchandeau S et al (2013) Emergence of a new lagovirus related to rabbit haemorrhagic disease virus. Vet Res 44:1–13

Le Pendu J, Abrantes J, Bertagnoli S et al (2017) Proposal for a unified classification system and nomenclature of lagoviruses. J Gen Virol 68:1658–1666

Leach K, Montgomery WI, Reid N (2015) Biogeography, macroecology and species traits mediate competitive interactions in the order Lagomorpha. Mammal Rev 45:88–102

Leach K, Montgomery WI, Reid N (2017) Characterizing biotic interactions within the order Lagomorpha using joint species distribution models at 3 different spatial scales. J Mammal 98:1434–1442

Lees AC, Bell DJ (2008) A conservation paradox for the 21st century: the European wild rabbit *Oryctolagus cuniculus*, an invasive alien and endangered native species. Mammal Rev 38:304–320

Liu SJ, Xue HP, Pu BQ et al (1984) A new viral disease in rabbit. Anim Husb Vet Med 16:253–255

Lloyd HG, MacCowan D (1968) Some observation on the breeding burrows of the wild rabbit *Oryctolagus cuniculus* on the island of Stockholm. J Zool 156: 540–549

Lockley RM (1961) Social structure and stress in the rabbit warren. J Anim Ecol 30:385–423

Lombardi L, Fernández N, Moreno S et al (2003) Habitat-related differences in rabbit (*Oryctolagus cuniculus*) abundance, distribution and activity. J Mammal 84: 26–36

Lombardi L, Fernández N, Moreno S (2007) Habitat use and spatial behaviour in the European rabbit in three Mediterranean environments. Basic Appl Ecol 8:453–463

Lozano J, Virgós E, Cabezas-Díaz S et al (2007) Increase of large game species in Mediterranean areas: is the European wildcat (*Felis sivestris*) facing a new threat? Biol Conserv 138:321–329

Macdonald DW (2007) The Encyclopedia of mammals, 3rd edn. Oxford University Press, Oxford

Manning PJ, Ringler DH, Newcomer CE (1994) The biology of the laboratory rabbit, 2nd edn. American College of Laboratory Animal Medicine/Academic, San Diego

Marques C, Mathias M (2001) The diet of the European wild rabbit, *Oryctolagus cuniculus* (L.), on different coastal habitats of Central Portugal. Mammalia 65: 437–449

Martínez-Gómez M, Juárez M, Distel H et al (2004) Overlapping litters and reproductive performance in the domestic rabbit. Physiol Behav 82:629–636

Mátrai K, Altbäcker V, Hahn I (1998) Seasonal diet of rabbits and their browsing effect on juniper in Bugac juniper Forest (Hungary). Acta Theriol 43:107e112

McKillop IG, Phillips KV, Ginella SGV (1992) Effectiveness of two types of electric fences for excluding European wild rabbits. Crop Prot 11:279–285

Mead-Briggs AR, Page RJC (1975) Records of Anoplocephaline cestodes from wild rabbits and hares collected throughout Great Britain. J Helminthol 49: 49–56

Mead-Briggs AR, Vaughan JA (1973) The incidence of Anoplocephaline cestodes in a population of rabbits in Surrey, England. Parasitology 70:103–118

Millán J, Casáis R, Delibes-Mateos M et al (2012) Widespread exposure to *Sarcoptes scabiei* in wild European rabbits (*Oryctolagus cuniculus*) in Spain. Vet Parasitol 183:323–329

Monterroso P, Garrote G, Serronha A et al (2016) Disease-mediated bottom-up regulation: an emergent virus affects a keystone prey, and alters the dynamics of trophic webs. Sci Rep 6:36072

Moreno S, Villafuerte R (1995) Traditional management of scrubland for the conservation of rabbits *Oryctolagus cuniculus* and their predators in Doñana National Park. Biol Conserv 73:81–85

Moreno S, Villafuerte R, Delibes M (1996) Cover is safe during the day but dangerous at night: the use of vegetation by European wild rabbits. Can J Zool 74:1656–1660

Moreno S, Villafuerte R, Cabezas S et al (2004) Wild rabbit restocking for predator conservation in Spain. Biol Conserv 118:183–193

Moreno S, Beltrán JF, Cotilla I et al (2007) Long-term decline of the European wild rabbit (*Oryctolagus cuniculus*) in South-Western Spain. Wildl Res 34: 652–658

Myers K (1958) Further observations on the use of field enclosures for the study of the wild rabbit, *Oryctolagus cuniculus* (L.). Wildl Res 3:40–49

Mykytowycz R, Fullagar PJ (1973) Effects of social environment on reproduction in the rabbit, *Oryctolagus cuniculus* (L.). J Reprod Fertil 19:503–522

Mykytowycz R, Ward MM (1971) Some reactions of nestlings of the wild rabbit, *Oryctolagus cuniculus* (L.), when exposed to natural rabbit odours. Forma Funct 4: 137–148

Navarro-Castilla A, Sánchez-González B, Barja I (2019) Latrine behaviour and faecal corticosterone metabolites as indicators of habitat-related responses of wild rabbits to predation risk. Ecol Indic 97:175–182

Palomares F (2003) The negative impact of heavy rains on the abundance of a Mediterranean population of European rabbits. Mammal Biol 68:224–234

Parer I (1992) Dispersal of the wild rabbit, *Oryctolagus cuniculus*, at Urana in New South Wales. Aust Wildl Res 9:427–441

Parer L, Libke JA (1985) Distribution of rabbit, *Oryctolagus cuniculus*, warrens in relation to soil type. Aust Wild Res 12:387–405

Parer I, Fullagar PJ, Malafant KW (1987) The history and structure of a large warren of the rabbit, *Oryctolagus cuniculus*, at Canberra, A.C.T. Aust Wildl Res 14: 505–513

Peacock DE, Sinclair RG (2009) Longevity record for a wild European rabbit (*Oryctolagus cuniculus*) from South Australia. Aust Mammal 31:65–66

Pelletier M, Cochard D, Boudadi-Maligne M et al (2015) Lower Pleistocene leporids (Lagomorpha, Mammalia) in western Europe: new data from the Bois-de-Riquet (Lézignan-la-Cèbe, Hérault, France). CR Palevol 14: 371–385

Penteriani V, Kuparinen A, Delgado MM et al (2013) Responses of a top and a meso predator and their prey to moon phases. Oecologia 173:753–766

Pinheiro A, Almeida T, Esteves PJ (2019) Genetic diversity of *IGHM* and *IGHE* in the Leporids revealed different patterns of diversity in the two European Rabbit Subspecies (*O. cuniculus algirus* and *O. c. cuniculus*). Animals 9:955

Piorno V, Villafuerte R, Branco M, Carneiro M, Ferrand N, Alves PC (2015) Low persistence in nature of captive reared rabbits after restocking operations. Eur J Wildlife Res 61:591–599

Rafati N, Blanco-Aguiar JA, Rubin CJ et al (2018) A genomic map of clinal variation across the European rabbit hybrid zone. Mol Ecol 27:1457–1478

Reglero M, Vicente J, Rouco C et al (2007) *Trypanosoma spp.* infection in wild rabbits (*Oryctolagus cuniculus*) during a restocking program in southern Spain. Vet Parasitol 149:178–184

Ríos-Saldaña CA, Delibes-Mateos M, Castro F et al (2013) Control of European rabbit in Central Spain. Eur J Wildl Res 59:573–580

Rödel HG, Bora A, Kaetzke P, Khaschei M, Hutzelmeyer H, Zapka M, von Holst D (2005) Timing of breeding and reproductive performance of female European rabbits in response to winter temperature and body mass. Can J Zool 83:935–942

Rödel HG (2005) Winter feeding behaviour of European rabbits in a temperate zone habitat. Mammal Biol 70: 300–306

Rödel HG, Dekker JJA (2012) Influence of weather factors on population dynamics of two lagomorph species based on hunting bag records. Eur J Wildl Res 58: 923–932

Rödel HG, Starkloff A (2014) Social environment and weather during early life influence loads of an intestinal parasite in a group-living small mammal. Oecologia 176:389–398

Rödel HG, Bora A, Kaetzke P et al (2004a) Over-winter survival in subadult European rabbits: weather effects, density dependence, and the impact of individual characteristics. Oecologia 140:566–576

Rödel HG, Bora A, Kaiser J et al (2004b) Density-dependent reproduction in the European rabbit: a consequence of individual response and age-dependent reproductive performance. Oikos 104:529–539

Rödel HG, Prager G, Stefanski V et al (2008a) Separating maternal and litter-size effects on early postnatal growth in two species of altricial small mammals. Physiol Behav 93:826–834

Rödel HG, Hudson R, von Holst D (2008b) Optimal litter size for individual growth of European rabbit pups depends on their thermal environment. Oecologia 155:677–689

Rödel HG, Starkloff A, Bautista A et al (2008c) Infanticide and maternal offspring defence in European rabbits under natural breeding conditions. Ethology 114:22–31

Rödel HG, von Holst D, Kraus C (2009a) Family legacies: short- and long-term fitness consequences of early-life conditions in female European rabbits. J Anim Ecol 78: 789–797

Rödel HG, Starkloff A, Seltmann MW et al (2009b) Causes and predictors of nest mortality in a European rabbit population. Mammal Biol 74:198–209

Rödel HG, Zapka M, Talke S et al (2015) Survival costs of fast exploration during juvenile life in a small mammal. Behav Ecol Sociobiol 69:205–217

Rödel HG, Valencak TG, Handrek A et al (2016) Paying the energetic costs of reproduction: reliance on postpartum foraging and stored reserves. Behav Ecol 27: 748–756

Rödel HG, Bautista A, Roder M et al (2017) Early development and the emergence of individual differences in behavior among littermates of wild rabbit pups. Physiol Behav 173:101–109

Rosell JM (2000) Enfermedades del conejo. Editorial Mundi-Prensa, Madrid

Rosell JM, de la Fuente LF (2012) Body weight and body condition of breeding rabbits in commercial units. J Anim Sci 90:3252–3258

Ross S, Zarrow MX, Sawin PB et al (1963) Maternal behaviour in the rabbit under semi-natural conditions. Anim Behav 11:283–285

Rouco C, Ferreras P, Castro F et al (2008) The effect of exclusion of terrestrial predators on short-term survival of translocated European wild rabbits. Wildl Res 3: 625–632

Rouco C, Villafuerte R, Castro F et al (2011a) Responses of naïve and experienced European rabbits to predator odour. Eur J Wildl Res 57:395–398

Rouco C, Villafuerte R, Castro F et al (2011b) Effect of artificial warren size on a restocked European wild rabbit population. Anim Conserv 14:117–123

Rouco C, Moreno S, Santoro S (2016) A case of low success of blind vaccination campaigns against myxomatosis and rabbit haemorrhagic disease on survival of adult European wild rabbits. Prev Vet Med 133:108–113

Rouco C, Abrantes J, Serronha A et al (2018) Epidemiology of RHDV2 (*Lagovirus europaeus*/GI.2) in free living European rabbits in Portugal. Transbound Emerg Dis 65:372–382

Rouco C, Aguayo-Adán JA, Santoro S et al (2019a) Worldwide rapid spread of the novel rabbit haemorrhagic disease virus (GI.2/RHDV2/b). Transbound Emerg Dis 66:1762–1764

Rouco C, Farfán MA, Olivero J et al (2019b) Favourability for the presence of wild rabbit warrens in motorway verges: implications for the spread of a native agricultural pest species. Ecol Indic 104:398–404

Rouco C, Abrantes J, Delibes-Mateos M (2020) Lessons from viruses that affect lagomorphs. Science 369:386

Ruiz-Aizpurua L, Guerrero-Casado J, Carpio AJ et al (2014) High rabbit abundance proves detrimental to the population growth rate in European rabbit (*Oryctolagus cuniculus* L.) extensive breeding enclosures. World Rabbit Sci 22:179–186

Santilli F, Bagliacca M, Paci G (2015) Density and habitat use of sympatric Brown hares and European rabbits in a Mediterranean farmland area of Tuscany (Central Italy). Ethol Ecol Evol 27:233–243

Saunders WB (2005) Rabbits. In: O'Malley B (ed) Clinical anatomy and physiology of exotic species. Structure and function of mammals, birds, reptiles, and amphibians. Saunders, Edinburgh, pp 173–195

Schaal B, Coureaud G, Langlois D et al (2003) Chemical and behavioural characterization of the rabbit mammary pheromone. Nature 424:68–72

Seltmann MW, Ruf T, Rödel HG (2009) Effects of body mass and huddling on resting metabolic rates of post-weaned European rabbits under different simulated weather conditions. Funct Ecol 23:1070–1080

Seltmann MW, Rangassamy M, Zapka M et al (2017) Timing of maternal nest building and perinatal offspring survival in a group-living small mammal. Behav Ecol Sociobiol 71:64

Sharples CM, Fa JE, Bell DJ (1996) Geographic variation in size in the European rabbit *Oryctolagus cuniculus* (Lagomorpha: Leporidae) in western Europe and North Africa. Zool J Linn Soc–Lond 117:141–158

Silvério D, Lopes AM, Melo-Ferreira J et al (2018) Insights into the evolution of the new variant rabbit haemorrhagic disease virus (GI.2) and the identification of novel recombinant strains. Transbound Emerg Dis 65:983–992

Simón MA, Gil-Sánchez JM, Ruiz G et al (2012) Reverse decline of the endangered Iberian lynx. Conserv Biol 26:731–736

Smith AT, Boyer AF (2007) *Oryctolagus cuniculus*. The IUCN red list of threatened species 2007:e.T41291A10414811. Available via https://www.iucnredlist.org/species/41291/10414811. Accessed 4 Nov 2020

Smith GC, Garthwaite DG, Prickett AJ (2006) Rabbit control in Great Britain. In: Feare DG, Cowan DP (eds) Advances in vertebrate pest management, vol 4. Filander Verlag, Fürth, pp 165–174

Smith AT, Johnston CH, Alves PC et al (2018) Lagomorphs: pikas, rabbits and hares of the world. Johns Hopkins University Press, Baltimore

Soriguer RC (1981) Biología y dinámica de una población de conejos (*Oryctolagus cuniculus* L.) en Andalucía Occidental. Doñana Acta Vertebr 8:1–379

Sroka J, Zwolinski J, Dutkiewicz J et al (2003) Toxoplasmosis in rabbits confirmed by strain isolation: a potential risk of infection among agricultural workers. Ann Agr Env Med 10:125–128

Sturdee AP, Chalmers RM, Bull SA (1999) Detection of Cryptosporidium oocysts in wild mammals of mainland Britain. Vet Parasitol 80:273–280

Tablado Z, Revilla E (2012) Contrasting effects of climate change on rabbit populations through reproduction. PLoS One 7:e48988

Tablado Z, Revilla E, Palomares F (2009) Breeding like rabbits: global patterns of variability and determinants of European wild rabbit reproduction. Ecography 32: 310–320

Tablado Z, Revilla E, Palomares F (2012) Dying like rabbits: general determinants of spatio-temporal variability in survival. J Anim Ecol 81:150–161

Thompson HV, King CM (eds) (1994) The European rabbit: history and biology of a successful colonizer. Oxford University Press, Oxford, UK

Trout RC, Langton S, Smith GC et al (2000) Factors affecting the abundance of rabbits (*Oryctolagus cuniculus*) in England and Wales. J Zool 252:227–238

Tyndale-Biscoe CH, Williams RM (1955) A study of natural mortality in a wild population of the rabbit, *Oryctolagus cuniculus* (L.). New Zeal J Sci Tech B 36:561–580

Vaquerizas PH, Delibes-Mateos M, Piorno V et al (2020) The paradox of endangered European rabbits regarded as pests: subspecies differences in trends matter. Endanger Species Res 43:99–102

Vaquerizas PH, Santoro S, Delibes-Mateos M, Castro F, Villafuerte R (2021) Keeping an eye on the use of eye-lens weight as a universal indicator of age for European wild rabbits. Scient Rep 11:8711

Velarde R, Cavadini P, Neimanis A et al (2017) Spillover events of infection of brown hares (*Lepus europaeus*) with rabbit haemorrhagic disease type 2 virus (RHDV2) caused sporadic cases of an European brown hare syndrome-like disease in Italy and Spain. Transbound Emerg Dis 64:1750–1761

Villafuerte R (1994) Riesgo de depredación y estrategias defensivas del conejo (*Oryctolagus cuniculus*) en el Parque Nacional de Doñana. PhD thesis, University of Córdoba, Córdoba

Villafuerte R, Delibes-Mateos M (2019) *Oryctolagus cuniculus*. The IUCN red list of threatened species 2019:e.T41291A45189779. Available via https://doi.org/10.2305/IUCN.UK.2019-2RLTS.T41291A45189779.en. Accessed 4 Nov 2020

Villafuerte R, Moreno S (1997) Predation risk, cover type, and group size in European rabbits in Doñana (SW Spain). Acta Theriol 42:225–230

Villafuerte R, Calvete C, Gortázar C et al (1994) First epizootic of rabbit hemorrhagic disease in free-living populations of *Oryctolagus cuniculus* at Doñana National Park, S.W. Spain. J Wildl Dis 30:176–179

Villafuerte R, Calvete C, Blanco JC et al (1995) Incidence of viral hemorrhagic disease in wild rabbit populations in Spain. Mammalia 59:651–659

Villafuerte R, Lazo A, Moreno S (1997) Influence of food abundance and quality on rabbit fluctuations: conservation and management implications in Doñana National Park (SW Spain). Rev Ecol Terre Vie 52: 345–356

Villafuerte R, Viñuela J, Blanco JC (1998) Extensive predator persecution caused by population crash in a game species: the case of red kites and rabbits in Spain. Biol Conserv 84:181–188

Villafuerte R, Ramírez E, Castro F et al (2017) Large-scale assessment of myxomatosis prevalence in European wild rabbits (*Oryctolagus cuniculus*) 60 years after first outbreak in Spain. Res Vet Sci 114:281–286

Viñuela J, Villafuerte R (2003) Predators and rabbits (*Oryctolagus cuniculus*) in Spain: a key conflict for European raptor conservation. In: Thompson DBA, Redpath SM, Fielding AH et al (eds) Birds of prey in a changing environment. The Stationery Office, Edinburgh, pp 511–526

Viñuela J, Villafuerte R, Le Court C (1994) Nesting dispersion of a black kite population in relation to location of rabbit warrens. Can J Zool 72:1680–1683

Virgós E, Cabezas-Díaz S, Malo A et al (2003) Factors shaping European rabbit abundance in continuous and fragmented populations of Central Spain. Acta Theriol 48:113–122

von Holst D, Hutzelmeyer H, Kaetzke P et al (1999) Social rank, stress, fitness, and life expectancy in wild rabbits. Naturwissenschaften 86:388–393

von Holst D, Hutzelmeyer H, Kaetzke P et al (2002) Social rank, fecundity and lifetime reproductive success in wild European rabbits (*Oryctolagus cuniculus*). Behav Ecol Sociobiol 51:245–254

Voza GV, Chabaud T, Landau I (2003) Coccidiosis of the wild rabbit (*Oryctolagus cuniculus*) in France. Parasite 10:51–57

Wallage-Drees JM (1986) Seasonal changes in the condition of rabbits, *Oryctolagus cuniculus* (L.), in a coastal sand dune habitat. Z Säugetierkd 51:26–36

Webb NJ (1993) Growth and mortality in juvenile European wild rabbits (*Oryctolagus cuniculus*). J Zool 230:665–677

Webb NJ, Ibrahim KM, Bell DJ et al (1995) Natal dispersal and genetic structure in a population of the European wild rabbit (*Oryctolagus cuniculus*). Mol Ecol 4:239–248

Wheeler S, King DR (1980) The use of eye-lens weights for aging wild rabbits, *Oryctolagus cuniculus* (L.), in Australia. Aust Wildl Res 7:79–84

Williams D, Acevedo P, Gortázar C et al (2007) Hunting for answers: rabbit (*Oryctolagus cuniculus*) population trends in northeastern Spain. Eur J Wildl Res 53:19–28

Williams F, Eschen R, Harris A et al (2010) The economic cost of invasive non-native species on Great Britain. Centre for Agriculture and Bioscience International (CABI), Surrey

Wood DH (1980) The demography of a rabbit population in an arid region of New South Wales, Australia. J Anim Ecol 49:55–79

Zhou L, Xiao Q, Bi J et al (2018) RabGTD: a comprehensive database of rabbit genome and transcriptome. Database–Oxford 2018:bay075

Eastern Cottontail *Sylvilagus floridanus* (J. A. Allen, 1890)

3

Sandro Bertolino, David E. Brown, Jacopo Cerri, and John L. Koprowski

Contents

S. Bertolino (✉)
Department of Life Sciences and Systems Biology, University of Turin, Torino, Italy
e-mail: sandro.bertolino@unito.it

D. E. Brown
Natural History Collections, Research Associate, Arizona State University, Tempe, AZ, USA

J. Cerri
Institute for Environmental Protection and Research, Livorno, Italy

J. L. Koprowski
Haub School of Environment & Natural Resources, University of Wyoming, Laramie, WY, USA
e-mail: jkoprows@uwyo.edu

K. Hackländer, P. C. Alves (eds.), *Primates and Lagomorpha*, Handbook of the Mammals of Europe,
https://doi.org/10.1007/978-3-030-34043-8_14

Common Names

English	Eastern cottontail, Florida cottontail
German	Östliches Baumwollschwanzkaninchen, Florida-Waldkaninchen
French	Lapin à queue blanche, lapin de Floride
Spanish	Conejo de Florida, Conejo castellano
Italian	Silvilago Orientale
Russian	Флори́дский кро́лик

Taxonomy, Systematics and Paleontology

The Eastern cottontail was first described by renowned naturalist Joel Asaph Allen in 1890. In recognition of the vast distribution of the species, no type specimen or locality was provided and the description referred to the "common gray rabbit" of the eastern United States. Eastern cottontails are closely related to other eastern North American congeners to include New England cottontails, *S. transitionalis*, and Appalachian cottontails, *S. obscurus* (Halanych and Robinson 1997; Ruedas 1998), which suggests that past vicariance events have led to speciation in species that are sympatric or parapatric in their current distribution. Given the wide distribution of *S. floridanus* across much of the United States, eastern Central America and northern South America, more than 30 subspecies have been described; however, a thorough assessment of phylogeography using modern molecular genetics is necessary (Lee et al. 2010; Schai-Braun and Hackländer 2016). Fossils of *S. floridanus* are known from Florida from 2.6 to 1.6 mya in the late Pliocene to early Pleistocene (Chapman et al. 1980). By 0.3 mya in the Irvingtonian, fossils document the presence of the species in the south central United States as well as the southeast. By the late Pleistocene, the Eastern cottontail fossils are found throughout most of its contemporary range (Nielsen and Berkman 2018).

Current Distribution

The Eastern cottontail has the largest distribution of the members of the genus *Sylvilagus*. It occurs from southern Canada, through eastern and south-central United States, Central America (central and eastern Mexico, Guatemala, Honduras, El Salvador, Nicaragua, Costa Rica), to the northern part of South America in Colombia and Venezuela (Chapman et al. 1980; AMCELA et al. 2008).

Widely introduced for hunting within and outside its native range, translocated populations are now also present in British Columbia, Washington, and Oregon. The species was also introduced in Cuba (Mancina et al. 2015) and in Europe into France, Italy, and Spain always for hunting (Masseti and De Marinis 2008; Bertolino et al. 2011a; Delibes-Mateos et al. 2018).

In Europe, the species seems to currently occur only in Italy (Lapini 1999), where it was introduced in 1966 in the Piedmont region (Mussa et al. 1996). Subsequently the species was introduced many other times, but data on these releases are not available (Angelici and Spagnesi 2008). One of these introductions was in Marettimo Island (Egadi Archipelago) during the 1980s, but it failed (Masseti and De Marinis 2008). Presently the Eastern cottontail is widespread in the northern and central parts of Italy (Bertolino et al. 2011a; Dori et al. 2019; Fig. 1).

Description

A medium to large cottontail, the total length of adults ranges from 395 to 477 mm with females tending to weigh slightly more than males. Skull measurement (N = 4): occipito-nasal length

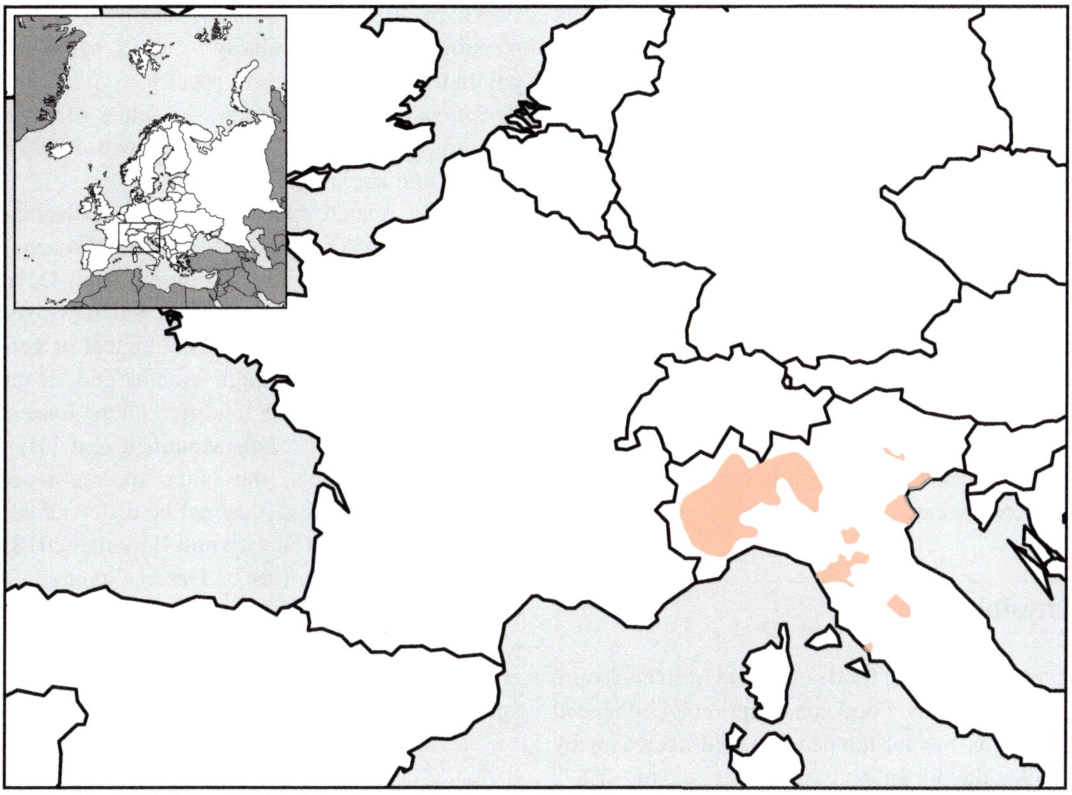

Fig. 1 *Sylvilagus floridanus* distribution in Europe. For data source please see text. (Map template: © Copyright Getty Images/iStockphoto)

73–79 mm; zygomatic width 35.0–38.2 mm; post-orbital constriction 10.7–13.9 mm; skull width 24.7–26.5 mm; nasal length 31.3–35.5 mm; mandible length 53.0–58.7; upper tooth row length 14.0–15.7 mm (Niethammer and Angelici 2003). Mean weights (800 to 1,500 grams) increase from north to south and from west to east (Nielsen and Berkman 2018).

The soft dorsal pelage varies from tan to gray grading into a white underside, especially under the short, fluffy tail (25–61 mm) giving the animal its common name (Fig. 2). The legs are washed in russet and white, and russet-orange fur typically colors the nape and head area between relatively long ears (55 to 67 mm in length). The hind feet are between 90 and 105 mm with weights ranging from 800 to 1,500 g (Nelson 1909; Nielsen and Berkman 2018).

The brain case is rounded with a broad and high interorbital region. Posterior projections of the superorbital process are transversely broad and generally fused to the braincase on the posterior ends. Anterior projections are slim, short, and nearly lacking or fused to the frontals. The presence of the premolar foramina is variable and can be single symmetrically or asymmetrically paired

Fig. 2 *Sylvilagus floridanus*. (Photo Aurelio Perrone)

between the second and fourth upper premolars (Nielsen and Berkman 2018).

Because of its wide distribution, the Eastern cottontail comes in contact with at least seven other species of cottontails, several of which are similar in appearance. Differentiation of species can be difficult and often depends on body and skull measurements along with color markings. For example, specimens of the Desert cottontail (*S. audubonii*) – exhibit a greater amount of white on the tail margins than Eastern cottontails (Hoffmeister 1986). A spot of white hairs on the forehead are used to differentiate Eastern cottontails from both New England cottontails (*S. transitionalis*) and Appalachian cottontails (*S. obscurus*), which have a black spot between the ears.

Physiology

Average weight of food consumed in an enclosure is 71–114 g/day. Food consumption is influenced by environmental temperature and decreases by 0.868 g/day for an increase in the range of −7.8 C to 27.2 C (Rose 1973). Based on this measurement, the annual energy requirement of an adult cottontail in the climatic condition of central United States (Illinois) is approximately 9.2×10^4 kcal which correspond to about 1.7% of the annual net productivity of aboveground vegetation of 0.4 ha (Rose 1973).

Metabolic rate of young exceeds that of larger animals. Assimilation and respiration are greater for larger rabbits but the rates of assimilation and respiration per gram of body weight decrease with increasing body size (Rose 1974). This higher metabolic rate in the younger cottontails enables them to develop rapidly and reach breeding condition at an early age (Rose 1974). The higher energy needs of young are satisfied with a diet rich in foods with more protein and available digestible energy.

Genetics

The diploid chromosome number is 42. Fossil evidence indicates that the species has been extant since the Late Pliocene, 2.6 mya, after which it evolved into nearly 30 subspecies (Nielsen and Berkman 2018). Interestingly, despite a similar physiogenomic similarity to sympatric populations of other species, *Sylvilagus floridanus* shows no genetic evidence of interbreeding (Findley et al. 1975; Hoffmeister 1986; Nielsen and Berkman 2018).

Closely related species are the Appalachian (*S. transitionalis*), New England (*S. obscurus*), Manzano Mountain (*S. cognatus*), and Davis Mountain (*S. robustus*) cottontails, which evolved from *floridanus* due to recent ecological or geographic isolation. Although cranial and dental characters can be used to differentiate Eastern cottontails from Manzano Mountain and Davis Mountain cottontails, the latter species is so recently separated that it cannot be differentiated by genetic markers (Nielsen and Berkman 2018). By way of contrast, Desert cottontails (*S. audubonii*) and Mountain cottontails (*S. nuttallii*) are as genetically distinct from Eastern cottontails as the Swamp rabbit (*S. aquaticus*) and Marsh rabbit (*S. palustris*).

Life History

Vital rates for Eastern cottontails vary widely across the substantial range of the species. The information below is from the native range of the Eastern cottontail. Breeding can occur throughout the year in southerly locations characterized by mild weather, whereas breeding is more seasonal where winters are severe and green emergent vegetation is limited, often initiating in February or March and ceasing in September but is delayed further at high elevation or latitude (Chapman et al. 1980). Gestation is about 28–30 days (range of 25 to 35 days) after which nearly hairless altricial young are born within a shallow burrow or depression lined with grass or other leafy vegetation often supplemented with fur. Three percent to 12% of embryos are resorbed. Litter size typically ranges between 3 and 6 young with a record litter of 12 young (Kirkpatrick 1960); first litter of the year is often smaller than subsequent litters. Mating occurs immediately after parturition and pregnancy is maintained while nursing of young continues for 3 or 4 weeks. Newborn young are about 30 to 45 g and 90 to 110 mm in total length but are particularly vulnerable to inclement weather and mortality can be complete following heavy rains or spells of cold weather. Young grow at the rate of 2.5 g/day (Kurta 1995). Eyes open at 6 to 7 days and young begin to leave the natal nest at 13 to 16 days, achieve adult body mass by 5 months and reproductive maturity can be reached well before the first year of life with 2 to 45% of young-of-the-year able to produce a litter (Chapman et al. 1980). As many as 8 litters can be produced per year in mild climates with 3–4 produced in more harsh climates (Baker 1983; Nielsen and Berkman 2018). Adult females produce between 6.2 and 35.0 young/year, which creates the potential for rapid population growth (Chapman et al. 1980). The large reproductive capacity of Eastern cottontails results in age-structured population that can be dominated by 80% young of the year by the end of the breeding season.

Sex ratios range from 40% to 79% males but typically do not vary significantly from an even 50:50 ratio among any age class of *S. floridanus* (Petrides 1951). Longevity typically does not exceed 3 years with 4-year old animals uncommon within populations (Baker 1983); however, animals may live to 9 years in captivity (Schwartz and Schwartz 1959). Annual turnover rates can reach 80% with annual mortality rates of adults ranging from 20 to 50% with slightly higher values for females than males (Chapman 1999; Bond et al. 2001a, b).

Habitat and Diet

The Eastern cottontail has considerable flexibility to use diverse habitats, in its native and introduced environments. This flexibility is probably one of the key factors behind the large geographical distribution in North America, and also explains it's the rapid colonization of North and Central Italy in less than forty years.

In North America, cottontails occur across different environments, from intensive agricultural environments to woodlands. In farmlands, traditional landscapes with small fields and permanent cover are the habitats where the species reaches the highest densities (Chapman and Litvaitis 2003). In these environments, home ranges (HR) rarely exceed 6 ha and core areas are usually below 1 ha (Trent and Rongstad 1974; Chapman and Flux 1990; Bond et al. 2001b). The presence of permanent cover is needed for antipredator behavior and to avoid extreme temperatures (Chapman and Litvaitis 2003). In intensive cereal croplands, cottontails can have extremely large home ranges, characterized by a strong seasonal variation. Mankin (1993), in an intensively farmed region of Illinois, reported average HR of about 33 and 9 ha, for male and female cottontails, which decreased about 2.5 times after crop harvesting, from November to May. In forested environments of North America, Eastern cottontails characterized intermediate stages of forest succession. While always dependent on ecotones and permanent cover, eastern cottontails are able to forage further from them, compared to other cottontail species (Smith and Litvaitis 2000). Eastern cottontails can also adapt to urbanized environments, such as urban parks. In these contexts, cottontails probably have larger home ranges, but they seem to rely on patches of permanent cover, like in their natural environments (Baker et al. 2015; Hunt et al. 2014).

Cottontail density in North American agricultural environments was 9 individuals/ha (Trent and Rongstad 1974). Densities in urban parks were high, from 9.4 to 16.3 individuals/ha (Hunt et al. 2014). In playa lake basins of Texas, Scribner and Warren (1990) reported very high densities, ranging from 8.2–11.6 individuals/ha in the pre-reproductive period, to 28.4 and 26.9 individuals/ha in the reproductive season. On islands, cottontail densities might reach 10–15 individuals/ha during fall (Chapman and Litvaitis 2003).

In Northern and Central Italy, cottontails adapted well to various agricultural environments. In their introduced range, cottontails selected permanent cover, like fallows of a poplar stand, for resting during the daytime. When foraging, at night, cottontails selected open habitats adjacent to permanent cover (Bertolino et al. 2011a, b, 2013; Vidus-Rosin et al. 2008, 2010a, b, 2011, 2012). In the Po plain, characterized by intensive agriculture, riverine vegetation seems to be important to guarantee cover and contributes to the dispersal of the species (Bertolino et al. 2011c). Interestingly, eastern cottontails were also able to establish in proximity to sand dunes in the Po estuary (De Barba 2015). Home ranges of cottontails in their introduced range were similar to those reported for North America, rarely exceeding 5–6 ha (Bertolino et al. 2011a, 2013). Cottontail densities at their introduction sites in Northern Italy vary greatly. In the Piemonte region, Silvano et al. (2000) reported average winter densities of about 0.25–0.28 individuals/ha and Bertolino et al. (2013) ranging from 0.2 ± 0.08 to 0.6 ± 0.4 individuals/ha. In the Lombardy region, Vidus-Rosin et al. (2008) reported densities of 0.6 ± 0.06 and 0.36 ± 0.06 individuals/ha, while Vidus-Rosin et al. (2010a, b) estimated average cottontail density as ranging from 0.47 to 0.62 individuals/ha, with densities of 0.28 ± 0.07 individuals/ha in the pre-breeding period and 1.3 ± 0.25 individuals/ha during the breeding season. Cerri (unpublished data), for Central Italy, estimated cottontail density as ranging from 0.07 ± 0.0 to 0.7 ± 0.1 individuals/ha.

In North America, cottontails have a broad diet, which varies according to available plants in the environment. Usually, cottontails exploit herbaceous plants during the growing season and woody species in winter. Available studies indicate that cottontails fed on alfalfa (*Medicago sativa*), Kentucky bluegrass (*Poa pratensis*), Canada bluegrass (*P. compressa*), timothy (*Phleum pratense*), quack grass (*Agropyron repens*), orchard grass (*Dactylis glomerata*), red clover (*Trifolium pratense*), and wild carrot (*Daucus carota*). Cottontails can also fed on woody plants such as apple (*Malus pumila*), sumac (*Rhus thypina*), red maple (*Acer rubrum*), blackberry (*Rubus allegheniensis*), red raspberry (*R. strigosus*), ray birch (*Betula populifolia*), aspen (*Populus tremuloides*), choke cherry (*Prunus virginiana*) and black cherry (*Prunus serotina*), dewberry (*Robus villosus*), willow (*Salix* spp.), black alder (*Iles beticillata*), maleberry (*Lyonia ligustrina*), and highblush blueberry (*Vaccinium corumbosum*) (Chapman et al. 1980). No diet study is available for cottontails in their introduced environment in Italy, but the fact that they were able to colonize sand dunes in the Po estuary might indicate their exploitation of Mediterranean sclerophylls and alophytes characterizing these environments. Cottontails are a major pest for croplands in North America, whereas no significant damage is known for their introduction range. In Italy, damage to crops is limited.

Behavior

Foraging is mostly nocturnal and it occurs in the surroundings of patches of permanent cover. Foraging is a nonsocial activity, but where densities are high, cottontails often forage in groups (Chapman and Litvaitis 2003).

Resource Competition

In its native environments, Eastern cottontails compete with other species of cottontails. Competition occurs mostly through aggression, where eastern cottontails are able to displace other cottontails from permanent cover and resting sites. This made them successful invaders at many introduction sites in the American continent (Leach et al. 2015a).

In Italy, cottontails do not seem to compete for habitat resources with native European hares (*Lepus europaeus*), as the two lagomorphs show different habitat preferences. Cottontails exploit permanent cover to a greater extent than hares, which can also use open areas as resting sites during the daytime. When foraging at night, the two species also select different sites, with hares relying on open areas where they can outrun predators, whereas cottontails prefer areas near permanent cover (Bertolino et al. 2013; Vidus-Rosin et al. 2011). However, no study assessed the selection of plant species by hares in areas of sympatry and allopatry with cottontails: recent studies showed that hares are more selective than previously thought (Schai-Braun et al. 2015), and this might lead to competition. Cottontails also exert apparent competition on hares, through increased predation from the red fox (*Vulpes vulpes*) (Cerri et al. 2017), and it is unclear whether such a relationship is mediated by habitat heterogeneity. No research about competition between the Eastern cottontail and the European rabbit (*Oryctolagus cuniculus*) is available, nor the interrelationships between cottontails and the Apennine hare (*Lepus corsicanus*) have been so far investigated where the two species potentially overlap in Central Italy.

Social Behavior

In semi-controlled environments, cottontails seem to have hierarchical relationships maintained through aggressive behavior in both sexes. In captivity, Brenner and Flemming (1979) observed that males and juveniles practiced paw-boxing and used such aggressive interactions to continuously maintain the hierarchy. Females also have hierarchies maintained through aggressive interactions (Bruch and Chapman 1983).

Eastern cottontails are not a social species. Nevertheless, females occasionally were reported to share their nest, but this does not seem to occur on a regular basis (Chapman and Litvaitis 2003).

Females create their nest on the soil, by using vegetable material, without burrowing, like the European rabbits. Two females occasionally can share their nest, but this is rare. In a suburban neighborhood in New York, a female was reported to have built their nest in a tree, at 2.5 m above the ground (McCallum et al. 2018). Eastern cottontails are polygamous.

Parasites and Diseases

In Northern America, cottontails carry a very high number of ectoparasites, including ticks. The most common ticks that are found on cottontails belong to the families Ixodidae, Pulicidae, and Leptopsyllidae. *Rickettsia rickettsii* is the main vector for the Rocky Mountain spotted fever (Chapman 2003). In Italy, cottontails were found to be infested by *Ixodes ricinus, Euhoplopsyllus glacialis, Ctenocephalides canis, Chtenophtalmus agyrtes* (Tizzani et al. 2002).

Cottontail in North America were also found to host a considerable number of endoparasites, like nematodes from the genera *Obeliscoides*, *Trichostrongylus*, *Longistriata,* and *Trichuris*. Cestodes include the genus *Mosgovoyia* and *Taenia* (Chapman and Litvaitis 2003). In Northern Italy, Eastern cottontails were found to carry eight species of coccidians from the genus *Eimeria*: seven species have a North American origin and were never reported in Italian lagomorphs, while only *Eimeria leporis* was previously described in *Lepus europaeus* and other congeners (Bertolino et al. 2010). Among nematodes, Tizzani et al. (2014) reported *Trichostrongylus affini*, a North-American species, and Tizzani et al. (2011) *Obeliscoides cuniculi*, a nematode which is usually found on European lagomorphs only. The Eastern cottontail seem therefore to be susceptible to cross-infection with the same endo- and ectoparasites that infest European leporids, namely, the European hare and the European rabbit.

In Italy, non-native eastern cottontails are suspected to play a role in the transmission of two major lagoviruses to native lagomorphs: the European Brown Hare Syndrome virus (EBHSV) and the Rabbit Haemorrhagic Disease virus (RHDV). Cottontails were found to be susceptible of infection with the EBHSV, both in laboratory trials and in the wild (Lavazza et al. 2015). It is therefore plausible to assume their role as a reservoir or as a carrier for this virus, which can seriously undermine hare populations. On the other hand, the scenario is more confused for RHDV. Cottontails might be infected by RHDV and by the RHDv2, a variant of the virus which appeared recently and also affected European hares, but do not develop the virulence (Lopes et al. 2018). This might indicate that eastern cottontails, despite not suffering from lagoviruses directly, can help lagoviruses to jump from one lagomorph species to another.

Eastern cottontails are one of the main vectors of tularemia in the United States. Their ectoparasites, notably ticks, are also responsible for the transmission of the Rocky Mountain spotted fever to humans. Moreover, cottontails can also carry *Staphylococcus aureus* and eastern encephalitis (Chapman and Litvaitis 2003). Eastern cottontails are also a reservoir for the West Nile virus in

North America (Tiawsirisup et al. 2005). In Italy, they were found to carry various species of dermatophyte fungi that can infect humans, notably *Microsporum canis, Microsporum gypseum,* and *Trichophyton mentagrophytes* (Gallo et al. 2005), as well as the main agent of the Toxoplasmosis, *Toxoplasma gondii* (Tizzani et al. 2002; Zanet et al. 2013). Tularemia was reported in Italy as well (Tizzani et al. 2002), altogether with brucellosis.

Population Ecology

Juvenile breeding in the first year of life is common among females: 52.4% of juvenile females were sexually active in Oregon and breeding juvenile females accounted for almost 4% of pregnancies in Maryland. Only a few juvenile males are sexually active, but it is supposed that early-born individuals might participate in late-season breeding. The breeding season varies greatly among regions, but usually starting about 28 days later than the first available fresh plants. In mild climates, like the southern United States, cottontails were reported to breed all year long. Litter size varies from 3 to 5 individuals, and about 13% of the eggs ovulated fail to implant or are reabsorbed as embryos. In their native range, Eastern cottontails are reported to produce up to 35–39 young in 7–8 litters and net annual productivity is about 3–20 individuals/female (Chapman and Litvaitis 2003).

Densities of Eastern cottontails in native range typically range between 2 and 10 individuals/ha but can exceed 20 individuals/ha and on islands can be greater than 10 individuals/ha (Chapman et al. 1980). Introduced Eastern cottontail densities in Italy are considerably lower, at 0.44 to 1.10 individuals/ha (Bertolino et al. 2011a). Population fluctuations can be considerable, but are generally irregular and population cycles are not known. The high densities and moderate body size of *S. floridanus* result in their frequent occurrence in the diet of mesocarnivorous mammals such as canids, felids, procyonids, and birds such as hawks, owls, and crows (Chapman et al. 1980). Road mortality is also considerable due to the nocturnal activity of the species. Eastern cottontail hunting is common in some areas in the native range, and even high hunting effort typically does not seem to create long-term declines in populations (Baker 1983). In Italy, the red fox (*Vulpes vulpes*) specialize on *S. floridanus* when available (Balestrieri et al. 2005), and predator populations appear to lag behind rabbit populations (Baker 1983). Niche overlap between foxes and badgers (*Meles meles*) decreased in areas with non-native Eastern cottontails (Balestrieri et al. 2005).

The interaction of fecal pellets and soil pH indicates a negative effect of Eastern cottontails that increases on acidic soils (Gheza et al. 2018). Bryophyte cover responds positively to fecal pellets with greater negative effects on lichens may be expected on calcareous soils (Gheza et al. 2018).

Eastern cottontails are sympatric in Italy with European hares and Apennine hares. Coexistence with European hares appears to occur due to spatial partitioning of niches (Bertolino et al. 2011b), while information on the interactions with Apennine hares are lacking. The species differentiated their habitat use for daily resting sites and during feeding activities. Natural herbaceous vegetation was used by both species for daily resting sites; however, the species differed in their use of large scale habitat types; hares used riverbanks whereas Eastern cottontails inhabited shrubby areas. *S. floridanus* foraged in natural herbaceous cover, poplar (*Populus*) plantations, meadows, and stubble fields, in proximity shelter from potential predators. *L. europaeus* fed in meadows, croplands, and natural herbaceous vegetation, but at distances ≥50 m from refuges, considerably further than Eastern cottontails (Bertolino et al. 2011a).

Habitat loss and climate change are expected to reduce Eastern cottontails in Mexico (Lorenzo et al. 2018). Response to climate change by members of the genus *Sylvilagus* are not well known with all predictions suggesting change but the direction and severity vary by study. Species distribution models suggest that the Eastern cottontail will shift significantly poleward in their native range (Leach et al. 2015b). These results would also suggest similar shifts in Italy and perhaps movement upward in elevation.

Conservation Status

The status Eastern cottontail is classified by the IUCN Red List as Least Concern (Nielsen and Lanier 2019). The species is widespread and it has been introduced widely outside its original range, where it sometimes presents a threat to sympatric species (AMCELA et al. 2008). Being an introduced species in Europe, it is not considered by the Bern Convention and EU Habitats Directive. It is also not listed in CITES. In contrast, Eastern cottontails were classified as alien invasive species in Europe since 1985 (Recommendation No. R (85) 14: https://rm.coe.int/090000168050a292) after the studies conducted in France by Chapuis et al. (1985). Nevertheless, this species is not included in the list of Invasive Alien Species of EU concern (Regulation No 1143/2014, https://environment.ec.europa.eu/topics/nature-and-biodiversity/invasive-alien-species_en).

Management

Despite being a widespread species in the native range, state federal-aid surveys and hunt questionnaire data indicate a declining trend in cottontail population levels, at least in part of the western United States (Nielsen and Berkman 2018). These declines cannot be attributed to hunting as hunt pressure has declined throughout most of Eastern cottontail's range. The locations of the states and regions in the United States most affected suggest that prolonged drought, coupled with land use change, may be the major cause of the decline with increased mortality due to increased predation rates as a possible factor.

In Europe, the Eastern cottontail is now reported only for Italy where populations are spreading. In Italy, the species can be hunted from September to January, according to different provinces. In some areas, the Eastern cottontail is also controlled for the damage it produces to crops and for the competition with the native European hare. After the amendment in 2015 of Art. 2 to Legge n. 157/1992 "Norme per la protezione della fauna selvatica omeoterma e per il prelievo venatorio" (Rules for the protection of homeothermic wildlife and hunting), the management of alien mammals and birds in Italy aims to reduce and, when possible, eradicate introduced populations. However, since the Eastern cottontail is not included in the European list of Invasive Alien Species, this species is not considered a management priority among introduced species, and local authorities implement control activities only occasionally.

Future Challenges for Research and Management

In Europe, research on the environmental impacts where the Eastern cottontail has been introduced remains limited. Population dynamic is not well understood in non-native populations. The mechanisms of apparent competition with the European hare are not clear (Cerri et al. 2017) and the role as possible vector of lagoviruses and other parasites should be better investigated. Studies on the possible competition with other native lagomorphs (i.e., Appenine hare, European rabbit) are completely lacking.

References

AMCELA (Mexican Association for Conservation and Study of Lagomorphs), Romero Malpica FJ, Rangel Cordero H (2008) *Sylvilagus floridanus*. The IUCN red list of threatened species 2008: e. T41299A10418836

Angelici FM, Spagnesi M (2008) *Sylvilagus floridanus* (J.A. Allen, 1890). In: Amori G, Contoli L, Nappi A (eds) Mammalia II. Erinaceomorpha, Soricomorpha, Lagomorpha, Rodentia. Fauna d'Italia Vol. XLIV. Edizioni Calderini e Il Sole 24 Ore, Milano, pp 305–313

Baker RH (1983) Michigan mammals. Michigan State University Press, Lansing

Baker MAA, Emerson SE, Brown JS (2015) Foraging and habitat use of eastern cottontails (*Sylvilagus floridanus*) in an urban landscape. Urban Ecosyst 18(3):977–987

Balestrieri A, Remonti L, Prigioni C (2005) Local feeding specialization of the red fox (*Vulpes vulpes*) in response to eastern cottontail (*Sylvilagus floridanus*) introduction (NW Italy). Hystrix 16:113–126

Bertolino S, Hofmannová L, Girardello M, Modry D (2010) Richness, origin and structure of an Eimeria community in a population of Eastern cottontail (*Sylvilagus floridanus*) introduced into Italy. Parasitology 137(8):1179–1186

Bertolino S, Perrone A, Gola L, Viterbi R (2011a) Population density and habitat use of the introduced Eastern cottontail (*Sylvilagus floridanus*) compared to the native European hare (*Lepus europaeus*). Zool Stud 50(3):315–326

Bertolino S, Cordero di Montezemolo N, Perrone A (2011b) Daytime habitat selection by introduced eastern cottontail *Sylvilagus floridanus* and native European hare *Lepus europaeus* in Northern Italy. Zool Sci 28(6):414–419

Bertolino S, Ingegno B, Girardello M (2011c) Modelling the habitat requirements of invasive eastern cottontail (*Sylvilagus floridanus*) introduced to Italy. Eur J Wildl Res 57(2):267–274

Bertolino S, Cordero di Montezemolo N, Perrone A (2013) Habitat use of coexisting introduced eastern cottontail and native European hare. Mamm Biol 78(4):235–240

Bond BT, Leopold BD, Burger LW Jr, Godwin KD (2001a) Movements and home range dynamics of cottontail rabbits in Mississippi. J Wildl Manag 65:1004–1013

Bond BT, Burger LW Jr, Leopold BD, Godwin KD (2001b) Survival of cottontail rabbits (*Sylvilagus floridanus*) in Mississippi and an examination of latitudinal variation. Amer Midl Nat 145:127–136

Brenner FJ, Flemming DP (1979) Behavioral determination of dominance in captive male cottontail rabbits (*Sylvilagus floridanus*). Proc Pennsylv Acad Sci 53(2):139–142

Bruch D, Chapman JA (1983) Social behavior of the eastern cottontail, *Sylvilagus floridanus* (Lagomorpha: Leporidae) in a woodland habitat, with descriptions of new behaviors. Proc Pennsylv Acad Sci 57:74–78

Cerri J, Ferretti M, Bertolino S (2017) Rabbits killing hares: an invasive mammal modifies native predator–prey dynamics. Anim Conserv 20(6):511–519

Chapman JA (1999) Eastern cottontail *Sylvilagus floridanus*. In: Wilson DE, Ruff S (eds) Smithsonian book of North American mammals. Smithsonian Institution Press, Washington, DC/London, pp 687–688

Chapman JA, Flux JE (eds) (1990) Rabbits, hares and pikas: status survey and conservation action plan. IUCN, Gland

Chapman JA, Litvaitis JA (2003) Eastern cottontail. In: Feldhamer GA, Thompson BC, Chapman JA (eds) Wild mammals of North America. Johns Hopkins University Press, Baltimore, pp 101–125

Chapman JA, Hockman JG, Ojeda CMM (1980) *Sylvilagus floridanus*. Mamm Spec 136:1–8

Chapuis JL, Forgeard F, Didillon MC (1985) Etude de *Sylvilagus floridanus* en région méditerranéenne dans des conditions de semi-liberté. Régime alimentaire au cours d'un cycle annuel par l'analyse micrographique des fèces. Gibier Faune Sauvage 2:59–104

De Barba H (2015) Presenza di *Sylvilagus floridanus* (JA Allen, 1890) nei diversi habitat del Delta del Po. Tesi di Laurea Magistrale in Scienze Forestali e Ambientali, Università degli Studi di Padova

Delibes-Mateos M, Castro F, Piorno V, Ramírez E, Blanco-Aguiar JA, Aparicio F, Mínguez LE, Ferreira CC, Rouco C, Ríos-Saldaña CA, Recuerda P, Villafuerte R (2018) First assessment of the potential introduction by hunters of eastern cottontail rabbits (*Sylvilagus floridanus*) in Spain. Wildl Res 45:571–577

Dori P, Scalisi M, Mori E (2019) "An American near Rome"... and not only! Presence of the eastern cottontail in Central Italy and potential impacts on the endemic and vulnerable Apennine hare. Mammalia 83:307–312

Findley JS, Harris AH, Wilson DE, Jones C (1975) Mammals of New Mexico. University of New Mexico Press, Albuquerque

Gallo MG, Tizzani P, Peano A, Rambozzi L, Meneguz PG (2005) Eastern cottontail (*Sylvilagus floridanus*) as carrier of dermatophyte fungi. Mycopathologia 160(2): 163–166

Gheza G, Assini S, Marini L, Nascimbene J (2018) Impact of an invasive herbivore and human trampling on lichen-rich dry grasslands: soil-dependent response of multiple taxa. Sci Total Environ 639:633–639

Halanych KM, Robinson TJ (1997) Phylogenetic relationships of cottontails (*Sylvilagus*, Lagomorpha): congruence of 12S rDNA and cytogenetic data. Mol Phylogen Evol 7:294–302

Hoffmeister DF (1986) Mammals of Arizona. University of Arizona Press and Arizona Game and Fish Department, Tucson

Hunt VM, Magle SB, Vargas C, Brown AW, Lonsdorf EV, Sacerdote AB, Sorley EJ, Santymire RM (2014) Survival, abundance, and capture rate of eastern cottontail rabbits in an urban park. Urban Ecosyst 17(2):547–560

Kirkpatrick CM (1960) Unusual cottontail litter. J Mammal 41:119–120

Kurta A (1995) Mammals of the Great Lakes region. University of Michigan Press, Ann Arbor

Lapini L (1999) *Sylvilagus floridanus* (Allen JA 1890). In: Mitchell-Jones AJ, Amori G, Bogdanowicz W, Kryötufek B, Reijnders PJH, Spitzenberger F, Stubbe M, Thissen JBM, Vohralik V, Zima J (eds) The atlas of the European mammals. Academic, London, pp 174–175

Lavazza A, Cavadini P, Barbieri I, Tizzani P, Pinheiro A, Abrantes J, Esteves PJ, Grilli G, Gioia E, Zanoni M, Meneguz PG, Guitton J-S, Marchandeau S, Chiari M, Capucci L (2015) Field and experimental data indicate that the eastern cottontail (*Sylvilagus floridanus*) is susceptible to infection with European brown hare syndrome (EBHS) virus and not with rabbit haemorrhagic disease (RHD) virus. Vet Res 46(1):13

Leach K, Montgomery WI, Reid N (2015a) Biogeography, macroecology and species' traits mediate competitive interactions in the order Lagomorpha. Mamm Rev 45: 88–102

Leach K, Kelly R, Cameron A, Montgomery WI, Reid N (2015b) Expertly validated models and phylogenetically-controlled analysis suggests responses to climate change are related to species traits in the order Lagomorpha. PLoS One 10:e0122267

Lee DN, Pfau RS, Ammerman LK (2010) Taxonomic status of the Davis Mountains cottontail, *Sylvilagus robustus*, revealed by amplified fragment length polymorphism. J Mammal 91:1473–1483

Lopes AM, Breiman A, Lora M, Le Moullac-Vaidye B, Galanina O, Nyström K, Marchandeau S, Le Gall-Reculé G, Strive T, Neimanis A, Bovin NV, Ruvoën-Clouet N, Esteves PJ, Abrantes J, Le Pendu J (2018) Host-specific glycans are correlated with susceptibility to infection by lagoviruses, but not with their virulence. J Virol 92(4):e01759–e01717

Lorenzo C, Sántiz EC, Bolaños-Citalán J, Navarrete-Gutiérrez D (2018) Influences of long-term changes in land cover on mammal populations: an example from Mexico. Oryx 53:321–328

Mancina C, Díaz H, García Tió E, Borroto-Páez R, Cervantes F (2015) Taxonomic identity of invasive rabbits in Cuba: first record of Eastern Cottontail, *Sylvilagus floridanus* (Mammalia: Lagomorpha). Check List 11(6):1–7

Mankin PC (1993) Agricultural land use and the eastern cottontail in Illinois. PhD Thesis, University of Illinois, Urbana-Champaign

Masseti M, De Marinis AM (2008) Prehistoric and historic artificial dispersal of lagomorphs on the Mediterranean Islands. In: Alves PC, Ferrand N, Hackländer K (eds) Lagomorph biology: evolution, ecology, and conservation. Springer, Berlin/Heidelberg, pp 13–26

McCallum ML, McCallum AA, McCallum DB Jr, McCallum JM, McCallum MS, McCallum SA (2018) Arboreal nesting by an Eastern cottontail in a suburban setting. Northeast Nat 25(1):N4–N6

Mussa PP, Meineri G, Bassano B (1996) Il Silvilago in Provincia di Torino. Habitat 61:5–11

Nelson EW (1909) The rabbits of North America. N Amer Fauna 29:9–287

Nielsen CK, Berkman LK (2018) *Sylvilagus floridanus* (JA Allen 1890) Eastern cottontail. In: Smith AT, Johnston CH, Alves PC, Hackländer K (eds) Lagomorphs: Pikas, rabbits and hares of the world. Johns Hopkins University Press, Baltimore, pp 137–140

Nielsen CK, Lanier HC (2019) *Sylvilagus floridanus*. The IUCN red list of threatened species 2019: e. T41299A45191626. https://dx.doi.org/10.2305

Niethammer J, Angelici FM (2003) *Sylvilagus floridanus* (J. A. Allen, 1890) – Baumwollschwanzkaninchen. In: Krapp F (ed) Handbuch der Saugetiere Europas. Band 3/II: Hasentiere. Lagomorpha. Aula-Verlag, Kempten, pp 291–295

Petrides GA (1951) The determination of sex and age ratios in the cottontail rabbit. Am Midl Nat 46:312–336

Rose GB (1973) Energy metabolism of adult cottontail rabbits, *Sylvilagus floridanus*, in simulated field conditions. Am Midl Nat 89:473–478

Rose GB (1974) Energy dynamics of immature cottontail rabbits. Am Midl Nat 91:473–477

Ruedas LA (1998) Systematics of *Sylvilagus* gray, 1867 (Lagomorpha: Leporidae) from southwestern North America. J Mammal 79:1355–1378

Schai-Braun SC, Hackländer K (2016) Family Leporidae (hares and rabbits). In: Wilson DE, Lacher TE Jr, Mittermeier RA (eds) Handbook of the mammals of the world, Volume 6. Lagomorphs and rodents I. Lynx Edicions, Barcelona, pp 62–149

Schai-Braun SC, Reichlin TS, Ruf T, Klansek E, Tataruch F, Arnold W, Hackländer K (2015) The European hare (*Lepus europaeus*): a picky herbivore searching for plant parts rich in fat. PLoS One 10(7): e0134278

Schwartz CW, Schwartz ER (1959) The wild animals of Missouri. University Missouri Press, Columbia

Scribner KT, Warren RJ (1990) Seasonal demography and movements of cottontail rabbits on isolated playa basins. J Wildl Manag 54:403–409

Silvano F, Acquarone C, Cucco M (2000) Distribution of the Eastern cottontail (*Sylvilagus floridanus*) in the province of Alessandria. Hystrix 12:75–78

Smith DF, Litvaitis JA (2000) Foraging strategies of sympatric lagomorphs: implications for differential success in fragmented landscapes. Can J Zool 78(12): 2134–2141

Tiawsirisup S, Platt KB, Tucker BJ, Rowley WA (2005) Eastern cottontail rabbits (*Sylvilagus floridanus)* develop West Nile virus viremias sufficient for infecting select mosquito species. Vector Borne Zoonotic Dis 5(4):342–350

Tizzani P, Lavazza A, Capucci L, Meneguz PG (2002) Presence of infectious agents and parasites in wild population of cottontail (*Sylvilagus floridanus*) and consideration on its role in the diffusion of pathogens infecting hares. The European Association of Zoo-and Wildlife Veterinarians (EAZWV) 4th scientific meeting, joint with the annual meeting of the European Wildlife Disease Association (EWDA). European Association of Zoo-and Wildlife Veterinarians (EAZWV), pp 245–248

Tizzani P, Menzano A, Catalano S, Rossi L, Meneguz PG (2011) First report of *Obeliscoides cuniculi* in European brown hare (*Lepus europaeus*). Parasitol Res 109(3):963–966

Tizzani P, Catalano S, Rossi L, Duignan PJ, Meneguz PG (2014) Invasive species and their parasites: eastern cottontail rabbit *Sylvilagus floridanus* and *Trichostrongylus affinis* (Graybill, 1924) from Northwestern Italy. Parasitol Res 113(4):1301–1303

Trent TT, Rongstad OJ (1974) Home range and survival of cottontail rabbits in southwestern Wisconsin. J Wildl Manag 38:459–472

Vidus-Rosin A, Gilio N, Meriggi A (2008) Introduced lagomorphs as a threat to "native" lagomorphs: the case of the Eastern cottontail (*Sylvilagus floridanus*) in northern Italy. In: Alves PC, Ferrand N, Hackländer K (eds) Lagomorph biology. Evolution, ecology and conservation. Springer, Berlin/Heidelberg, pp 153–164

Vidus-Rosin A, Meriggi A, Serrano-Perez S (2010a) Density and habitat requirements of introduced eastern cottontail *Sylvilagus floridanus* in northern Italy. Acta Theriol 55(2):139–151

Vidus-Rosin A, Montagna A, Meriggi A, Serrano Perez S (2010b) Density and habitat requirements of sympatric hares and cottontails in northern Italy. Hystrix 20(2): 101–110

Vidus-Rosin A, Meriggi A, Cardarelli E, Serrano-Perez S, Mariani MC, Corradelli C, Barba A (2011) Habitat overlap between sympatric European hares (*Lepus europaeus*) and Eastern cottontails (*Sylvilagus floridanus*) in northern Italy. Acta Theriol 56(1):53–61

Vidus-Rosin A, Lizier L, Meriggi A, Serrano-Perez S (2012) Habitat selection and segregation by two sympatric lagomorphs: the case of European hares (*Lepus europaeus*) and Eastern cottontails (*Sylvilagus floridanus*) in northern Italy. Acta Theriol 57(4): 295–304

Zanet S, Palese V, Trisciuoglio A, Alonso CC, Ferroglio E (2013) *Encephalitozoon cuniculi*, *Toxoplasma gondii* and *Neospora caninum* infection in invasive eastern cottontail rabbits *Sylvilagus floridanus* in Northwestern Italy. Vet Parasitol 197(3–4):682–684

Cape Hare *Lepus capensis* Linnaeus, 1758

4

Massimo Scandura, Anna Maria De Marinis, and Antonio Canu

Contents

M. Scandura (✉) · A. Canu
Department of Veterinary Medicine, University of Sassari, Sassari, Italy
e-mail: scandura@uniss.it

A. M. De Marinis
Italian Institute for Environmental Protection and Research (ISPRA), Ozzano dell'Emilia, Italy
e-mail: annamaria.demarinis@isprambiente.it

K. Hackländer, P. C. Alves (eds.), *Primates and Lagomorpha*, Handbook of the Mammals of Europe,
https://doi.org/10.1007/978-3-030-34043-8_10

Common Names

English	Cape hare, Sardinian hare
German	Kaphase
French	Lièvre du Cap
Spanish	Liebre de El Cabo
Italian	Lepre del Capo, Lepre sarda
Russian	Капский заяц

Taxonomy, Systematics and Paleontology

Evolutionary relationships and systematics within the polytypic genus *Lepus* are still not well understood and have been under continuous revision (see introductory chapter in this volume, Hackländer et al. 2008; Schai-Braun and Hackländer 2016; Smith et al. 2018). The validity of many hare species is controversial, and their number has ranged over the years from a dozen to over 30 depending on the authors (Flux and Angermann 1990; Smith et al. 2018). Cape hare *Lepus capensis* is one of the species that has been most influenced by these reassessments and its taxonomy is still under debate (Schai-Braun and Hackländer 2018).

Cape hares are listed among the faunal remains of the Late Upper Palaeolithic site of Kehf el Hammar in Northwest Morocco and are recorded from Epipalaeolithic sites of the Eastern Mediterranean; they were also found at Late Pleistocene–Early Holocene excavation sites in the Maghreb (among food remains found in Capsian excavation sites of ca. 10,000–7,500 years B.P.; Suchentrunk et al. 2006).

The taxonomy of the Cape hare is in urgent need of revision: the original descriptions of subspecies are seldom reliable as they are mostly based on a few morphological traits and a small number of individuals (Smith et al. 2018). As the range of morphological variation among areas and age classes is very high and has not been extensively studied, there is lack of agreement among taxonomists and some authors suppose that a number of subspecies may go extinct before their formal identification (Johnston et al. 2019). Moreover, it has been shown that such variation is clinal; therefore, the distinction in subspecies might be arbitrary and unreasonable (Smith et al. 2018). Flux and Angermann (1990) reported as many as 80 subspecies of Cape hare, although currently some of them are recognized as good species (e.g., *L. granatensis* and *L. tolai*). Hoffmann (1993) reported 50 synonyms just for Africa and Near East, disregarding forms occurring further east as their taxonomic status was particularly uncertain. According to Harrison and Bates (1991), there were 8 *L. capensis* subspecies in Arabia only, while Hoffmann and Smith (2005) reported 12 subspecies in total, partitioned into four groups (*L. c. arabicus* in the Near East; *L. c. capensis, L. c. aquilo*, *L. c. carpi*, and *L. c. granti* in South Africa; *L. c. aegyptius*, *L. c. hawkeri*, *L. c. isabellinus*, and *L. c. sinaiticus* in East Africa; *L. c. atlanticus*, *L. c. schlumbergeri*, and *L. c. whitakeri* in Northwest Africa), suggesting that they might represent as many distinct species of *capensis*-type hares. *L. capensis* has been described from the area of Cape of Good Hope in South Africa, and the description of the nominal subspecies *L. c. capensis* is based on an adult male sampled near Piketburg, in the Western Cape Region (Ben Slimen et al. 2008a). More recently, Smith et al. (2018) reported 13 subspecies: *capensis*, *aquilo*, *carpi*, *granti*, *aegyptius*, *hawkeri*, *isabellinus*, *sinaiticus*, *arabicus*, *atlanticus*, *mediterraneus*, *schlumbergeri*, and *whitakeri*.

Conversely, other authors were not in favor of splitting *L. capensis* in a number of taxa, but rather supported the hypothesis of conspecificity of *L. capensis* and *L. europaeus* advanced by Petter (1959, 1961) on the basis of morphological traits like body size and dental characters. This classification was accepted for about two decades (resulting in numerous common names for this species), until later morphological studies (Angermann 1983) led to the acceptance of separate species status for *L. capensis* and *L. europaeus* (e.g., Flux and Angermann 1990; Hoffmann 1993). However, Ben Slimen et al. (2008a) and Suchentrunk et al. (2008), by analyzing neutral and nonneutral genetic markers in various Cape and brown hare populations of

Africa, Europe, and Middle East, re-evaluated Petter's hypothesis. In fact, the degree of genetic differentiation they observed between populations belonging to the two different taxa was comparable to that observed within taxa, being consistent with an isolation-by-distance model. Ben Slimen et al. (2008a) thus concluded suggesting conspecificity for *L. capensis* and *L. europaeus* and speculating that brown hares and African Cape hares might be connected via the Near East by gene flow along a network of intergrading populations with regionally distinct gene pools.

However, this view has not found large consensus in the scientific community so far, nor support by subsequent studies. Recently, Lado et al. (2019) performed one of the most comprehensive studies on *L. capensis*, by analyzing several nuclear and mitochondrial markers in Cape hares sampled across Africa and Near East. The authors revealed that *L. capensis sensu lato* is not monophyletic and a high genetic divergence exists between African and Asian populations. Five major phylogeographical groups were inferred: Near East, Arabia, East Africa, South Africa, and Northwest Africa. The authors concluded that such high intraspecific differences likely resulted from fragmentation of ancestral ranges, divergence in allopatry, and interspecific introgression, suggesting that *L. capensis s.l.* could be a heterogeneous evolutionary entity representing several cryptic species, deserving a taxonomic revision. Accordingly, Soria-Boix et al. (2019) showed that three cryptic species occur in Northwest Africa, representing monophyletic and sharply differentiated mitochondrial DNA lineages.

The genetic position of the only European Cape hare population (occurring in Sardinia) was disregarded in the phylogeographic study conducted by Lado et al. (2019). However, Cape hares are not native to Sardinia, but supposedly stem from a small number of North African founders carried by seafarers (possibly from Tunisia or nearby regions of Lybia or Algeria; Suchentrunk et al. 2006; Canu et al. 2012; Soria-Boix et al. 2019). The finding of a tibia of hare in a late Bronze Age excavation site (Serra Niedda nuragic settlement) represents the oldest record for Sardinia, even though its dating was not certain (Wilkens and Delussu 2003).

Unsurprisingly, the taxonomic position of the Sardinian Cape hare population is uncertain. Amori et al. (1996) and Mitchell-Jones et al. (1999) classified it as a subspecies of *L. capensis* and *L. c. mediterraneus*, and this classification is the one that receives the greatest consensus today (e.g., Schai-Braun and Hackländer 2016). In the work by Alves et al. (2003), where both mitochondrial and nuclear genes were analyzed, the Sardinian hare was grouped with *L. capensis* from Morocco and separated from South African *L. capensis*. Canu et al. (2012) found divergence among Sardinian hares, Tunisian Cape hares, and Italian brown hares in terms of skull shape and both nuclear and mitochondrial genetic markers. Today the Sardinian hare is grouped with hares from the coastal regions of North Africa, to which it shows the highest morphological and genetic similarity (Alves et al. 2003; Ben Slimen et al. 2006; Scandura et al. 2007; Canu et al. 2012). This monophyletic group has been reported to be phylogenetically more related to *L. europaeus* and *L. saxatilis*, than to geographically distant South African *L. capensis* populations (e.g., Scandura et al. 2007; Koutsogiannouli et al. 2012). Thus, some authors suggested that hares from North Africa and Sardinia could represent a distinct taxon for which the name *L. mediterraneus* Wagner, 1841 was proposed (Alves et al. 2003; Pierpaoli et al. 1999; Scandura et al. 2007; Soria-Boix et al. 2019), but the debate remains open awaiting more comprehensive phylogeographic studies.

The taxonomic position of Chinese hare populations, previously ascribed to *L. capensis,* is under debate as well, and most recent studies suggest that they should not be included in *L. capensis*. Both phylogenetic and morphological investigations have indeed disclosed the huge divergence of these populations from African Cape hares and their similarity with *L. tolai* and *L. oiostolus* (Alves et al. 2008; Cheng et al. 2012).

A further taxonomic issue concerns the relationship between *L. capensis* and *L. victoriae* (occurring in most of the sub-Saharan Africa), given that in some areas, like Kenya and Somalia, specimens appear undistinguishable and with intermediate characteristics between the two species (Schai-Braun and Hackländer 2016), although some diagnostic traits were suggested in the past (Suchentrunk et al. 2007).

As consequence of the historical, but also current, lack of knowledge in the taxonomy of the genus that involves this species, much information on morphological and life history traits, physiology, and behavior, formerly attributed to *L. capensis*, actually refers to populations that are currently ascribed to other hare species. Therefore, in the following sections, available data were reconsidered on the basis of the most recent taxonomic studies.

Current Distribution

Due to the above-mentioned taxonomic uncertainties, even tracing the distribution of the Cape hare is challenging. In the past the species was considered to cover a wide range including Africa and a large portion of Asia (to China and Mongolia in the east, Flux and Angermann 1990). However, following recent taxonomic revisions, the sub-range east to Iraq should be no longer ascribed to *L. capensis* (Smith et al. 2018).

In Africa, its range is split into a southern and a northern sub-range. The former encompasses a region from the Cape Town area in the south to Namibia, Botswana, Zimbabwe, and Mozambique in the north. The northern sub-range includes Tanzania and Kenya, in the south-east, extending north to Sudan and Eritrea, and from here west, along the sub-Saharan (Sahel) region to Senegal and Mauritania, and north to Egypt, extending, across the whole Maghreb region, west to the Atlantic coast of Morocco.

The West Asian sub-range includes the Mediterranean region from the Sinai Peninsula north to Syria and western Iraq and several scattered populations in the Arabian Peninsula.

In Europe, the species occurs only in Sardinia and in the surrounding islands of Asinara, San Pietro, and Sant'Antioco (Fig. 1). These populations are the result of multiple introductions in historical times (Masseti and De Marinis 2008). As attains to the Sardinian population, its distribution can be considered continuous from coasts and lowlands to the major reliefs (present in nearly 90% of the island, Meriggi et al. 2010).

Although even recent reviews on *L. capensis* still report that this lagomorph occurs on Cyprus (e.g., Schai-Braun and Hackländer 2018; Johnston et al. 2019), already Kasapidis et al. (2005) and later Ashrafzadeh et al. (2018) revealed that the population of this island is actually represented by *L. europaeus* genetically related to the Anatolian populations.

Description

The Cape hare is a medium-sized mammal (Fig. 2a), showing a geographical variation in body size resulting from genetic as well as ecological factors, such as spatial and temporal variation in food availability. Females are larger and heavier than males, due to positive relationship between size and reproductive capacity (Flux et al. 2016). Only the body weight variation of males seems to be in accordance with the Bergmann's rule.

Body measurements for the Sardinian population are: body mass 1.5–2.5 kg; head-body length 400–500 mm; hind foot length 93–103 mm; ear length 97–114 mm; tail length 80–100 mm; mandible length 57.4–66.5 mm; and zygomatic width 37.1–42.1 mm (Miller 1912; Toschi 1965; Canu et al. 2012).

Sex identification based on phenotypical characteristics is not possible from the distance, but only through a close examination of a living or dead specimen. This is because of the high similarity between genital signs of the two sexes, due to the presence of internal testes in males when they are not breeding. Sexing young individuals or those in reproductive stasis may be even more complicated (Trocchi and Riga 2005). Molecular sexing has been proposed based on the co-amplification of specific fragments of the Y chromosomal SRY gene and a portion of the Transferrin gene (Mengoni et al. 2015).

The skull (Fig. 3) is characterized by a preorbital region with long and broad nasals

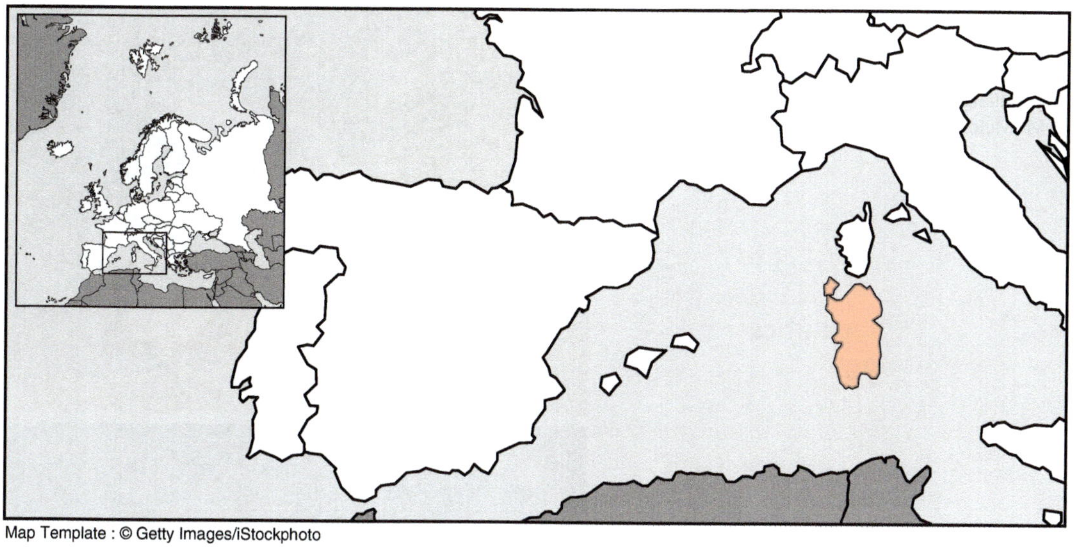

Fig. 1 Distribution map of the Cape hare in Europe, based on the IUCN Red List of Threatened Species, version 2021-3. (Map Template : © Getty Images/iStockphoto)

projecting usually beyond the incisors in dorsal view, an interorbital region with well-developed zygomatic arches, supraorbital processes with deep anterior notches, mesopterygoid fossa usually much wider than palatal bridge, and a postorbital region with elongated braincase and large auditory bullae (cf. Corbet 1978). As in other species of the genus *Lepus*, premolar foramen is absent (Fostowicz-Frelik and Meng 2013).

No significant differences in skull morphology were found between sexes (Riga et al. 2001; Suchentrunk et al. 2007; Canu et al. 2012).

A comparative craniometrical analysis of Sardinian and North African hares revealed the peculiar skull shape of Sardinian hares (Canu et al. 2012). After the introduction of the species in Sardinia from Northern Africa in historical times, the long-lasting isolation and the consequent genetic drift resulted in a skull differentiation, perhaps due to an adaptation to local environmental conditions (Canu et al. 2012). Sardinian hare skulls proved to be smaller than Moroccan and larger than Tunisian ones but were within the range of Algerian and Mauritanian populations (Canu et al. 2012). This confirms the role of environment on the large skull geographical variation in this species (Table 1).

Protocols based on morphological and morphometric parameters of the cranial and postcranial bones were developed to identify scrub (*L. saxatilis*) and Cape hares, even when they occur in fragmentary form as in archaeological faunal remains (Scott 2018).

Various glands (chin, orbital, inguinal, and anal) have an important role in the communication.

There is such a wide range of variation in pelage color throughout the species distribution that is impossible to provide a description that fits all these variations, if not in very general terms or referring to one or a few populations. In Sardinia, the Cape hare has tawny upper parts graying of black with rump tending toward grayish; under parts white in color extending from posterior portion of the chest to the inguinal patches in a central area wide not more than 50 mm; sides light tawny in color that fades into the white of the under parts; head like upper parts but darker with a clear rust nuchal patch; chin opaque white or light tawny; chest light tawny; ears with rounded dark tips and fringed with long pale hairs on inner surface and a narrow but strongly contrasting whitish rim on outer surface; clear whitish eye rings; limbs light tawny with inner surface lighter in color,

Fig. 2 (**a** and **b**) Cape hare in Sardinia. (**a**) Standing. (© Courtesy of V. Cadoni). (**b**) Running. (© Courtesy of M. Meloni)

especially in fore limbs, but without any evident line of demarcation; and tail black above and white below (Miller 1912; Toschi 1965; Trocchi and Riga 2005).

Throughout the species range, coat color can vary from exceptionally pale to almost black in appearance (Flux and Angermann 1990; Skinner and Chimimba 2005; Ben Slimen et al. 2007) with nuchal patch more or less contrasting with the color of the upper parts and the junction between the upper and under parts more or less marked. Individuals living in arid and semi-arid habitats are paler in color than those living in grassland habitats with higher mean annual rainfall (Skinner and Chimimba 2005), revealing that body coloration tends to match the background, like observed in other lagomorphs (Stoner et al. 2003). Variation in pelage color seems to be clinal.

The dorsal guard hairs are shielded. They are black in color fading into the grayish of the shaft with a tawny bar in the central part of the shield. The medullary pattern is multicellular column-shaped, typical of the Leporidae family; the shield cross section has one or two concave sides like a dumb bell (De Marinis and Agnelli 1993); cuticular scale pattern changes considerably along the length of the hair and overlap between hare species (Perrin and Campbell 1980; Teerink 1991; Fattorini et al. 2018). The analysis of hair structure has an important role in the studies of food habits of predators. Hair samples of *L. capensis* can be distinguished from those of *L. europaeus*, *L. timidus*, and *L. corsicanus* on the basis of morphological characters and morphometric parameters of the dorsal guard hairs, like hair length, pale bar width and distance between pale bar and hair tip (Fattorini et al. 2018). The width of the pale bar is typically smaller than in other species (on average 3.5 mm).

The molt process may influence the general appearance of the coat (Skinner and Chimimba 2005). An analysis of molting chronology

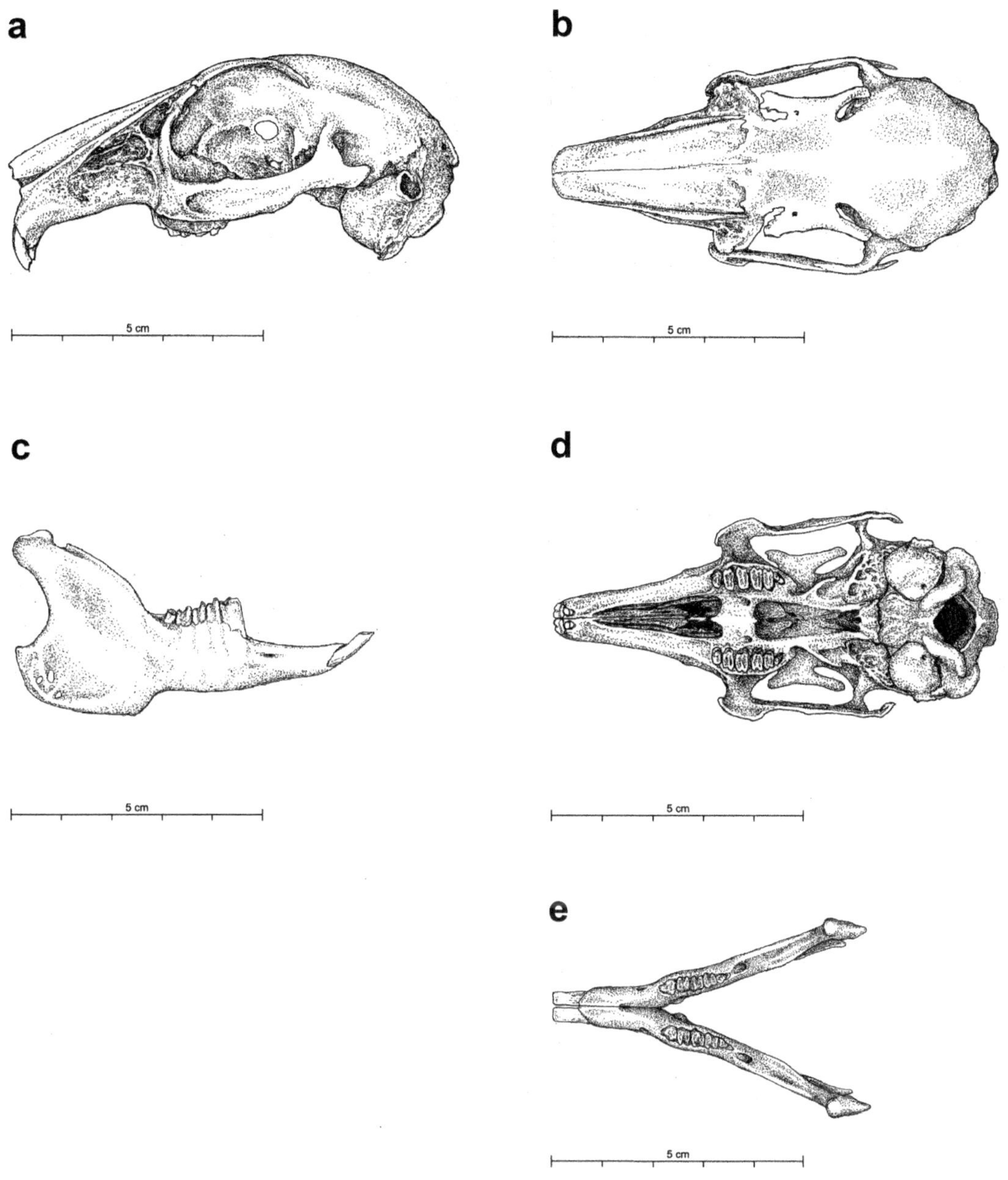

Fig. 3 Skull of *Lepus capensis*. (**a, b, d**) lateral, dorsal and ventral view of skull, (**c**) lateral view of left mandible (view from outside), (**e**) dorsal view of mandible. (© Franz Müller with kind permission)

involving seasonal timing and duration of molt in relation to age, sex, body condition, reproduction status, and population has not yet been carried out in this species.

Permanent dentition has formula I 2/1, C 0/0, P 3/2, M 3/3. Deciduous teeth are rarely seen as they are generally shed before birth or shortly after (within approx. 1 month).

Anterior upper incisors are squarish in cross-section along the entire length of the teeth with enamel groove in vestibular view consistently relatively shallow, always of simple configuration

Table 1 Variation in condylobasal length of Cape hare skulls across the species range

Locality	*N*	Average value (range), in mm	Reference
Sardinia	18	75.6 (72.5–78.8)	Canu et al. (2012)
Morocco	16	78.4 (72.7–83.8)	Canu et al. (2012)
Mauritania	3	72.1 (69.2–75.3)	Canu et al. (2012)
Algeria	11	72.2 (68.0–74.7)	Niethammer (1963)
Algeria	13	73.6 (70.8–76.5)	Canu et al. (2012)
Tunisia	33	71.0 (65.9–76.4)	Canu et al. (2012)
Egypt	16	78.8 (75.3–82.3) males	Osborn and Helmy (1980)
	8	76.8 (74.9–78.9) females	
Kenya	20	75.2 (71.8–75.2) males	Suchentrunk et al. (2007)
	27	76.6 (74.4–79.1) females	
South Africa	20	78 (75.2–81.6)	Palacios et al. (2008)

and filled with cement (Robinson 1986); this shape is probably associated with the texture of the vegetation eaten (Flux and Angermann 1990). Upper incisor cross-sections do not change with age (Robinson 1986). The similarity in incisor patterns among specimens from different localities may be indicative of the reliability of this character for diagnostic purposes (Robinson 1986). Cheek teeth are composed by two transverse lobes; premolars are molarized and their morphology was described in detail by Vismara (2012).

The pattern of spatial divergence in the occlusal surface generally reflects the gene pool differentiation among populations and between hare species (e.g., Robinson 1986; Azzaroli-Puccetti 1987; Palacios and Fernandez 1992; Suchentrunk et al. 1994). Therefore, nonmetric dental characters have been widely used for separating extant species and identifying fossil specimens (Suchentrunk and Flux 1996). Three dichotomous keys have been developed on the basis of morphological characters on the occlusal surface of P^2, P^3, and P_3 for the identification of *L. capensis*, *L. corsicanus*, *L. timidus*, and *L. europaeus* (Vismara 2012).

As in other leporids, skulls can be indicative of the age of the individual, according to the progressive ossification of the sutures. In particular, the *sutura frontalis, sut. coronaria, sut. sagittalis,* and *sut. parietotemporalis* are used for age estimation (Palacios et al. 2008; see also Palacios 1989). The most commonly applied method to distinguish young and adult hares is that described by Stroh (1931), consisting in checking the presence of the epiphyseal protrusion of the ulna, approximately 1 cm away from the wrist, through palpation, after dissection or using radiography. The time at which the ossification is completed is similar to that of the brown hare, though the two species differ considerably in body weight. This method can be applied under field conditions and on living animals.

Eye lens growth is continuous throughout life and is independent of gender (Trocchi et al. 2003a). Eye lens dry weight allows to distinguish the hares younger than 1 year from those older than 1 year. The reference curve for aging Cape hares by means of eye lens dry weights, including confidence interval, was constructed according to a nonlinear regression of dry weight on real age of hares bred in captivity in Sardinia (Trocchi and Riga 2005). In the first 2 months of life, the weight increases linearly and it is possible to estimate the age in days and also the date of birth, knowing the date of death (Trocchi and Riga 2005). The cut-off point for the weight of an eye lens belonging to an adult hare corresponds to 240 mg (Trocchi and Riga 2005). After the first year of life, the age estimation is progressively less reliable due to individual variation in lens weights (Trocchi et al. 2003a).

Annual layers formed in the cementum of the teeth root cannot be used for age determination because of hares having rootless and evergrowing teeth (Klevezal and Kleinenberg 1967). The periosteal growth lines which are formed in mandibles and postcranial bones of many lagomorphs (Klevezal 1996), as well as tibia, phalanges, and metatarsals, are suitable for age determination.

Physiology

Cape hare has particular physiological adaptations to life in extreme arid environments. Individuals maintained on dry alfalfa hay in captive conditions (Kronfeld and Shkolnik 1996) showed the capacity to digest efficiently high-fiber plant material (80% of the food eaten is digested), to use highly concentrated salt solutions (almost twice the osmotic concentration of sea water) as a water source, and to excrete the excess of salt in a concentrate urine. Moreover, the hares showed low values of oxygen consumption and water turnover rates.

In relation to their cursorial locomotion, Cape hares have a large heart (Smith et al. 2018), and the voluntary muscles are adapted to high endurance requirements, with many capillaries per mm^2 and more myoglobin (Flux and Angermann 1990).

Like in other hare species, the temperature is controlled by panting and through heat radiation from the ears. The level of blood circulation through the ears is under the control of the central nervous system and up to 40% of heat loss can be via the ears.

Genetics

The diploid chromosome number is $2n = 48$, like in all other hare species (Flux and Angermann 1990).

Phylogenetic reconstructions within the genus *Lepus* have long been hampered by poor genetic sampling, the rapid radiation of the genus, and the lack of phylogenetically informative cytogenetic and mitochondrial DNA characteristics (Melo-Ferreira et al. 2012; Schai-Braun and Hackländer 2016). Recently, Ferreira et al. (2021) used a phylogenomic approach based on exome sequences to understand the complex evolutionary relationships among 15 hare species. Major adaptive radiation of the genus occurred approximately in the last 4 million years, and regional speciation was even more recent (less than 1 mya for some species group), creating the conditions for reticulate evolution. The genomic analysis has identified groups of closely related species (e.g., a group of African hares including *L. capensis*, *L. fagani*, *L. habessinicus*, and *L. starcki*) and highlighted the extensive and recurrent introgressive hybridization between and within these lineages (Ferreira et al. 2021). Such introgression has seemingly characterized hares evolution and the consequent shared polymorphisms have likely played a role in the group radiation, enhancing local adaptation. Nonetheless, deep and comprehensive investigations on African hares are still lacking (see Happold 2013; Lado et al. 2019).

Table 2 shows genetic diversity indices at nuclear (microsatellites) and mitochondrial (control region) markers that are reported in the scientific literature for Cape hare populations (although some of them should be considered rough approximations, given the low sample size). Given their uncertain taxonomical position, East Asian populations formerly attributed to Cape hare are not included in the table.

Focusing on the only European population of *L. capensis*, inhabiting Sardinia, Canu et al. (2012) showed the presence of an unstructured population considering nuclear markers, while mtDNA analyses indicated the presence of a gradient of increasing diversity from the northern to the southern part of the island. This is in line with the general lack of concordance between nuclear and mitochondrial markers in hare populations (e.g., Ben Slimen et al. 2008b; Lado et al. 2019), with higher structuring in the latter, likely caused by female philopatry and differences in effective population size and mutation rates between the two types of markers.

As expected for an insular population, nuclear genetic diversity in Sardinian hares (He = 0.59–0.61, Table 2) was generally lower than in conspecific populations of Africa; nevertheless, it was in the range of other related species as *L. europaeus* (0.46–0.71 in Ben Slimen et al. 2008a). On the other hand, mitochondrial genetic diversity in Sardinian Cape hare was surprisingly

Table 2 Genetic diversity in Cape hare populations. (A) Nuclear DNA (microsatellites), (B) mitochondrial DNA (control region). *N* sample size, N_A average number of alleles per locus, H_e expected heterozygosity, H_o observed heterozygosity, N_m number of investigated microsatellite loci. n.d. - not determined.

(A) Nuclear DNA

Area	*N*	N_A	H_e	H_o	N_m	Source
Sardinia	92	8.00	0.59	0.52	13	Mengoni (2011)
Sardinia	85	7.17	0.61	0.52	12	Canu et al. (2012)
Tunisia	26	9.75	0.80	0.54	12	Canu et al. (2012)
South Africa	9	5.82	0.71	0.59	11	Ben Slimen et al. (2008a)
Morocco	17	6.91	0.69	0.57	11	Ben Slimen et al. (2008a)
Tunisia	26	7.73	0.69	0.62	11	Ben Slimen et al. (2008a)
Egypt	6	3.90	0.64	0.61	11	Ben Slimen et al. (2008a)
South Israel	6	4.09	0.60	0.50	11	Ben Slimen et al. (2008a)
South Africa (all subspecies)	66	10.54	0.76	0.60	13	Suchentrunk et al. (2009)
South Africa (subspecies *capensis*)	13	4.23	0.57	0.56	13	Suchentrunk et al. (2009)
South Africa (subspecies *centralis*)	12	6.08	0.68	0.64	13	Suchentrunk et al. (2009)
South Africa (subspecies *kalaharicus*)	11	4.69	0.62	0.61	13	Suchentrunk et al. (2009)
South Africa (subspecies *vernayi*)	7	5.31	0.68	0.59	13	Suchentrunk et al. (2009)
South Africa (subspecies *ochropus*)	7	5.38	0.68	0.65	13	Suchentrunk et al. (2009)
South Africa (subspecies *ermeloensis*)	11	4.23	0.61	0.62	13	Suchentrunk et al. (2009)

(B) Mitochondrial DNA

Area	*N*	Nr. of haplotypes	Haplotype diversity	Nucleotide diversity	Source
Sardinia	21	9	0.87	0.014	Pierpaoli et al. (2003)
Sardinia	49	19	0.82	0.013	Canu et al. (2012)
Sardinia	86	27	0.83	0.012	Mengoni (2011)
Tunisia	54	10	0.74	0.009	Ben Slimen et al. (2006)
Morocco	6	5	0.93	0.013	Ben Slimen et al. (2006)
South Africa	9	4	0.75	0.019	Ben Slimen et al. (2006)
Tunisia	133	n.d.	0.97	0.027	Ben Slimen et al. (2017)

high and in line with African populations (haplotype diversity = 0.82–0.87; nucleotide diversity = 0.012–0.014). This is mainly due to the occurrence of two sharply differentiated mitochondrial haplogroups in the island (E1 and E2 in Scandura et al. 2007), both shared with North African Cape hares. Nevertheless, all the mitochondrial haplotypes found in Sardinia were private, and Sardinian hares turned out to diverge from North African ones considering nuclear markers as well (Canu et al. 2012), showing an endemic genetic diversity, as expected after two millennia of independent evolution.

Introgressive hybridization has been reported in many hare species (e.g., Alves et al. 2008; Melo-Ferreira et al. 2012; Ferreira et al. 2021), being more likely in cases of secondary contact of poorly differentiated gene pools. The Cape hare does not represent an exception. Lado et al. (2019) detected by coalescent simulations a possible mtDNA introgression from *L. europaeus* into Near Eastern populations of *L. capensis*, supporting the gene flow between *L. capensis* and *L. europaeus* hypothesized by Ben Slimen et al. (2008a, see section "Taxonomy, Systematics, and Paleontology"). However, no evident sign of recent introgression from *L. europaeus* was found in the Sardinian Cape hare population at nuclear and mitochondrial loci, despite the two species could have come into contact on the island as result of uncontrolled restocking practices for hunting (Canu et al. 2012). Introgression was also documented in Chinese hares that were previously ascribed to *L. capensis* (Wu et al. 2011; Lado et al. 2019).

Table 3 Life history traits of Cape hares in Sardinia. *NA* not available

Trait	Mean (min–max)	Reference
Gestation (days)	42	Angelici and Spagnesi (2008)
Weight at birth (g)	NA (61–127)	Trocchi and Riga (2005)
Litter size (*n*)	1.4 (1–3)[a]	De Marinis et al. (2007)
Litters per year (*n*)	2.6 (1–4)[a]	De Marinis et al. (2007)
Weaning (days)	24	Angelici and Spagnesi (2008)
Age at first reproduction (months)	4[b]	Beccu and Fassò (1993)
Female body mass at first reproduction (kg)	1.5[b]	Beccu and Fassò (1993)
Maximum longevity (years)	8[b]	Angelici and Spagnesi (2008)

[a]Through the analysis of placental scars
[b]In captivity

Life History

Cape hares do not show a strong seasonality in reproduction. Male testes do not vary in size, indicating a constant activity through the year. Similarly, females can be pregnant in any moment of the year (Flux 1981), as observed in other species living in the Mediterranean region, for example, *L. corsicanus* (De Marinis et al. 2007), *L. europaeus* (Antoniou et al. 2008), and *L. granatensis* (Alves et al. 2002).

Gestation lasts approximately 42 days and the interval between births averages 39 days (Jones et al. 2009).

Leverets are born precocial in ground depressions, as in other hare species. Females place newborns under grass to hide them from predators (Caravaggi 2018). Leverets stay close to the birthplace for the first days after birth and are nursed once per day with highly concentrated milk; gradually they enlarge their movements and return to the birthplace only for nursing, relying on odor cues from its own body independently of any maternal contribution to the nursing site odor (Stavy et al. 1985). The growth rate is approximately 10 g/day and the adult size is reached in 4–5 months (Flux and Jarvis 1970). Weaning age and weaning mass range, respectively, from 24 to 33 days (Angelici and Spagnesi 2008; Jones et al. 2009) and from 915 to 943.4 g (Ernest 2003; Jones et al. 2009). Sexual maturity is reached at 5–7 months of age (Jones et al. 2009). The number of young/year is on average 11.6 (Flux 1981), equivalent to 55% of adult female body mass, that is, a rather high percentage among lagomorphs (Happold 2013).

Breeding activity and reproductive parameters change by locality, probably in response to changes in rainfall and forage (Happold 2013). The reproductive activity of Cape hare in Sardinia (see Table 3) increases in winter and has a regressive phase in summer, as revealed by the analysis of placental scars in a sample of wild hares (De Marinis et al. 2007). Females had on average 2.6 l/year (ranging from 1 to 4) with a mean litter size of 1.4 (ranging from 1 to 3) or an average of 3.6 young/year (ranging from 1 to 6). A 7-month female was found to be pregnant while a 4-month individual showed resorption of embryos and a *corpus luteum* (De Marinis et al. 2007). A 22% prenatal mortality, due to the resorption of embryos, was observed in pregnant females; this phenomenon allows to reduce the litter size in unfavorable conditions and was recorded also in other leporids (Pépin et al. 1981; Alves et al. 2002; De Marinis et al. 2005). In captivity, the reproductive output of Sardinian hares appears higher than in natural conditions: females have on average 4 l/year, up to a maximum of 8 l/year, with an average of 8.5 young/year (ranging from 1 to 20). Captive females can reproduce from the year of birth up to 6 years of age, with an average interval between consecutive births of 55.4 days. The reproductive activity lasts throughout the year and shows an increase in spring unlike what observed in the wild.

Habitat and Diet

The Cape hare is extremely adaptable and lives in a wide range of habitats. Across its range, it occurs in grasslands, Acacia (*Acacia* sp.) and miombo (*Brachystegia* sp.) savannah, Sahel and Sudan savannah, steppe, semi-desert, desert, and on alpine meadows of mountain valleys up to 2400 m (Smith et al. 2018). Pastures overgrazed by livestock often represent a favored habitat, but Cape hares can also take advantage of bush clearing and extensive savannah fires, moving into burned areas as soon as the grass begins to sprout (Flux and Angermann 1990).

In Sardinia, the Cape hare occurs in all environments, from the sea level to the mountains. It is common in habitats characterized by small, cultivated plots and arable land mixed with bushy areas (Mediterranean scrub) and grasslands (Amori et al. 2008). According to preliminary results obtained by Murgia et al. (2003), cultivated areas seem to be preferred to wooded areas and low scrubland.

Cape hares are herbivorous. There is little information available on their diet, but it is reported to vary according to habitat (Flux and Angermann 1990). In fact, the Cape hare is an opportunistic species which selects different grass species according to their relative availabilities (Smith et al. 2018). Proportions of grass species in the diet may differ significantly among sites, while seasonal differences in diet are apparently less important. Cape hares are able to adapt even to conditions that are prohibitive for other species. For example, in the Negev Desert (Israel), the main source of water for Cape hares during the prolonged dry summer are the succulent plants that contain sap with high concentrations of electrolytes. At the end of summer, the Cape hare diet in this region consists mainly of emaciated desert shrubs, high in plant cell-wall constituents, or xerophilic succulent plants, high in salt content (Kronfeld and Shkolnik 1996). In times of food shortage, coprophagy (ingestion of hard feces) may help individuals to survive to extended fasting periods (Schai-Braun and Hackländer 2016).

Behavior

Cape hares are typically nocturnal. They usually graze during the night but, in cloudy days, it may happen to find them feeding at daylight (Smith et al. 2018). Resting takes usually place in depressions of the ground. Occasionally, in desert environments, Cape hares are observed to use burrows to shelter from the sun (Dixon 1975). When disturbed, they run into open habitats and use their speed to escape predators. Nonetheless, they appear slower than *L. europaeus* (Angelici and Spagnesi 2008).

Home ranges of Cape hares vary depending on habitat features and on the period of the year (Flux and Angermann 1990). Data on space use by Sardinian hares are limited and refer to three radio-collared individuals (two males and one female, Onida et al. 1995). This preliminary study revealed that male home ranges (minimum convex polygon) were relatively small in October (1.6–23 ha) but increased considerably and overlapped in the following 2 months (75–106 ha), shrinking again in January (0.8–3.2 ha). The female home range was instead small and stable during the same period (3–6 ha). Onida et al. (1995) speculated that the increase observed in male home ranges in November–December could be related to a phase of low reproductive activity leading to a relaxation of intrasexual competition. Home ranges of different individuals are likely to overlap at feeding grounds.

The Cape hare is a solitary animal that only occasionally lives in small groups (Flux 1981). During the estrus, females may be followed by several males for a short period. Groups of maximum five to six individuals are reported (Trocchi et al. 2003b).

Specific information on the mating behavior of *L. capensis* is lacking. According to the timing of parturition, breeding activity can take place year-round. Males can become bolder during the rut, occasionally showing aggressive interactions with other males, as observed in several leporids (Nowak 1999).

Little is known about the escape behavior in Cape hares. It was reported that they use to stop and do not run when they hear a shot or are dazzled by a spotlight (Flux 1981).

Parasites and Diseases

Cape hares may carry a number of external parasites and, in particular, across their wide range, they harbor a large number of tick species. In many cases, more than 50 ticks/individuals were recorded (e.g., Al Rammhi 2013). No information on tick presence is available for Sardinian hares.

The adults of some of the tick species infesting hares as larvae or nymphs are important vectors of disease or toxins to domestic livestock or humans, whereas others are of minor concern (Horak et al. 2005). For example, among the tick species detected on Cape hares, adults of *Rhipicephalus warburtoni* cause paralysis in sheep and goats, while adults of *Hyalomma truncatum* secrete a toxin that is the cause of sweating sickness in domestic cattle, sheep, and goats (Horak et al. 2005). As many mammals, Cape hares may be infested by such ticks as *Ixodes* spp., vectors of *Borrelia burgdorferi*, a spiral bacterium which is transferred to humans by tick bites and cause the Lyme borreliosis (Anderson and Magnarelli 1993). Ticks of the genus *Hyalomma* carried by Cape hares may host the Crimean-Congo hemorrhagic fever virus (CCHFV), which is the etiological agent of the most widespread tick-borne viral infection worldwide (Shayan et al. 2015).

As regards endoparasites, the most common are represented by tapeworms (Cestoda). In Sardinia, larvae of the nematode *Graphidium strigosum* and of the protozoa *Eimeria leporis* and *E. semisculpta* were reported at high frequencies (90–100%) in captive Cape hares (A. Scala, pers. comm.). The nematode *Passalurus ambiguus* and the trematode *Fasciola hepatica* were also found at lower frequencies (Arru et al. 1967; A. Scala, pers. comm.).

Cape hares are affected by the Rabbit hemorrhagic disease virus 2 (RHDV2; Lagovirus, genotype GI.2), which is a pathogenic calicivirus that mostly affects wild rabbits (*Oryctolagus cuniculus*) and, more sporadically, other hare species like *L. europaeus*, *L. corsicanus*, and *L. timidus* (Puggioni et al. 2013). This virus has been detected in France for the first time in 2010 in domestic and wild rabbits and the following year in Italian *L. europaeus* and Sardinian Cape hares. Since then, it has spread throughout Europe, replacing the circulating RHDV/RHDVa strains in most European countries. RHDV2 causes acute and lethal forms of hepatitis, with gross damages to liver, spleen, lungs, trachea, kidneys, and central nervous system (Rocchi and Dagleish 2018). While the first RHDV2 isolates from 2010 to 2011 were only mildly pathogenic, highly pathogenic RHDV2 strains have emerged during the virus evolution and have become prevalent in the field (Capucci et al. 2017). Conversely, there is no evidence of sensitivity by Cape hares to a similar form of hepatitis, the European Brown Hare Syndrome, caused by the calicivirus EBHSV (highly related to RHDV), which affects *L. europaeus*, *L. corsicanus* and *L. timidus* (Frölich and Lavazza 2008; Lavazza and Guberti 2007).

The impact of pathogens on the present status of the Sardinian hare population has not been investigated.

Population Ecology

The population dynamics of *L. capensis* is not well known and deserves further studies. In Sardinia, the population size seems to increase from May to December, as a consequence of reproductive activity (Trocchi et al. 2003b). In winter, this activity declines, leading to a stabilization of the population size until the next spring.

No specific studies have been done on the effects of competition with other sympatric lagomorphs, like *O. cuniculus* in Sardinia or related *Lepus* species in Africa (e.g., *L. saxatilis* and *L. victoriae*). Similarly, relationships with predators are unclear. Carnivores are the main predators of Cape hares, whereas eagles cannot represent a big threat due to their diurnal habits. Red fox (*Vulpes vulpes*), wild and stray/feral cats (*Felis lybica and F. catus*), stray/feral dogs (*Canis lupus familiaris*), and pine martens (*Martes martes*) are the typical predators of *L. capensis* in Sardinia (Angelici and Spagnesi 2008). However, the amount of mortality due to natural predation, as well as its impact on the population dynamics of the species, is unknown.

A relevant cause of juvenile mortality is reported to be linked to agricultural activities, as in the first weeks of life, hares, if threatened, tend to become immobile and are crushed by agricultural vehicles.

Conservation Status

The Cape hare is considered fairly common across its broad geographic range and its status was classified as Least Concern by the International Union for the Conservation of Nature (IUCN), both globally (Johnston et al. 2019) and in Europe (Temple and Terry 2009), though a general decreasing trend was observed across the species' range (Smith et al. 2018). The species "*Lepus capensis* (*europaeus*)" appears in the Appendix III (protected species) of the Bern Convention, but seems to be referred to an undistinguished taxon. Instead, the Cape hare is not included in the Annexes of the EU Habitats Directive.

As attains to the Sardinian population, a declining trend was reported (Meriggi et al. 2010), although there are signals of an overall improvement in status, despite the great geographic heterogeneity. Notwithstanding this population was introduced by humans to the island, it represents an ancient introduction and is accordingly classified by the Italian legislation as "parautochthonous," that is, to be managed like a native species (Loy et al. 2019). Its conservation value derives from its being, together with the less common wild rabbit, the only lagomorph inhabiting Sardinia playing a major and exclusive ecological role.

Threats to the species in the island are represented by predation, diseases (e.g., RHDV2), poaching, overhunting, habitat fragmentation, fires, and pesticides used in agriculture. A further possible threat can be represented by the introduction of continental hares belonging to different species (namely *L. europaeus*) because of the possible ecological, sanitary, and genetic impact. Fortunately, no striking evidence of recent introduction/introgression has been detected so far by genetic inspection of hunted or captive-reared specimens (Canu et al. 2012). Although climate change is considered a possible threat to some lagomorph species (Leach et al. 2015), it is not expected to represent a concern for *L. capensis*, since this species is well adapted to live in semi-desert and even desert environments. In fact, an increase of this species in Africa was predicted under a climate change model (Olwoch et al. 2009).

Management

Hares are usually game species particularly appreciated by hunters. As such they raise special interests and populations are usually managed to safeguard and possibly enhance their availability over time. However, a programmed form of hunting has never found application in Sardinia, yet prescribed by a national law, because of the opposition of local stakeholders (basically hunters). This lack, together with the scarcity of population data, has so far hampered an effective management of Cape hares in Sardinia.

Sardinian hares are typically hunted in autumn, between the end of September and October. In the last 10 years, the number of available days for hunting dropped from five to zero (back to two at the time of writing), as consequence of precautionary initiatives aimed to cope with the declining trend of the species reported between the 1990s and 2010, yet with strong local variations (Trocchi et al. 2003b; Luchetti and Sacchi 2007; Meriggi et al. 2010).

A more recent investigation, based on spring spotlight census in hunting grounds, suggests a stabilization at medium-low densities in most areas of Sardinia (2–10 hares/km^2; Cossu et al. 2018). Nonetheless, spring densities higher than 50 hares/km^2 were occasionally unveiled both in protected (Meriggi et al. 2010) and in nonprotected areas (Cossu et al. 2018). The density values recently observed in Sardinia are similar to those previously recorded in the south of the island (1.6–7.3 hares/km^2 in 1994–1998; Murgia et al. 2003), as well as in other parts of the species' range. For instance, densities of 4.7–24.8 hares/km^2 were observed in South Africa (Happold 2013), whereas values in the range 2.1–8.0 hares/km^2 were estimated in three protected areas of Tunisia, spanning from semi-desert to arid Mediterranean environments (Mohsen et al. 2017).

Hunting takes place in organized hunting grounds (covering around 10% of the island surface) or in open public and private lands. The traditional form of hunting consists in flushing the animal out with the help of one or a few trained dogs. Hares are usually hunted in combination with other resident game species: the barbary partridge (*Alectoris barbara*), sharing the same habitats and ubiquitous in the island, and occasionally the wild rabbit, patchily distributed. No annual quota is defined, but a maximum number of heads per hunter and per hunting day are set (currently 1 hare/day).

In consequence of the high variation in local densities, hunting yields may differ a lot among areas. Meriggi et al. (2010) reported that on average 1.6 hares/km^2 were harvested in hunting grounds (the only areas for which hunting bag data are currently available), with very few areas harvesting >5 hares/km^2. In the hunting season 2012/2013, a total harvest of 4890 hares was reported by hunters in Sardinia (91% of data coverage).

With the aim to increase local densities, areas are specifically set to favor the natural reproduction of Cape hare, wild rabbit, and barbary partridge (named "Zone Temporanee di Ripopolamento e Cattura"). In these areas, hunting is forbidden and annual activities (population counts, habitat improvements, etc.) are carried out to increase the productivity and natural dispersal to surrounding areas and to define a number of animals that could be captured and transferred to external depleted areas. Unfortunately, though existing in good numbers in Sardinia, these areas are seldom managed in a proper way and the conditions for restocking with wild-captured heads are persistently lacking.

Cape hares are bred in captivity in a very limited number of public and private farms. Breeding protocols used for the *L. europaeus* (L. Mandas, pers. comm.) were initially applied but resulted ineffective and were progressively adapted to the biological requirements of the Cape hare. Nonetheless, a few releases of captive-reared hares have taken place in the past.

The interest towards this species makes it a valuable resource. Actually, possible economic implications in Sardinia are constrained by a number of factors, namely: (1) poor management of the species, (2) prohibition of selling the meat of hunted hares, (3) prohibition of exporting Cape hares to the mainland, (4) low number of breeding farms, and (5) restrictions to the release of captive-reared hares. Therefore, the only economic repercussions are limited to those linked to a naturalistic enjoyment or to the hunting activity (weapons, equipment, hunting dogs, hunting tourism).

The species in Sardinia has a relatively low impact on human activities. Damages to crops are negligible, as well as those caused by traffic accidents. Nevertheless, some economically important diseases can be transmitted to livestock by tick infesting hares (see section "Parasites and Diseases").

Future Challenges for Research and Management

As remarked by many authors (e.g., Smith et al. 2018), knowledge on the Cape hare is largely incomplete.

Notwithstanding the insights provided by recent studies, taxonomic uncertainties remain that would take advantage from new comprehensive studies on the phylogeography of the *L. capensis* complex. They would deserve a targeted sampling in some areas that have not been sufficiently considered in previous investigations (e.g., sub-Saharan and East African populations).

As concerns the European population of Cape hares, researches on its ecology, behavior, and population dynamics urge, so as to assist a conservative management of the Sardinian population.

In particular, it will be crucial to focus on:

- Factors that are responsible of the low densities observed in many areas of Sardinia (even in presence of a high habitat suitability)
- Relative contribution of the different mortality factors in free-living populations
- Spatial behavior (especially habitat selection and dispersal patterns)
- Food habits in different ecological conditions (agricultural landscape, Mediterranean shrubland, high-elevation pastures)

Hunting needs to be reorganized to make possible the definition of sustainable annual quotas, on the basis of local abundances and recruitment rates estimated annually on the basis of standardized protocols. Hunting bag data must be regularly collected and analyzed, in order to provide information on relevant population parameters (sex and juvenile/adult ratios). Measures should be undertaken to favor natural reproduction and dispersal (e.g., by an efficient management of protected areas and the "Zone Temporanee di Ripopolamento e Cattura"), yet short-range (e.g., <50 km) translocations of live-trapped hares may be considered to restock heavily depleted areas. Instead, the recourse to captive-reared individuals should be discouraged.

Controls should be enforced to limit poaching, while predator populations should be monitored and, in presence of high densities, an active control of predators should be carried out.

Finally, care should be taken to inform and educate stakeholders to a sustainable use of the species, aimed to safeguard its viability over time.

Acknowledgments We are grateful to A. Scala and A. Varcasia for the valuable information provided.

References

Al Rammhi HM (2013) Tick infestation of hares (*Lepus capensis*) in Al-Qasim district – Babylon, Iraq. Euphrates J Agric Sci 5:8–14

Alves PC, Gonçalves H, Santos M, Rocha A (2002) Reproductive biology of the Iberian hare, *Lepus granatensis*, in Portugal. Mamm Biol 67:358–371

Alves PC, Ferrand N, Suchentrunk F, Harris DJ (2003) Ancient introgression of *Lepus timidus* mtDNA into *L. granatensis* and *L. europaeus* in the Iberian Peninsula. Mol Phylogenet Evol 27:70–80

Alves PC, Melo-Ferreira J, Freitas H, Boursot P (2008) The ubiquitous mountain hare mitochondria: multiple introgressive hybridization in hares, genus *Lepus*. Philos Trans R Soc Lond Ser B Biol Sci 363:2831–2839

Amori G, Angelici FM, Prigioni C, Vigna Taglianti A (1996) The mammal fauna of Italy: a review. Hystrix 8:3–7

Amori G, Contoli L, Nappi A (2008) Mammalia II. Erinaceomorpha, Soricomorpha, Lagomorpha, Rodentia. Calderini, Sole24Ore, Bologna

Anderson JF, Magnarelli LA (1993) Epizootiology of Lyme disease-causing borreliae. Clin Dermatol 11: 339–351

Angelici F, Spagnesi M (2008) *Lepus capensis* Linnaeus, 1758. In: Amori G, Contoli L, Nappi A (eds) Fauna d'Italia. Mammalia II. Erinaceomorpha, Soricomorpha, Lagomorpha, Rodentia, 2nd edn. Calderini, Bologna, pp 248–253

Angermann R (1983) The taxonomy of Old World *Lepus*. Acta Zool Fenn 174:17–21

Antoniou A, Kotoulas G, Magoulas A, Alves PC (2008) Evidence of autumn reproduction in female European hares (*Lepus europaeus*) from southern Europe. Eur J Wildl Res 54:581–587

Arru E, Deiana S, Nuvole A (1967) Distomatosi epatica nei Leporidi selvatici in Sardegna. Atti Soc Ital Sc Vet 21: 762–766

Ashrafzadeh MR, Djan M, Szendrei L, Paulauskas A, Scandura M, Bagi Z, Ilie DE, Kerdikoshvili N, Marek P, Soós N, Kusza S (2018) Large-scale mitochondrial DNA analysis reveals new light on the phylogeography of Central and Eastern-European brown hare (*Lepus europaeus* Pallas, 1778). PLoS One 13(10):e0204653

Azzaroli-Puccetti ML (1987) On the hares of Ethiopia and Somalia and the systematic position of *Lepus whytei* Thomas, 1984 (Mammalia, Lagomorpha). Atti Acc Naz Lincei 19:1–19

Beccu E, Fassò C (1993) Allevamento sperimentale della Lepre sarda (*Lepus capensis mediterraneus* W.). In: Cossu S, Onida P, Torre A (eds). Atti del 1° Convegno regionale sulla fauna selvatica in Sardegna, Oristano, pp 403–406

Ben Slimen H, Suchentrunk F, Memmi A, Sert H, Kryger U, Alves PC, Ben Ammar Elgaaied A (2006) Evolutionary relationships among hares from North Africa (*Lepus* sp. or *Lepus* spp.), Cape hares (*L. capensis*) from South Africa, and brown hares (*L. europaeus*), as inferred from mtDNA PCR-RFLP and allozyme data. J Zool Syst Evol Res 44:88–99

Ben Slimen H, Suchentrunk F, Shahin AB, Ben Ammar Elgaaied A (2007) Phylogenetic analysis of mtCR-1 sequences of Tunisian and Egyptian hares (*Lepus* sp. or spp., Lagomorpha) with different coat colours. Mamm Biol 72:224–239

Ben Slimen H, Suchentrunk F, Stamatis C, Mamuris Z, Sert H, Alves PC, Kryger U, Shahin AB, Elgaaied AB

(2008a) Population genetics of cape and brown hares (*Lepus capensis* and *L. europaeus*): a test of Petter's hypothesis of conspecificity. Biochem Syst Ecol 36: 22–39

Ben Slimen H, Suchentrunk F, Ben Ammar Elgaaied A (2008b) On shortcomings of using mtDNA sequence divergence for the systematics of hares (genus *Lepus*): an example from Cape hares. Mamm Biol 73:25–32

Ben Slimen H, Schaschl H, Knauer F, Suchentrunk F (2017) Selection on the mitochondrial ATP synthase 6 and the NADH dehydrogenase 2 genes in hares (*Lepus capensis* L., 1758) from a steep ecological gradient in North Africa. BMC Evol Biol 17(1):46

Canu A, Cossu A, Foddai R, Iacolina L, Ben Slimen H, Suchentrunk F, Apollonio M, Scandura M (2012) Differentiation under isolation and genetic structure of Sardinian hares as revealed by craniometric analysis, mitochondrial DNA and microsatellites. J Zool Syst Evol Res 50(4):328–337

Capucci L, Cavadini P, Schiavitto M, Lombardi G, Lavazza A (2017) Increased pathogenicity in rabbit haemorrhagic disease virus type 2 (RHDV2). Vet Rec 180(17):426

Caravaggi A (2018) Lagomorpha life history. In: Vonk J, Shackelford TK (eds) Encyclopedia of animal cognition and behavior. Springer, Cham, pp 1–9

Cheng C, Ge D, Xia L, Zhou C, Yang Q (2012) Morphometrics study on the so called "Cape hare" (Lagomorpha: Leporidae: *Lepus*) in China. Acta Theriol Sin 32:275–286

Corbet GB (1978) The mammals of the palaearctic region. A taxonomic review. British Museum (Natural History)/Cornell University Press, London/Ithaca

Cossu A, Luccarini S, Marcia P, Apollonio M (2018) Indagine per la valutazione della consistenza di lepre sarda (*Lepus capensis mediterraneus*) e pernice sarda (*Alectoris barbara*) sul territorio della regione Sardegna. Unpublished report for Regione Autonoma della Sardegna

De Marinis AM, Agnelli P (1993) Guide to the microscope analysis of Italian mammals hairs: Insectivora, Rodentia and Lagomorpha. Boll Zool 60(2):225–232

De Marinis AM, Trocchi V, Fallico A, Toso S (2005) Fertility and breeding season of the European rabbit (*Oryctolagus cuniculus*) in Sicily. In: Hadjisterkotis E (ed) Proceedings of the 25th international congress of the International Union of Game Biologists (IUGB) and the 9th international symposium Perdix, pp 86–91

De Marinis AM, Trocchi V, Mangiafico S, Fassò C, Mallia E (2007) Strategie riproduttive in tre specie di Lepre (*Lepus* sp. pl.) in Italia. In: De Filippo G, De Riso L, Riga F (eds) Conservazione di *Lepus corsicanus* De Winton, 1898 e stato delle conoscenze. IGF Publishing, Napoli, pp 75–81

Dixon JEW (1975) A note on the burrows used by the hare *Lepus capensis*. Modoqua 9:45–46

Ernest SKM (2003) Life history characteristics of placental nonvolant mammals: ecological archives E084-093. Ecology 84(12):3402

Fattorini N, Burrini L, Morao G, Ferretti F, Romeo G, Mori E (2018) Splitting hairs: how to tell hair of hares apart for predator diet studies. Mamm Biol 89:84–89

Ferreira MS, Jones MR, Callahan CM, Farelo L, Tolesa Z, Suchentrunk F, Boursot P, Mills LS, Alves PC, Good JM, Melo-Ferreira J (2021) The legacy of recurrent introgression during the radiation of hares. Syst Biol 70(3):593–607

Flux JEC (1981) Reproductive strategies in the genus *Lepus*. In: Myers K, MacInnes CD (eds) Proceedings of the World Lagomorph Conference. University of Guelph, Guelph, pp 155–174

Flux JEC, Angermann R (1990) The hares and jackrabbits. In: Chapman JA, Flux JEC (eds) Rabbits, hares and pikas: status survey and conservation action plan. The World Conservation Union, Gland, pp 61–94

Flux JEC, Jarvis JUM (1970) Growth rates of two African hares, *Lepus capensis*. J Mammal 51:798–799

Flux JEC, Flux MM, Knauer F, Suchentrunk F (2016) Body weight variation in Cape hares (*Lepus capensis*) from Kenya. Mamm Biol 81(Suppl):6

Fostowicz-Frelik L, Meng J (2013) Comparative morphology of premolar foramen in lagomorphs (Mammalia: Glires) and its functional and phylogenetic implications. PLoS One 8(11):e79794

Frölich K, Lavazza A (2008) European brown hare syndrome. In: Alves PC, Ferrand N, Hackländer K (eds) Lagomorph biology: evolution, ecology, and conservation. Springer, Berlin/Heidelberg, pp 253–261

Hackländer K, Ferrand N, Alves PC (2008) Overview of Lagomorph research: what we have learned and what we still need to do. In: Alves PC, Ferrand N, Hackländer K (eds) Lagomorph biology: evolution, ecology, and conservation. Springer, Berlin/Heidelberg, pp 381–394

Happold DCD (2013) Mammals of Africa, volume III: rodents, hares and rabbits. Bloomsbury Publishing, London, p 784

Harrison DL, Bates PJJ (1991) The mammals of Arabia. Harrison Zoological Museum, Sevenoaks

Hoffmann RS (1993) Order Lagomorpha. In: Wilson DE, Reeder DM (eds) Mammal species of the world. Smithsonian Institution Press, Washington, DC, pp 807–827

Hoffmann RS, Smith AT (2005) Order Lagomorpha. In: Wilson DE, Reeder DM (eds) Mammal species of the world, 3rd edn. Johns Hopkins University Press, Baltimore, pp 185–211

Horak IG, Fourie LJ, Braack LEO (2005) Small mammals as hosts of immature ixodid ticks. Onderstepoort J Vet Res 72:255–261

Johnston CH, Robinson TJ, Child MF, Relton C (2019) *Lepus capensis*. IUCN Red List Threat Species 2019:e. T41277A45186750

Jones KE, Bielby J, Cardillo M, Fritz SA, O'Dell J, Orme CDL, Safi K, Sechrest W, Boakes EH, Carbone C, Connolly C, Cutts MJ, Foster JK, Grenyer R, Habib M, Plaster CA, Price SA, Rigby EA, Rist J, Teacher A, Bininda-Emonds ORP, Gittleman JL,

Mace GM, Purvis A (2009) PanTHERIA: a species-level database of life history, ecology, and geography of extant and recently extinct mammals. Ecology 90:2648

Kasapidis P, Suchentrunk F, Magoulas A, Kotoulas G (2005) The shaping of mitochondrial DNA phylogeographic patterns of the brown hare (*Lepus europaeus*) under the combined influence of Late Pleistocene climatic fluctuations and anthropogenic translocations. Mol Phylogenet Evol 34:55–66

Klevezal GA (1996) Recording structures of mammals. Balkema Publishers, Amsterdam

Klevezal GA, Kleinenberg SE (1967) Age determination of mammals from annual layers in teeth and bones. Israel Program for Scientific Translations. Translated from Russian by J. Salkind

Koutsogiannouli EA, Moutou KA, Stamatis C, Mamuris Z (2012) Analysis of MC1R genetic variation in *Lepus* species in Mediterranean refugia. Mamm Biol 77:428–433

Kronfeld N, Shkolnik A (1996) Adaptation to life in the desert in the brown hare (*Lepus capensis*). J Mammal 77:171–178

Lado S, Alves PC, Islam MZ, Brito JC, Melo-Ferreira J (2019) The evolutionary history of the Cape hare (*Lepus capensis sensu lato*): insights for systematics and biogeography. Heredity 123:634–646

Lavazza A, Guberti V (2007) Malattie di *Lepus corsicanus* e programmi di monitoraggio e gestione sanitaria. In: De Filippo G, De Riso L, Riga F, Trocchi V, Troisi SR (eds) Conservazione di *Lepus corsicanus* De Winton, 1898 e stato delle conoscenze. IGF Publishing, Napoli, pp 83–88

Leach K, Kelly R, Cameron A, Montgomery WI, Reid N (2015) Expertly validated models and phylogenetically-controlled analysis suggests responses to climate change are related to species traits in the order Lagomorpha. PLoS One 10(4):e0122267

Loy A, Aloise G, Ancillotto L, Angelici FM, Bertolino S, Capizzi D, Castiglia R, Colangelo P, Contoli L, Cozzi B, Fontaneto D, Lapini L, Maio N, Monaco A, Mori E, Nappi A, Podestà MA, Sarà M, Scandura M, Russo D, Amori G (2019) Mammals of Italy: an annotated checklist. Hystrix 30(2):87–106

Luchetti S, Sacchi O (2007) Distribution and density of Mediterranean hare (*Lepus capensis mediterraneus*) and wild rabbit (*Oryctolagus cuniculus*) in Sardinia (Italy). Hystrix 18(Suppl):120

Masseti M, De Marinis AM (2008) Prehistoric and historic artificial dispersal of Lagomorphs on the Mediterranean Islands. In: Alves PC, Ferrand N, Hackländer K (eds) Lagomorph biology: evolution, ecology, and conservation. Springer, Berlin/Heidelberg, pp 13–26

Melo-Ferreira J, Boursot P, Carneiro M, Esteves PJ, Farelo L, Alves PC (2012) Recurrent introgression of mitochondrial DNA among hares (*Lepus* spp.) revealed by species-tree inference and coalescent simulations. Syst Biol 61(3):367

Mengoni C (2011) Phylogeny and genetic diversity of Italian species of hares (genus *Lepus*). PhD thesis, University of Bologna

Mengoni C, Mucci N, Randi E (2015) Sex identification in four leporid species *(Lepus corsicanus, Lepus europaeus, Lepus timidus* and *Lepus capensis mediterraneus*). Conserv Genet Resour 7(3):619–621

Meriggi A, Giordano M, Medda M, Milanesi P, Nelli L, Sacchi O, Vidus A, Ziliani U (2010) Aggiornamento Carta Faunistica della Regione Sardegna – Relazione conclusiva sulle ricerche su Pernice sarda e Lepre sarda. Unpublished report for the Regione Autonoma della Sardegna

Miller GS (1912) Catalogue of the mammals of western Europe (Europe exclusive of Russia) in the collection of the British Museum. British Museum (Natural History), Department of Zoology, London

Mitchell-Jones AJ, Amori G, Bogdanowicz W, Krystufek B, Reijnders PJH, Spitzenberger F, Stubbe M, Thissen JBM, Vohralik V, Zima J (1999) The atlas of European mammals. T & AD Poyser Natural History, London

Mohsen C, Yamna K, Mohsen J, Touhami K (2017) Estimating Cape hare occupancy and abundance in southern Tunisia using spotlighting data. Afr Zool 52(3): 137–146

Murgia C, Murgia A, Deiana AM (2003) Aspetti dell'ecologia della lepre sarda (*Lepus capensis mediterraneus*) in aree della Sardegna meridionale. Rend Sem Fac Sci Univ Cagliari 73(1):19–25

Niethammer J (1963) Nagetiere und Hasen aus der zentralen Sahara (Hoggar). Z Säugetierkd 28:350–369

Nowak RM (1999) Walker's mammals of the world, 6th edn. Johns Hopkins University Press, Baltimore

Olwoch JM, Reyers B, Van Jaarsveld AS (2009) Host–parasite distribution patterns under simulated climate: implications for tick-borne diseases. Int J Clim 29:993–1000

Onida P, Cossu S, Torre A, Meriggi A, Cesaris C, Sacchi O, Fassò C (1995) Primo studio finalizzato alla gestione della Lepre Sarda, *Lepus capensis mediterraneus*, nella provincia di Oristano. IVRAM, University of Pavia, Azienda Foreste Demaniali, Regione Autonoma della Sardegna

Osborn DJ, Helmy I (1980) The contemporary land mammals of Egypt (including Sinai). Fieldiana Zool 5:84–93

Palacios F (1989) Biometric and morphologic features of the species of the genus *Lepus* in Spain. Mammalia 73: 227–264

Palacios F, Fernandez J (1992) A new subspecies of hare from Majorca (Balearic Islands). Mammalia 5(6):71–85

Palacios F, Angelone C, Alonso G, Reig S (2008) Morphological evidence of species differentiation within *Lepus capensis* Linnaeus, 1758 (Leporidae, Lagomorpha) in Cape Province, South Africa. Mamm Biol 73:358–370

Pépin D, Meunier M, Angibault J (1981) Étude de la reproduction du lièvre (*Lepus europaeus*) dans le bassin parisien. Bull Mens Off Natl Chasse 11:3–26

Perrin MR, Campbell BS (1980) Key to the mammals of the Andries Vosloo Kudu Reserve (Eastern Cape), based on their hair morphology, for use in predator scat analysis. S Afr J Wildl Res 10(1):1–14

Petter F (1959) Eléments d'une révision des lièvres africains du sous-genre *Lepus*. Mammalia 23:41–67

Petter F (1961) Eléments d'une révision des lièvres européens et asiatique du sous-genre *Lepus*. Z Säugetierkd 26:30–40

Pierpaoli M, Riga F, Trocchi V, Randi E (1999) Species distinction and evolutionary relationships of the Italian hare (*Lepus corsicanus*) as described by mitochondrial DNA sequencing. Mol Ecol 8:1805–1817

Pierpaoli M, Riga F, Trocchi V, Randi E (2003) Hare populations in Europe: intra and interspecific analysis of mtDNA variation. C R Biol 326:S80–S84

Puggioni G, Cavadini P, Maestrale C, Scivoli R, Botti G, Ligios C, Le Gall-Reculé G, Lavazza A, Capucci L (2013) The new French 2010 rabbit hemorrhagic disease virus causes an RHD-like disease in the Sardinian Cape hare (*Lepus capensis mediterraneus*). Vet Res 44:96

Riga F, Trocchi V, Randi E, Toso S (2001) Morphometric differentiation between the Italian hare (*Lepus corsicanus* De Winton, 1898) and the European brown hare (*Lepus europaeus* Pallas, 1778). J Zool 253(2):241–252

Robinson T (1986) Incisor morphology as an aid in the systematics of the South African Leporidae (Mammalia: Lagomorpha). S Afr J Zool 21:297–302

Rocchi MS, Dagleish MP (2018) Diagnosis and prevention of rabbit viral haemorrhagic disease 2. In Pract 40:11–16

Scandura M, Iacolina L, Ben Slimen H, Suchentrunk F, Apollonio M (2007) Mitochondrial CR-1 variation in Sardinian hares and its relationships with other old world hares (genus *Lepus*). Biochem Genet 45:305–323

Schai-Braun S, Hackländer K (2016) Family Leporidae (hares and rabbits). In: Wilson DE, Lacher TE Jr, Mittermeier RA (eds) Handbook of the mammals of the world, vol 6, Lagomorph and Rodents. Lynx Edicions, Barcelona, pp 62–149

Schai-Braun SC, Hackländer K (2018) *Lepus capensis* Linnaeus, 1758 Cape hare. In: Smith AT, Johnston CH, Alves PC, Hackländer K (eds) Lagomorphs: pikas, rabbits, and hares. John Hopkins University Press, Baltimore, pp 176–179

Scott K (2018) A comparative morphological and morphometric study of the cranial and post-cranial osteology of South African hares – Cape hare (*Lepus capensis*) and Scrub hare (*Lepus saxatilis*) and its application in archaeozoology. PhD thesis, University of South Africa

Shayan S, Bokaean M, Shahrivar MR, Chinikar S (2015) Crimean-Congo hemorrhagic fever. Lab Med 46:180–189

Skinner JD, Chimimba CT (2005) The mammals of the southern African subregion. Cambridge University Press, Cambridge, UK/Cape Town

Smith AT, Johnston CH, Alves PC, Hackländer K (eds) (2018) Lagomorphs: pikas, rabbits, and hares of the world. Johns Hopkins University Press, Baltimore

Soria-Boix C, Donat-Torres MP, Idrissi HR, Urios V (2019) Evolutionary ecology of hares (*Lepus* spp.) from northwest Africa: the existence of cryptic species and description of a new species (*Lepus saharae* sp. nov.). Evol Ecol Res 20:537–555

Stavy M, Goldblatt A, Terkel J (1985) Home odor preferences in young hares (*Lepus capensis syriacus*): effects of age and role of maternal presence. Dev Psychobiol 18(2):125–139

Stoner CJ, Bininda-Emonds ORP, Caro T (2003) The adaptive significance of coloration in lagomorphs. Biol J Linn Soc 79(2):309–328

Stroh G (1931) Zwei sichere Altersmerkmale beim Hasen. Tierärztl Wochenschr 12:180–181

Suchentrunk F, Flux J (1996) Minor dental traits in East African Cape hares and savanna hares (*Lepus capensis* and *Lepus victoriae*): a study of intra- and interspecific variability. J Zool 238(3):495–451

Suchentrunk F, Willing R, Hartl GB (1994) Nonmetrical polymorphism of the first lower premolar (P-3) in Austrian brown hares (*Lepus europaeus*) – a study on regional differentiation. J Zool 232:79–91

Suchentrunk F, Ben Slimen H, Stamatis C, Sert H, Scandura M, Apollonio M, Mamuris Z (2006) Molecular approaches revealing prehistoric, historic, or recent translocations and introductions of hares (genus *Lepus*) by humans. Hum Evol 21:151–165

Suchentrunk F, Flux JEC, Flux MM, Ben Slimen H (2007) Multivariate discrimination between East African Cape hares (*Lepus capensis*) and savanna hares (*L. victoriae*) based on occipital bone shape. Mamm Biol 72:372–383

Suchentrunk F, Ben Slimen H, Sert H (2008) Phylogenetic aspects of nuclear and mitochondrial gene-pool characteristics of South and North African Cape hares (*Lepus capensis*) and European hares (*Lepus europaeus*). In: Alves PC, Ferrand N, Hackländer K (eds) Lagomorph biology. Ecology, evolution, and conservation. Springer, Berlin/Heidelberg, pp 65–85

Suchentrunk F, Ben Slimen H, Kryger U (2009) Molecular evidence of conspecificity of South African hares conventionally considered *Lepus capensis* L., 1758. Mamm Biol 74(5):325–343

Teerink BJ (1991) Hair of West-European mammals: atlas and identification key. Cambridge University Press, Cambridge, UK

Temple HJ, Terry A (2009) European mammals: Red List status, trends, and conservation priorities. Folia Zool 58:248–269

Toschi A (1965) Fauna d'Italia. Mammalia. Lagomorpha – Rodentia – Carnivora – Ungulata – Cetacea. Calderini, Bologna

Trocchi V, Riga F (2005) I Lagomorfi in Italia. Linee guida per la conservazione e la gestione. Ministero delle Politiche Agricole e Forestali – Istituto Nazionale Fauna Selvatica. Doc Tec 25:1–128

Trocchi V, Fassò C, Riga F (2003a) Determinazione dell'età nella Lepre sarda *Lepus capensis mediterraneus* mediante il peso del cristallino. Hystrix 14(Suppl):135

Trocchi W, Riga F, Randi E (2003b) *Lepus* ("*capensis*") *mediterraneus* Wagner 1841 – Sardischer Hase. In: Krapp F (ed) Handbuch der Säugetiere Europas, vol 3/II. Aula-Verlag, Wiebelsheim, pp 104–116

Vismara P (2012) Il genere *Lepus* in Italia: chiavi diagnostiche morfo-odontologiche e contributi paleobiogeografici. PhD thesis, University of Milan

Wilkens B, Delussu F (2003) Wild and domestic mammals in holocenic Sardinia. In: Legakis A, Sfenthourakis S, Polymeni R, Thessalou-Legaki M (eds) The new panorama of animal evolution. Proceedings of the 18th international congress of zoology. Pensoft, Sofia/Moscow, pp 303–308

Wu Y, Xia L, Zhang Q, Yang Q, Meng X (2011) Bidirectional introgressive hybridization between *Lepus capensis* and *Lepus yarkandensis*. Mol Phylogenet Evol 59:545–555

Broom Hare *Lepus castroviejoi* Palacios, 1977

5

Fernando Ballesteros and Paulo C. Alves

Contents

F. Ballesteros (✉)
Fundación Oso Pardo (Brown Bear Foundation), Santander, Spain
e-mail: fballes0@gmail.com

P. C. Alves (✉)
CIBIO, Centro de Investigação em Biodiversidade e Recursos Genéticos, InBIO, Universidade do Porto, Vairão, Portugal

BIOPOLIS Program in Genomics, Biodiversity and Land Planning, CIBIO, Vairão, Portugal

Departamento de Biologia, Faculdade de Ciências, Universidade do Porto, Porto, Portugal

Estação Biológica de Mértola (EBM), Mértola, Portugal
e-mail: pcalves@fc.up.pt

K. Hackländer, P. C. Alves (eds.), *Primates and Lagomorpha*, Handbook of the Mammals of Europe,
https://doi.org/10.1007/978-3-030-34043-8_11

Common Names

English	Broom hare
German	Ginsterhase
French	Lièvre Cantabrique
Spanish	Liebre de piornal
Italian	Lepre cantabrica
Russian	Веник заяц

Taxonomy, Systematics and Paleontology

The broom hare (*Lepus castroviejoi* Palacios, 1977) is an Iberian endemism. Hares reflect the high biodiversity and abundance of endemism usually observed in the Iberian Peninsula, since three of the five species of hares naturally occurring in Europe are present in this region, being two of them endemic. The broom hare was previously considered as a form of the Iberian hare (*L. granatensis*), and it was not described as a new species until 1977 (Palacios 1977). The identity of the species was not accepted at first by all researchers (Angermann 1983; Corbet 1986), although the first comparative biochemical and genetic data contributed to confirm its clear separation from the other Iberian hares (Palacios 1979; Bonhomme et al. 1986; Pérez-Suárez et al. 1994). Subsequent genetic studies confirmed the specific identity of the broom hare as a divergent species from *L. granatensis* and the European hare (*L. europaeus*) (Alves et al. 2003; Alves and Niethammer 2003; Estonba et al. 2006), including recent studies using whole exome (Ferreira et al. 2021) and whole genome (Giska et al. 2019) sequences. Nevertheless, interestingly both mitochondrial and nuclear DNA analysis have shown that the broom hare is closely related (sister species) to the Apennine hare (*L. corsicanus*), sharing not only genetic similarities (Alves et al. 2003, 2008a; Ferreira et al. 2021) but also phenotypic features (Palacios 1996) and ecological niche properties (Acevedo et al. 2014). No hybrids have been found with the Iberian and the European hare, despite occurring in parapatry. However, the mitochondrial DNA is quite similar to the lineage observed in the mountain hare (*L. timidus*), suggesting ancestral hybridization with this species (Alves et al. 2008b). Also, analyses based on nuclear DNA data suggest minor levels of admixture from the Iberian hare, which is thus either rare or ancient (Melo-Ferreira et al. 2009, 2012). No subspecies or geographic variation has been described, something to be expected considering its small distribution range.

No paleontological information exists about *L. castroviejoi* or its possible ancestors. Nevertheless, the strong genetic, morphological, and ecological similarity between *L. castroviejoi* and *L. corsicanus* suggest the existence of a common ancestral of these two species. This common ancestral most likely had a wider distribution throughout Europe during the late Pleistocene, before the range isolation in Iberian (*L. castroviejoi*) and Italian (*L. corsicanus*) Peninsulas. Therefore, it is possible that paleontological remains of this ancestral may occur in different regions, but so far was not identified, or was confound with another hare species, due taxonomic uncertainties.

Current Distribution

The broom hare distribution is restricted to the Cantabrian Mountains (Northwest of Spain), where it occupies a small distribution range of about 5000 km^2 (230 km long, 25–40 km wide) in the Spanish provinces of Lugo, León, Asturias, Palencia, and Cantabria (Fig. 1). The species is found from Sierra de los Ancares, between Lugo and León, up to Sierra de Peña Labra, between

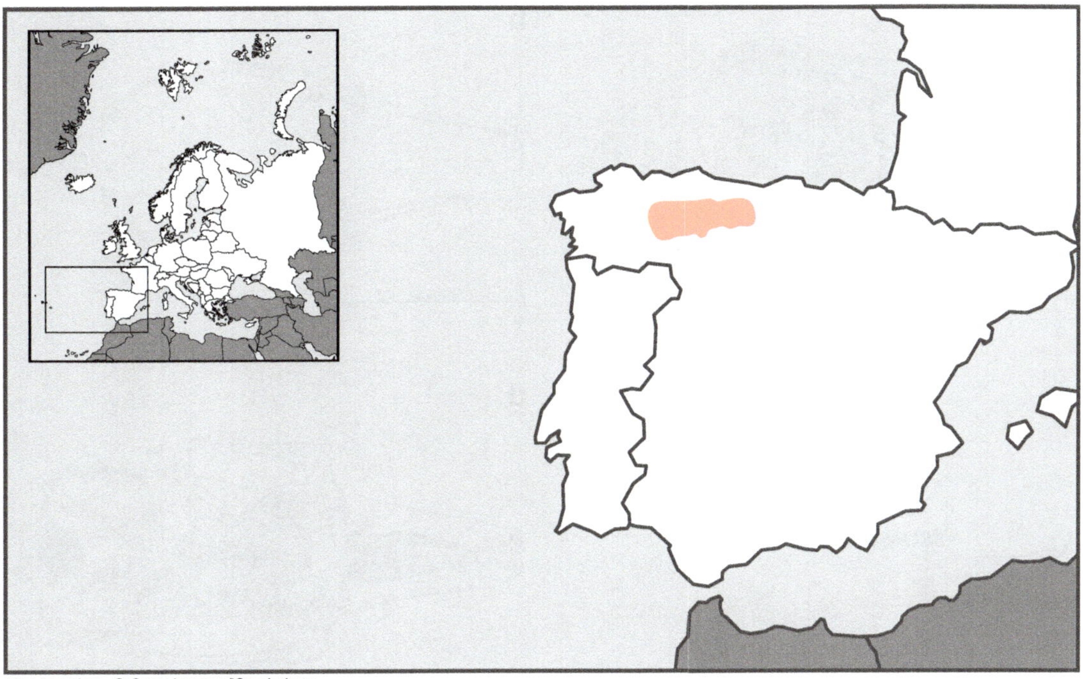

Map template: © Getty Images/iStockphoto

Fig. 1 Distribution map of *Lepus castroviejoi* in Europe (in apricot). This distribution is essentially based on the IUCN Red List of Threatened Species (Ballesteros and Smith 2019). (Map template: © Getty Images/iStockphoto)

Cantabria and Palencia (Palacios and Meijide 1979). It occurs at elevations ranging from 1000 to 1900 m above sea level, with a highly fragmented distribution, as it occupies specialized patches of habitat within pastures and shrubland, sparse in a diverse mountain landscape (Ballesteros 2007, 2018). The range shows a differential distribution between northern and southern slopes of the Cantabrian Mountains, because of the differences in orography and favorable habitat availability, with deep valleys and high altitudinal variations in the northern slope and broader highlands in the southern one. In some peripheral areas of its range, the broom hare is very scarce or has even disappeared (e.g., Peña Manteca or Sierra del Aramo in Asturias, Áliva or Peña Labra in Cantabria, Ballesteros 2009). In the southwest of Asturias, southeast of Galicia, and a large part of the southern limit of its distribution range in León, the broom hare contacts with the Iberian hare. In east of Cantabria and northeast of Palencia, broom hare also contacts with the European hare.

Description

The broom hare has an intermediate size between those of the smaller Iberian hare and the bigger European hare, the two other hare species occurring in the Iberian Peninsula. Data of 26 hares hunted in Asturias show that weight vary between 1450 and 3300 g, with an average value of 2424 g (sd $=$ 400.6), and a greater average weight in females (2577 g, sd $=$ 397.2, n $=$ 12) compared to males (2293 g, sd $=$ 367.3, n $=$ 14). Some biometric measurements presented by Palacios (1989) are: head and body length 503 $\pm$ 9.0 mm (n $=$ 5), length of the tail 87 $\pm$ 9.3 mm (n $=$ 5), foot length 140 $\pm$ 4.7 mm (n $=$ 10), and ear length 93 $\pm$ 1.9 mm (n $=$ 8).

The cranial and leg bone measurements have also intermediate values between those of the Iberian and European hares. It has clear cranial differences with the other two species and according to the description of Palacios (1977),

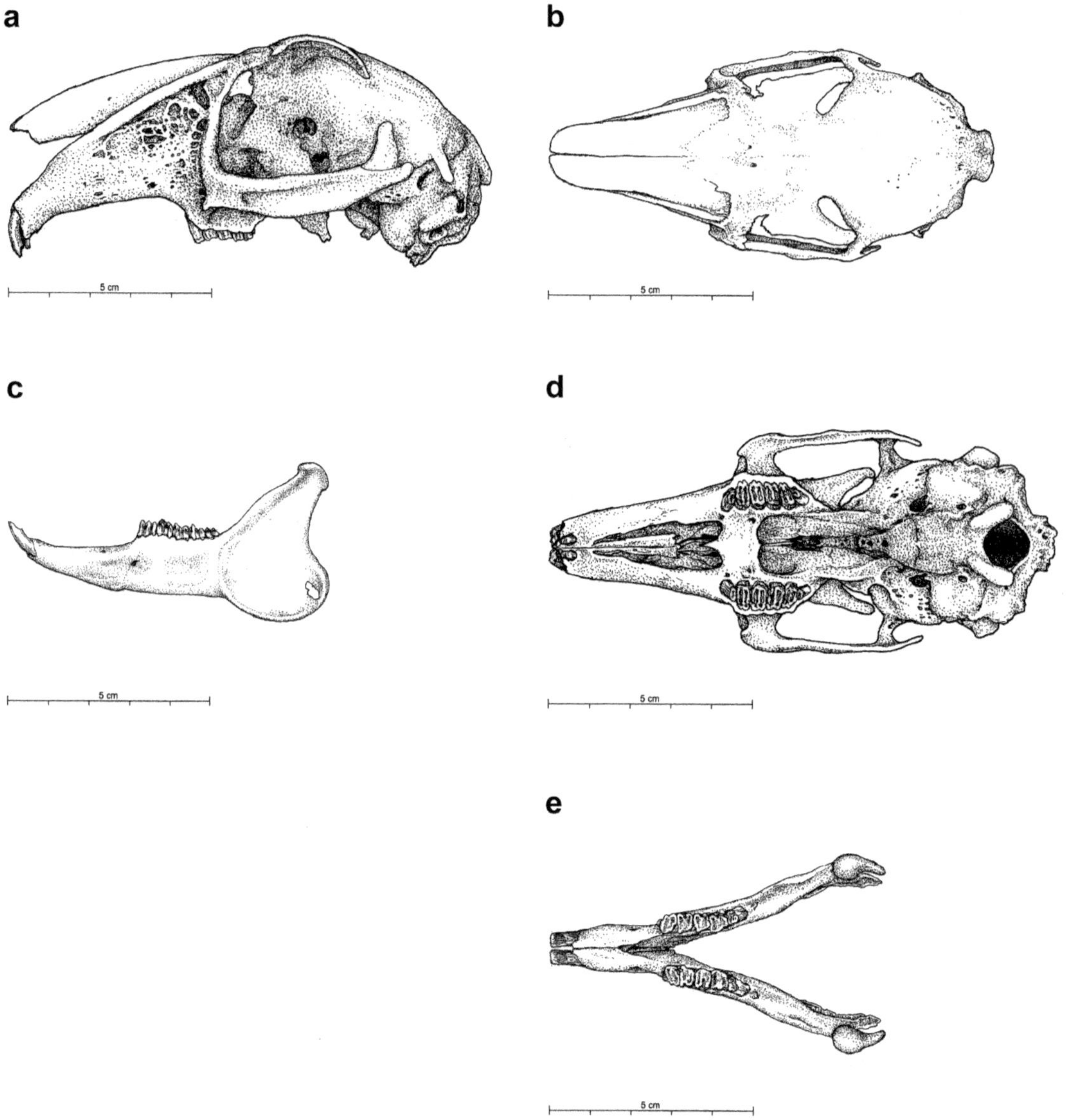

Fig. 2 Skull of *Lepus castroviejoi*. (**a**, **b**, **d**) Lateral, dorsal, and ventral view of the skull, (**c**) lateral view of the left mandible (view from outside), and (**e**) dorsal view of the mandible. (© Franz Müller with kind permission)

some of its most distinctive cranial features are the open V-shaped nasofrontal suture, the small orbital process with very short oral branch, the coronary suture, the small dimples in the upper bones of the skull, and the tiny palatal foramina. It has a definitive dentition composed of 28 teeth and its dental formula is I^2, C^0, P^3, M^3/I_1, C_0, P_2, and M_3 (Fig. 2).

Dorsal and lateral pelage is grayish brown, with white ventral pelage and white and black tail (Fig. 3). The ventral white area is wider and clearer than in the European hare, sometimes reaching the forelegs, although not as extensive as in the Iberian hare, which often extends through the chest. The contrast between the color of the back and the ventral area is clear, unlike the diffused transition of the European hare. The facial design is very characteristic, as it has a lighter grayish band on the face, between the base of the ears and the lower part of the cheeks, which stands

Fig. 3 Adult broom hare (*Lepus castroviejoi*) in its habitat in the Cantabrian mountains. (© Photo J.M. Fernández Díaz-Formentí)

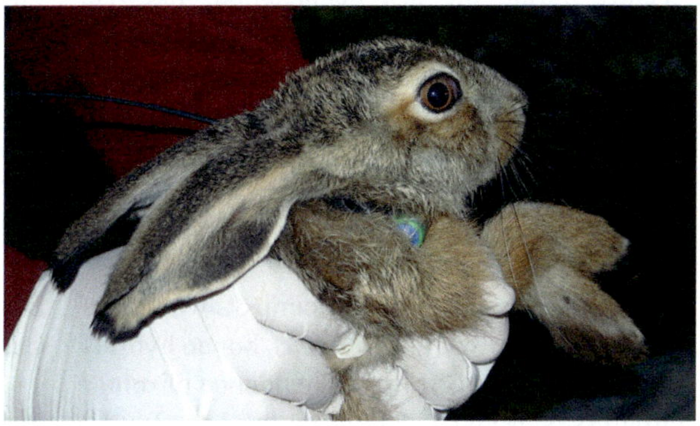

Fig. 4 Portrait of an adult broom hare (*Lepus castroviejoi*) showing its distinctive facial pattern (© Photo F. Ballesteros)

out against the brown color of the area located under the eye and the snout (Fig. 4). It has the ears and tail comparatively shorter than the Iberian and the European hare, and the hind foot longer, which could be related to adaptations to life in colder environments with more snow in winter. There are no sexual or seasonal variations in pelage coloration. There is no specific information on age determination in this hare species.

Physiology

There is no specific information about the physiology of the broom hare.

Genetics

It has the same number of chromosomes as other species of the genus *Lepus* ($2n = 48$) and similar chromosome morphology (Palacios 1979).

Genetic analyses have shown that the broom hare is closely related and a sister taxa to the Apennine hare (Alves et al. 2003, 2008a; Wu et al. 2005; Melo-Ferreira et al. 2012). This genetic similarity has been recently corroborated by a phylogenomic study that analyzed whole exome sequences (61.7 Mb) of 15 species of hares. These analyses showed that absolute genetic divergence (d_{xy}) between broom and Corsican hare is only 0.17%, the lowest value estimated among pairs of the analyzed species (Ferreira et al. 2021). The extremely low divergence between these two species suggests that they are very closely related and encourage a revision of species-level designations. Altogether, the genetic data suggest that these two species must have shared a recent common ancestor with a greater distribution throughout Europe during the late Pleistocene. Subsequent climate changes have caused their spatial reduction and their permanence only in two refuges in the Iberian and Italian peninsulas. Competition with other *Lepus* species could have maintained broom and Corsican hares in their respective small and allopatric ranges until the present.

The introgression of *L. timidus* mitochondrial DNA into *L. castroviejoi* has been described, which has replaced the native mitochondrial DNA (Melo-Ferreira et al. 2007). The different phylogenetic schemes constructed from nuclear and mitochondrial DNA markers suggest that the broom hare may have been affected by two events of mitochondrial DNA introgression from *L. timidus* throughout its history: the first also affected *L. corsicanus*, and the second, more recently during the last deglaciation, likewise affected the other two Iberian species of hares (*L. granatensis* and *L. europaeus*) (Melo-Ferreira et al. 2012). This repetition of two introgressions of mitochondrial DNA at two very different moments in history could be a unique case among mammals (Melo-Ferreira and Alves 2009).

The range of *L. castroviejoi* contacts with those of *L. granatensis* and *L. europaeus*, but no hybrids have been detected. However, evidence of minor introgression from *L. granatensis* has been found, which must either be rare or ancient (Melo-Ferreira et al. 2009, 2012). Despite the genetic similarity with *L. corsicanus*, the broom hare shows quite genetic differentiation in relation to all the other hare species inhabiting in Europe. In an electrophoretical analysis, Bonhomme et al. (1986) found a very low value of heterozygosity in the broom hare (1.4%), much lower than the values found in the same study in the Iberian and the European hare (between 7% and 8%). Subsequent studies using microsatellites have confirmed this low value of genetic diversity of *L. castroviejoi*, the lowest so far observed in hares. Alzaga (2010) found an average mean number of alleles (N_A) of 1.932, mean observed heterozygosity (H_o) of 0.275, and mean expected heterozygosity (H_e) of 0.339. A recent study conducted in 2019 across the entire range of the broom hare, including genetic noninvasive sampling (fecal pellets) and using 16 microsatellites, confirm the low genetic diversity observed in this species, with lower values of NA, Ho and He, than the reported for the other hare species occurring in Europe (Alves et al. unp. data). The extremely low levels of genetic diversity observed in the broom hare may reflect the small population size, its limited distribution area, as well as putative habitat fragmentation and inbreeding (Alves and Niethammer 2003). Thus, the low genetic diversity observed in this species brings an additional concern for its conservation.

Life History

There is no specific information about the reproduction or life history of the broom hare. Courtship behavior has been observed in April and May (Ballesteros 2009). Sex ratio in some samples seems to be biased in favor of males. Palacios (1977) described a sex ratio of 0.55 females per male ($n = 14$), and some later studies have found 0.86 females per male in Asturias ($n = 26$, Ballesteros 2003) and 0.84 females per male in Cantabria ($n = 11$, Ceballos et al. 2009).

The temporary distribution of hare sightings and tracks detections made throughout the year by rangers of a protected area (Picos de Europa National Park) from 1988 to 2008 shows a greater abundance and activity of the hares in the summer and autumn, with a peak of abundance of observations in September, which can reflect that most of the hares reproduce during the spring and summer, and thus most juveniles are incorporated in the normal activity at the beginning of the autumn (Ballesteros and Palacios 2009).

Habitat and Diet

The broom hare occupies a very characteristic habitat of the high elevation areas from the Cantabrian Mountains. These areas are typically occupied by abundant open pastures or small herbaceous clearings of variable surface, surrounded by broom formations (*Cytisus scoparius, C. multiflorus, Genista polygaliphylla, G. obtusirramea, G. legionensis,* and other similar leguminous shrubs), heathers (*Daboecia cantabrica, Erica*

Fig. 5 Typical habitat of broom hare (*Lepus castroviejoi*) in the upper part of the Cantabrian Mountains with a mosaic of open pastures and formations of broom, heather, and other shrubs. (© Photo F. Ballesteros)

aragonensis, E. arborea, E. australis, E. vagans), thorny bushes (*Genista occidentalis, Ulex cantabricum*), subalpine juniper (*Juniperus communis*) and forests of beech (*Fagus sylvatica*), oaks (*Quercus robur, Q. petraea, Q. pyrenaica*), birch (*Betula celtiberica*), and other species, located at altitudes varying between 1000 and 1900 m above sea level (Ballesteros 2018). Within this landscape, the broom hares occupy areas with quite specific characteristics, mainly intercalation of brooms and heaths with wide mountain pastures ("brañas" and "majadas" in the typical Spanish denomination) in the hills or at the top of the mountains (Fig. 5). The most relevant human traditional land use in these areas is the extensive cattle and horse breeding. In the past, and especially in the southern slope of the mountain range, the dominant livestock has been sheep, with huge herds of transhumant sheep that throughout history have been shaping a landscape dominated by large pastures and grasslands. The use of small grain cultures and orchards near mountain villages by the broom hare has been described in the past by Palacios and Meijide (1979), although these cultures are currently practically absent in the Cantabrian Mountains.

Herbaceous clearings and pastures between scrublands, which constitute the habitat and the basis of the broom hare diet, are replacement communities of the potential vegetation that may have an origin linked to the last glaciations, and the performance of the big herbivores of the late Pleistocene but have been widely managed and improved by anthropogenic actions as clearing, livestock grazing, fire, and eventual tillage. The broom hare is, therefore, occurring in dynamic communities in mosaic and in vigorous equilibrium with the shrub and forest formations (San Miguel and Perea 2009).

Habitat suitability models have determined that broom hares select areas characterized by a high percentage of broom and heather scrublands, high altitude and slope, and with limited human accessibility (Acevedo et al. 2007). They avoid extensive forests, extensive areas of continuous scrubland, and the more humanized valley bottoms. The broom hare selects positively the small clearings existing inside the broom formations as feeding zones, while the large pastures and scrublands are used, but in smaller proportion than that found in the habitat. A very positive selection has been also detected by the small areas of heathland or broom traditionally burned for maintenance of pastures for cattle grazing. The regrowth of these newly burned heath or broom formations is attractive to hares, which select these plots and concentrate on them (Ballesteros 2009). During drought periods, when pastures provide little fresh food, broom hares tend to remain inside broom formations, where they feed on the herbaceous vegetation growing in the fresh patches under broom plants.

An environmental favorability analysis for the three hare species occurring in the Iberian Peninsula found that topography, and more specifically altitude, is a very relevant factor, and considered together with land uses, shows a high explanatory power in the broom habitat model (Acevedo et al. 2009). This could indicate the environmental peculiarities of their distribution range, characterized by Atlantic mountain areas with a mosaic of livestock grazed pastures and shrubs (Ballesteros 2018). Climatological variables are also important to explain environmental favorability for all hare species. The environmental favorability for broom hare is marginal in Spain, though the areas that are suitable for the species have high values of favorability. The favorable zones are restricted to the Cantabrian Mountains, but also in small areas of northern Galicia and in the Pyrenees. Nevertheless, there are some areas of high favorability without the species occurrence, especially in the southern part of its current range, which means that there is adequate territory for a potential population expansion.

Although *L. castroviejoi* is herbivorous and it feeds on grasses and other herbaceous plants, there is no detailed scientific information about its diet. Feeding areas overlap extensively with that of domestic animals, especially cows, but also sheep and horses (Ballesteros 2018). There is a strong niche overlap between *L. castroviejoi* and the endangered population of the gray partridge (*Perdix perdix*) in the Cantabrian Mountains (Acevedo et al. 2007).

Behavior

The broom hare is mainly nocturnal. They began their activity one or few hours before sunset, moving to their feeding places (Ballesteros and Palacios 2009). At dusk and during the night, they come to feed in open pastures or clearings in scrublands, where they tend to stay relatively close to the broom or heath formations as an antipredatory strategy (medium distance to protecting shrubs is about 30 m, Ballesteros 2018). Broom has been the preferred refuge vegetation. Observations during nocturnal census surveys suggest some levels of gregarious distribution, both for its specific habitat selection and social interactions, establishing lax groups in the small clearings of herbaceous vegetation positively selected.

Home range estimation based on two radio-tracked female broom hares indicate a size of 53.2 and 60.3 ha, after 4 months of monitoring, although both hares spent most of the time in a preferred (core) area of about 7–8 ha (Ballesteros and Palacios 2009). It seems that broom hares living in a high quality and very diverse mountainous habitat could use smaller home ranges than that of other hares living in more extensive agricultural lands.

Parasites and Diseases

No comprehensive information is available about parasites and diseases in the broom hare. Regarding the ectoparasites, the occurrence of *Rhipicephalus* sp., *Ixodes* sp., and *Haemaphysalis* sp. have been reported (Alzaga 2010). The known endoparasites include Anoplocephalidae, *Graphidium strigosum*, *Trichostrongylus* sp., *Eimeria* sp., *Dicrocoelium dendriticum*, *Mosgovoyia pectinata*, and *Trichuris leporis* (Vila et al. 1999; Alzaga 2010). A recent study has detected a widespread presence of *Leishmania infantum* in *L. granatensis* and *L. europaeus* in Spain, although a low number of samples of *L. castroviejoi* prevented the confirmation of this parasite in broom hares (Ruiz-Fons et al. 2013). So far, no information on the existence of the European brown hare syndrome, tularemia, rabbit hemorrhagic disease, and neither the recent form of hare myxomatosis has been reported in the broom hare. Nevertheless, since there are no systematic studies on diseases, neither a monitoring program in this species, the lack of information does not mean the absence of diseases.

Population Ecology

As in many other issues, information about broom hare population ecology is scarce and it is necessary to increase the level of knowledge on this topic, to better improve conservation actions. Nocturnal censuses with spotlights have estimated average densities of 4.83/100 ha in Liébana (Cantabria), 6.89/100 ha in Picos de Europa National Park (León), or between 8.83 and 23.32/100 ha in Somiedo (Asturias) (Acevedo et al. 2009; Ballesteros and Palacios 2009). Abundance of broom hare is often evaluated using Kilometric Abundance Indexes (KAI) obtained in night surveys with spotlights. The average KAI value was 0.73 hares/km in Asturias in 1997 (varying between 0.21 and 1.51 in different areas), or 0.40 and 0.46 in Picos de Europa National Park in 2006 and 2007, respectively (Ballesteros and Palacios 2009). No big fluctuations or cycles in relative abundance or density have been documented, but the analysis of 565 hare locations collected in a protected area (Picos de Europa National Park) between 1988 and 2018 has shown a high variability in the number of observations each year, although the survey effort applied in this period has been more or less similar. It is foreseeable that part of the interannual variability may be due to a variation in abundance as a consequence of the alternation of years of good and poor reproduction (Ballesteros and Palacios 2009 and unpublished data).

Given the dispersed distribution of its suitable habitat, it seems that broom hares function as a metapopulation with different small local populations in good habitats, connected by dispersal through unfavorable areas (forests, rocks, human settlements, etc.). Thus, the demography of each local population of the broom hare will depend on the local factors conditioning fertility and survival, but also on the possibility of dispersion of individuals from nearby populations. The habitat quality models applied in the northern part of the Cantabrian Mountains indicate that there are good conditions for dispersion among populations (Acevedo et al. 2007), and the occasional observation of hares in unusual habitats between local populations supports these findings.

Adult and juvenile broom hares may be preyed by carnivores such as red foxes (*Vulpes vulpes*), Iberian wolves (*Canis lupus*), European wildcats (*Felis silvestris*), pine martens (*Martes martes*), as well as raptors such as golden eagles (*Aquila chrysaetos*) and goshawks (*Accipiter gentilis*), all of which are well established in broom hare range.

Under the available climate scenarios, significant impacts are expected in the distribution of favorable habitats for the broom hare. In particular, the higher incidence of droughts in summer and autumn can reduce the productivity of the population. However, the greatest risks for the future conservation of the species are mostly associated with the socioeconomic and ecological changes in the mountain areas it occupies.

Conservation Status

The broom hare is listed in the IUCN Red List as Vulnerable (Ballesteros and Smith 2019). The adjustment to the IUCN criteria is not easy to do, because the systematic lack of information in its range and population trends, necessary to verify the certainty of the assumptions, although the very low genetic variability, the small distribution range, the foreseeable small population size and the structure as a metapopulation with fragmented local populations, constitute significant risk factors for its conservation. In the Spanish National Red List, the species is also listed as Vulnerable (Ballesteros 2007), although it is not included in the Spanish Catalog of Endangered Species or the Spanish List of Wildlife Species of Special Protection Regime. The broom hare is not included in the European Union Habitats Directive or the Bern Convention, although due to its threatened status and endemic characteristics it would be reasonable for the species to be included in these conservation conventions.

The global population size and detailed distribution range are unknown. The species is distributed in metapopulations occupying specific habitat patches through their distribution range. In a study conducted through surveys by rangers, 70.83% of local populations were considered stable, compared to 16.67% of declining and 4.17% of increasing populations (Acevedo et al. 2007). The peripheral populations seem to be suffering a more pronounced regression and even some local disappearance (Ballesteros 2009). Nevertheless, this information has over 10 years, and no recent data is currently available.

Management

Much of the distribution range of the broom hare is included in natural protected areas of the regional autonomous communities and many are also included in the Natura 2000 network of the European Union, which greatly protects the habitat of the species against existing or future threats. But even within this preventive protection, socioeconomic ongoing changes in the rural environment together with climate change will generate complex management challenges for the conservation of this species.

Although *L. castroviejoi* is listed as Vulnerable in Spain, it remains considered as a hunting species in some regions. Despite some local bans, currently *L. castroviejoi* is legally hunted across most of its range, with a low extraction level and controlled management plans. Legal hunting is usually developed by a group of hunters and several specialized dogs that capture a single specimen in each hunting event. In addition, the broom hare may be easily visually confounded by hunters with other hare species in some regions, namely when in contact areas with *L. granatensis*. Moreover, poaching using vehicles and spotlights at night is considered a major threat for this species. Due the general low population size, and the fragmented distribution, even considering the limited number of hunted specimens, the effects of legal and illegal hunting on the evolution of the population should be carefully evaluated. As a precautionary principle, the hunting of this species should be avoided or strictly controlled throughout its distribution range, at least until more information is obtained about population size and trends in each area and the effect induced by hunting in its populations. Population reinforcements or translocations have not been made in the distribution area of the broom hare, neither other hare species been introduced in its range. Translocation of other species of hares must be avoided in the broom hare distribution range but also in the nearby areas where it contacts with the distribution of the Iberian or European hare. This species does not generate any conflict with human activities.

In addition to legal hunting and poaching vulnerability, there is uncertainty about the possible risk for broom hare conservation associated with global change driven by climate change and land use evolution in the mountain rural environment. In recent decades, changes in livestock breeds, species, and management systems have taken place in the Cantabrian Mountains associated with human depopulation, and flocks of sheep and goats have almost completely disappeared, been replaced in many areas by cows and horses, that have even increased their number (Blanco-Fontao et al. 2011). The generalized change in this livestock uses, which implies both reduction of grazing pressures in some highest areas and concentration of pressures in other areas, contributes in some parts of the Cantabrian Mountains to a reduction or modification of subalpine meadows and the expansion of scrubs and forest (San Miguel and Perea 2009), together with changes in plant species composition, structure, and nutritional value of grasslands. The possible effects of this scenario on the broom hare have not been evaluated yet, although this trend will increase predictably in the future and it is foreseeable that it influences the availability and use of habitat by hares (San Miguel and Perea 2009).

Future Challenges for Research and Management

The broom hare has an enormous scientific interest, since it constitutes a unique model for the study of phylogeography in the southern Europe refuges and for the understanding of the reticulated evolution of hares. Furthermore, regardless of its taxonomic status, *L. castroviejoi* is an endangered evolutionary significant unit (ESU), leaving in an isolated and very restricted region, which must be actively conserved. Besides, it is surprising the low level of knowledge about the biology and ecology of this species, as well as the lack of basic information and monitoring of population size and trends. Recommended measures must include the increase in research and the implementation of a population, disease, and genetic monitoring program. In addition, studies should be implemented to determine the effects of climate change, habitat change and hunting, as well as improvement in the protection against poaching. The establishment of a global hunting plan for all regional reserves and private hunting areas, and the establishment of a habitat management strategy considering grazing management, fire prevention, habitat restoration, preservation of most relevant habitats, and improvement of corridors are also a priority for the conservation and management of this endemic and highly range limited species (Ballesteros 2018). Basic information should also be collected on their reproduction, population dynamics, dispersal and connectivity between populations, as well as the possible effect of emerging diseases present on other species of hares. It also important to study the relationships and competition with the other hare species inhabiting in Iberia, namely the *L. granatensis* in the southern range, since land use and climate change may favor *L. granatensis*, and thus limiting even more the range for *L. castroviejoi*.

Considering the uniqueness of the broom hare, its interest as a potential indicator of Cantabrian Mountain grassland and meadows biodiversity, and of climate change (since it is adapted to high altitudes and lower temperatures), it is advisable to establish a specific monitoring program about broom hare habitat in different parts of its distribution range. There is some basic information about preferred habitats, but there is a great lack of knowledge on a microhabitat scale, diet, landscape use, and relations with other wild herbivores and livestock. In general, hares avoid herbaceous areas that are heavily grazed by livestock, due to the decrease in food and shelter availability, although a moderate grazing pressure by wild and domestic ungulates usually generates habitat quality improvement for the hares. The relations between the broom hare abundance and livestock pressure and management requires a more detailed analysis and has a special importance in the current scenarios in which mountain livestock is the object of economic support and territorial management policies. The broom hare is currently one of the most unknown species of the European vertebrate fauna, so research and monitoring are essential to ensure its conservation.

References

Acevedo P, Alzaga V, Cassinello J, Gortazar C (2007) Habitat suitability modelling reveals a strong niche overlap between two poorly known species, the broom hare and the Pyrenean grey partridge, in the north of Spain. Acta Oecol 31:174–184

Acevedo P, Alzaga V, Gortazar C (2009) Modelos predictivos para el estudio de la distribución de las especies: aplicación para la conservación de la liebre de piornal. In: Ballesteros F, Palacios B (eds) Situación y conservación de la liebre del piornal *Lepus castroviejoi* en la Cordillera Cantábrica. Ministerio de Medio Ambiente y Medio Rural y Marino, Madrid, pp 49–74

Acevedo P, Melo-Ferreira J, Real R, Alves PC (2014) Evidence for niche similarities in the allopatric sister species *Lepus castroviejoi* and *Lepus corsicanus*. J Biogeogr 41:977–986

Alves PC, Niethammer J (2003) Die iberischen Hasen. In: Niethammer J, Krapp F (eds) Handbuch der Säugetiere Europas, vol 3/II. Aula-Verlag

Alves PC, Ferrand N, Suchentrunk F, Harris JD (2003) Ancient introgression of *Lepus timidus* mtDNA into *L. granatensis* and *L. europaeus* in the Iberian Peninsula. Mol Phylogenet Evol 27:70–80

Alves PC, Melo-Ferreira J, Branco M, Suchentrunk F, Ferrand N, Harris J (2008a) Evidence for genetic similarity of two allopatric European hares (*Lepus corsicanus* and *L. castroviejoi*) based on nuclear genes sequences. Mol Phylogenet Evol 46:1191–1197

Alves PC, Melo-Ferreira J, Freitas H, Boursot P (2008b) The ubiquitous mountain hare mitochondria: multiple introgressive hybridization in hares, genus *Lepus*. Philos Trans R Soc Lond Biol Sci 363:2831–2839

Alzaga V (2010) Ecología de la parasitación de las liebres (*Lepus* spp.) de la Península Ibérica. PhD thesis, Instituto de Investigación en Recursos Cinegéticos (CSIC-UCLM-JCCM)

Angermann R (1983) Taxonomy of old world *Lepus*. Acta Zool Fenn 174:17–21

Ballesteros F (2003) Liebre de piornal, *Lepus castroviejoi* (Palacios, 1976). Galemys 15(1):3–13

Ballesteros F (2007) *Lepus castroviejoi* (Palacios 1977). In: Palomo LJ, Gisbert J, Blanco JC (eds) Atlas y Libro

Rojo de los Mamíferos Terrestres de España. Ministerio de Medio Ambiente, Madrid, pp 479–483

Ballesteros F (2009) Descripción, biología y estado de conservación de la liebre de piornal. In: Ballesteros F, Palacios B (eds) Situación y conservación de la liebre del piornal *Lepus castroviejoi* en la Cordillera Cantábrica. Ministerio de Medio Ambiente y Medio Rural y Marino, Madrid, pp 31–48

Ballesteros F (2018) *Lepus castroviejoi* Palacios, 1977; Broom Hare. In: Smith AT, Johnston CH, Alves PC, Hackländer K (eds) Lagomorphs: pikas, rabbits, and hares of the world. Johns Hopkins University Press, Baltimore, pp 179–181

Ballesteros F, Palacios B (2009) La liebre de piornal en el Parque Nacional de los Picos de Europa. In: Ballesteros F, Palacios B (eds) Situación y conservación de la liebre del piornal *Lepus castroviejoi* en la Cordillera Cantábrica. Ministerio de Medio Ambiente y Medio Rural y Marino, Madrid, pp 87–102

Ballesteros F, Smith AT (2019) *Lepus castroviejoi*. The IUCN red list of threatened species 2019: e. T11797A503908. https://doi.org/10.2305/IUCN.UK. 2019-2.RLTS.T11797A503908.en

Blanco-Fontao B, Quevedo M, Obeso JR (2011) Abandonment of traditional uses in mountain areas: typological thinking versus hard data in the Cantabrian Mountains (NW Spain). Biodivers Conserv 20(5):1133–1140

Bonhomme F, Fernández J, Palacios F, Catalán L, Machordon A (1986) Caractérisation biochimique du complex d'espéces du genre *Lepus* en Espagne. Mammalia 50:495–506

Ceballos I, Serdio A, Lucio AJ (2009) La liebre de piornal en Cantabria: situación y gestión de sus poblaciones. In: Ballesteros F, Palacios B (eds) Situación y conservación de la liebre del piornal *Lepus castroviejoi* en la Cordillera Cantábrica. Ministerio de Medio Ambiente y Medio Rural y Marino, Madrid, pp 69–78

Corbet GB (1986) Relationships and origins of the European lagomorphs. Mamm Rev 16:105–110

Estonba A, Solís A, Iriondo M, Sanz-Martín MJ, Perez-Suarez G, Markov G, Palacios F (2006) The genetic distinctiveness of the three Iberian hare species: *Lepus europaeus, L. granatensis*, and *L. castroviejoi*. Mamm Biol 71:52–59

Ferreira MS, Jones M, Callahan C, Farelo L, Tolesa Z, Suchentrunk F, Boursot P, Mills LS, Alves PC, Good JM, Melo-Ferreira J (2021) The legacy of recurrent introgression during the radiation of hares. System Biol 70:593

Giska I, Farelo L, Pimenta J, Seixas FA, Ferreira MS, Marques JP, Miranda I, Letty J, Jenny H, Hackländer K, Magnussen E, Melo-Ferreira J (2019) Introgression drives repeated evolution of winter coat color polymorphism in hares. Proc Natl Acad Sci U S A 116:24150–24156

Melo-Ferreira J, Boursot P, Randi E, Kryukov A, Suchentrunk F, Ferrand N, Alves PC (2007) The rise and fall of the mountain hare (*Lepus timidus*) during Pleistocene glaciations: expansion and retreat with hybridization in the Iberian Peninsula. Mol Ecol 16: 605–618

Melo-Ferreira J, Alves PC (2009) Evolutionary history, taxonomy and genetics of the broom hare. In: Ballesteros F, Palacios B (eds) Situación y conservación de la liebre del piornal *Lepus castroviejoi* en la Cordillera Cantábrica. Ministerio de Medio Ambiente y Medio Rural y Marino, Madrid, pp 17–27

Melo-Ferreira J, Alves PC, Freitas H, Ferrand N, Boursot P (2009) The genomic legacy from the extinct Lepus timidus to the three hare species of Iberia: contrast between mtDNA, sex chromosomes and autosomes. Molecular ecology 18(12):2643–2658

Melo-Ferreira J, Boursot P, Carneiro M, Esteves PJ, Farelo L, Alves PC (2012) Recurrent introgression of mitochondrial DNA among hares (*Lepus* spp.) revealed by species-tree inference and coalescent simulation. Syst Biol 61:367–381

Palacios F (1977) Descripción de una nueva especie de liebre (*Lepus castroviejoi*) endémica de la cordillera Cantábrica. Doñana Acta Vertebrata 3:205–223

Palacios F (1979) Análisis cromosómico, carga de DNA y electroforesis de las liebres españolas. Doñana Acta Vertebrata 6:203–215

Palacios F (1989) Biometric and morphologic features of the species of the genus *Lepus* in Spain. Mammalia 53: 227–263

Palacios F (1996) Systematics of the indigenous hares of Italy traditionally identified as *Lepus europaeus* Pallas 1778 (Mammalia: Leporidae). Bonn Zool Beitr 56:59–91

Palacios F, Meijide M (1979) Distribución geográfica y hábitat de las liebres en la Península Ibérica. Instituto Nacional para la Conservación de la Naturaleza, Madrid

Pérez-Suárez G, Palacios F, Boursot P (1994) Speciation and paraphyly in Western Mediterranean hares (*Lepus castroviejoi*, *L. europaeus*, *L. granatensis* and *L. capensis*) revealed by mitochondrial DNA phylogeny. Biochem Genet 32:423–436

Ruiz-Fons F, Ferroglio E, Gortázar C (2013) *Leishmania infantum* in free-ranging hares, Spain, 2004–2010. Euro Surveill 18:30

San Miguel A, Perea R (2009) Los pastos de puerto cantábricos y su conservación. In: Ballesteros F, Palacios B (eds) Situación y conservación de la liebre del piornal *Lepus castroviejoi* en la Cordillera Cantábrica. Ministerio de Medio Ambiente y Medio Rural y Marino, Madrid, pp 103–128

Vila T, Casanova J, Miquel J, Feliu C (1999) Sobre las Helmintofaunas de *Lepus castroviejoi* y *L. capensis* (Lagomorpha; Leporidae) Ibérica. In: Hernández S et al (eds) Paper presented at the VI Congreso Ibérico de Parasitología. Universidad de Córdoba, Córdoba, 21–24 September 1999, p 20

Wu C, Wu J, Bunch TD, Li Q, Wang Y, Zhang YP (2005) Molecular phylogenetics and biogeography of *Lepus* in Eastern Asia based on mitochondrial DNA sequences. Mol Phylogenet Evol 37:45–61

Corsican Hare *Lepus corsicanus* de Winton, 1898

6

Francesco Maria Angelici

Contents

F. M. Angelici (✉)
Fondazione Italiana per la Zoologia dei Vertebrati (FIZV),
Rome, Italy

K. Hackländer, P. C. Alves (eds.), *Primates and Lagomorpha*, Handbook of the Mammals of Europe,
https://doi.org/10.1007/978-3-030-34043-8_7

Common Names

English	Apennine hare, Corsican hare, Italian hare
German	Apenninenhase, Korsika-Hase
French	Lièvre des Apennins, Lièvre Corse
Spanish	Liebre des Apeninos, Liebre de Córcega
Italian	Lepre appenninica, lepre corsa, lepre italica (improper name)
Russian	Корсиканский заяц

Taxonomy and Systematics

The Apennine hare, *Lepus corsicanus*, was originally described by de Winton (1898) based on a specimen from Corsica, hence the scientific name. Latter, it was downgraded as subspecies of *L. europaeus*, *L. e. corsicanus*, and in this way confirmed for a long time, by various authors (e.g., Miller 1912; Ellermann and Morrison-Scott 1951; Toschi 1965; Corbet 1978) (Fig. 1).

Toschi (1965) believed that this subspecies was seriously threatened or even extinct, following hybridization and genetic introgression with other conspecific European hares. This transformation had occurred, according to Toschi (1965), following the introduction of thousands of *L. europaeus* from various origins occurring every year in mainland Italy. However, some Apennine hares were casually captured in Calabria (Southern Italy) in the years 1974–1975 (M. Spagnesi, pers. comm.), in Central Italy from 1981 (Angelici and Luiselli 1998), and in Sicily from 1982 to 1983 (Lo Valvo et al. 1997), although these were still considered as a subspecies of *L. europaeus*.

Fig. 1 The Apenine hare, *Lepus corsicanus* (adult). (Photo by A. Calabrese)

Since 1983, the possible specific distinction from *L. europaeus* of some Apennine hare populations of the Latium-Abruzzo Apennines (Central Italy) was hypothesized for the first time, based on simple observations of the coat (in particular the distinctive coloring of the neck) and of the linear measures of the body (Angelici 1989). The first scientific work supporting this evidence, developed entirely independently, was presented in 1989 at an international congress (Palacios et al. 1989); in this work, old specimens of *L. corsicanus* from Central-Southern Italy and Corsica were examined, prior to a period of intense introduction of allochtonous European hares. On the basis of univariate morphological and morphometric comparisons between European hares of different origins, it was concluded that the hares from Central-Southern Italy and Corsica could be considered distinct from *L. europaeus*, and most likely isolated reproductively. Subsequently Palacios (1996) in a more complete and exhaustive work, strengthened this hypotheysis by comparing skulls and teeth of *L. europaeus* and *L. corsicanus*, and highlighting some distinctive characteristics, in particular in the articular process of the mandible, in the upper incisors and in the occlusive surfaces of the premolars (cf Riga et al. 2003). Angelici and Luiselli (1998), analyzing a much larger sample of hares from Central-Southern Italy and Sicily, collected since 1982, confirmed, on the basis of morphometric measurements of the head and body, that the two species were well distinct, with very few or no linear overlaps (see also Angelici and Luiselli 2007). Subsequently, two important works carried out with two very different methodologies, the first with a genetic approach, using mitochondrial DNA, the second by multivariate analysis of the morphometric characters, again confirmed the specific validation of *L. corsicanus,* compared to *L. europaeus* and others species of the genus *Lepus* (Pierpaoli et al. 1999; Riga et al. 2001). It should be noted that, surprisingly, Hoffmann (1993), in his list of the world's Lagomorphs, already considered *L. corsicanus* as a distinct species, probably accepting the first morphological distinct characteristics suggested by Palacios et al. (1989). New genetic evidences, using traditional and genomic approaches, interesingly suggest that *L. corsicanus* and the broom hare *L. castroviejoi* are sister species (Suchentrunk et al. 2007; Alves and Melo-Ferreira 2007; Alves et al. 2008; Ferreira et al. 2021), as already hypothesized using biogeographical evidences by Angelici and Luiselli (2007).

Lepus corsicanus, based on current knowledge, is an endemic species of Central-Southern Italy and Sicily (Amori et al. 1996, 1999; Angelici 1998, 1999), and it was introduced in Corsica in historical times (Vigne 1992).

Paleontology

No leporid has ever been part of the fossils originating in Corsica, where *L. corsicanus* was introduced, perhaps in periods just before the sixteenth century CE, as evidenced by some historical evidence (Vigne 1988, 1990, 1992). In his description of Corsica, for example, the historian Polybius (third to second century BCE) clearly notes the absence of hares (Stories XII: 3,8-4,6) (Shuckburgh 1889). Currently there are no paleontological studies that specifically consider *L. corsicanus*; many *Lepus* fossil records have been classified only at the gender level or has been attributed to *L. europaus* or *L. timidus* (Riga et al. 2003). All fossil or sub-fossil material collected in Italy (e.g., Caloi and Palombo 1989; Iacoangeli 1997) will have to be re-examined, in light of the validation of *L. corsicanus* as a separate species. Some skulls examined by F. Riga and V. Trocchi of *Lepus* specimens from Grotta Romanelli (Salento, Apulia, Southern Italy) were identified as *L. corsicanus* and not *L. europaeus*, as previously believed (Riga et al. 2003). Moreover, other similar observations on findings of typical fanlets of the Leccese Stone Formation (Iacoangeli 1997) from the Upper Pleistocene of Melpignano (Salento, Apulia, Southern Italy) are attributable to *L. corsicanus*, rather than to *L. europaeus*. Interessingly, true fossil remains of *L. corsicanus* were recently found in Apulia (South-East Italy), attributable to the late Pleistocene (Kotsakis et al. 2020).

The Apennine hare may have been present originally in the Central-Southern Italian peninsula as a relic endemic population, diverging from the other hares widespread in Northern Italy and, perhaps, of more ancient origin (Masseti and De Marinis 2007). Moreover, these territories still host some mammalian taxa that differ substantially from their northern counterparts, such as *Talpa romana*, *Rupicapra pyrenaica*, *Vulpes vulpes*, together with a probable endemic taxon of *Chionomys nivalis* and perhaps also of *Capreolus capreolus* (cf Frati et al. 1998; Lorenzini et al. 2002; Masseti 2003).

Another problem concerns the origin of the Apennine hare in Sicily. On the basis of some ancient literature it was thought that it has been introduced from mainland Italy (Riga et al. 2003). At present it is believed that hares were already present in Sicily before the arrival of the Greeks; there are, in fact, both historical-linguistic and fossil evidence (Tagliacozzo 1993), as well as phylogenetic data, which would probably trace the presence of *L. corsicanus* in Sicily within a period between 45,000 and 121,000 years ago (Pierpaoli et al. 1999). Apart from these data, still preliminary, there are very likely historical findings of *L. corsicanus*, dating back to the eleventh and twelfth century CE, found in Alia (Palermo) (cf. Riga et al. 2003). The first certain data of Apennine hare for Sicily date back to the thirteenth century CE (Sarà 2005) on the basis of its discovery among the remains of the Norman Castro of Calathamet. However, it cannot be excluded that a revision of the Sicilian materials already traditionally referred to *L. europaeus* or *Lepus* sp. may allow a new attribution to *L. corsicanus*. Based on the determinations of the old museum finds made by Palacios et al. (1989), Palacios (1996), and subsequently by Riga et al. (2001), it was ascertained that the range of the species, up to the period 1930–40, included large part of peninsular Italy, with northern limits given roughly by the island of Elba on the Tyrrhenian side and by the province of Foggia (Apulia) on the Adriatic side, in addition to Sicilia and Corsica (Riga et al. 2003). However, new records in northern areas were discovered, both on the Adriatic side and on the Tyrrhenian side (Angelici and Spagnesi 2008a). Moreover, it should be highlighted that these findings, which are based in few specimens, derive from random captures and observations, and not from a comprehensive study. Therefore it cannot be excluded that *L. corsicanus* is present in other areas not represented by the mentioned findings, therefore the Northern limit of the range of the species might be, in reality, more toward the North.

Current Distribution

L. corsicanus is currently present in Central-Southern Italy (native), Sicily (native), and Corsica (historically introduced) (Fig. 2). Recently it has been reintroduced in the Isle of Elba (Tuscany, Central Italy) (Calamai 2019), where it was present until the end of the nineteenth CE (Scarselli et al. 2016) (Fig. 2). The northernmost records, to date, are located at some places in the Tuscan Maremma on the Tyrrhenian side, and in Umbria and the Southern Marche (Angelici and Spagnesi 2008a), while further east there are small populations on the Apennines of Abruzzo and Molise, and some fragmented populations on the Gargano (Northern Apulia) and even further South always in Apulia. Apart for Isle of Elba, *L. corsicanus* does not appear to be present in the Tuscan Archipelago, where *L. europaeus* is often present (Riga et al. 2003). The Apennine hare is well represented in the regions of Latium, Campania, Basilicata, and Calabria, where it exists with more substantial nuclei of occurrence; however it does not seem to always have a panmictic continuity, since the majority of the peninsular nuclei are relegated to inaccessible areas, with difficult access, or in historically protected areas, where restockings or introductions with *L. europaeus* have not occurred (Angelici 1998; Guglielmi et al. 2008).

In Sicily, the Corsican hare still seems fairly widespread on the whole island and is currently the only species of hare present, though tens of thousands of *L. europaeus* individuals have been introduced, at least up to the decade 1990–2000. Fortunately *L. europaeus* have not established a stable population in Sicily neither apparently have hybridized with the autochthonous Apennine hare (cf. Lo Valvo et al. 1997). However, in Sicily, although the species is still relatively frequent and apparently distributed continuously, there is a tendency to a fragmentation of the range, due to the strong degradation of the habitat and the very high increase of human infrastructure and settlements. According to Sarà (1998), the Apennine hare in Sicily is in drastic decline, with the exception of protected areas such as national parks and natural reserves, where it is more protected from poaching and where it can still find intact habitats. According to Lo Valvo (2007), on the contrary, the species is widespread on the whole island, but with variable local densities.

In Corsica, the species appeared, according to Vigne (1988) in the fifteenth century CE, as the genus *Lepus* is not native of Sardo-Corso massif (Azzaroli 1980; Esu and Kotsakis 1980) and would be introduced at the time of the Maritime Republics of Pisa and Genoa. However, it cannot be excluded that the introduction into Corsica is earlier, starting from the eleventh century CE, or even in ancient historical times (Vigne 1992; Scalera and Angelici 2003). *L. corsicanus* was considered extremely rare in Corsica and was considered extinct from 1984 (Dubray and Roux 1984), or from 1990 (Angelici 1999; Angelici and Luiselli 2001). However, the species was rediscovered in an inaccessible area in the South of the island in 2000 (Scalera and Angelici 2003), and was subsequently found also in other locations in Central and Northern Corsica (Riga et al. 2003) and currently it seems fairly widespread (Pietri 2007, 2015).

The current distribution of *L. corsicanus* can be interpreted, from a biogeographical point of view, as a residual distribution of an ancestral hare species that was previously more widespread, and subsequently drastically reduced its range by the subsequent invasion of *L. europaeus*.

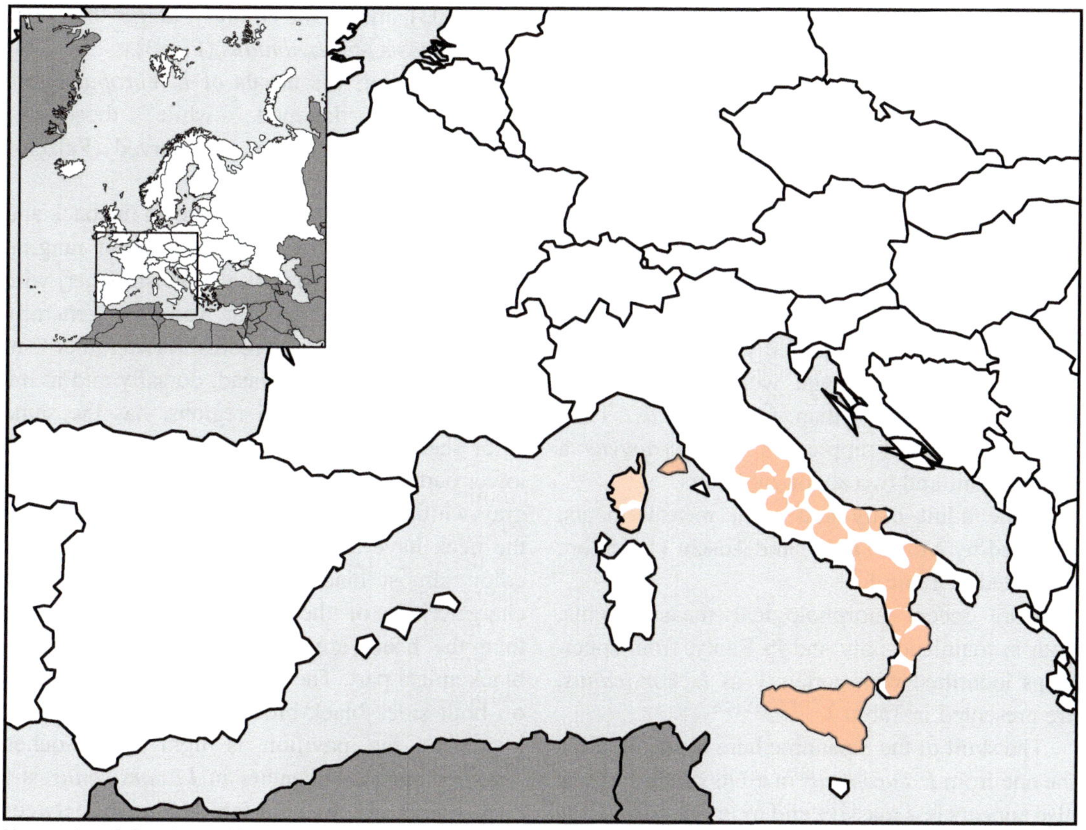

Fig. 2 Geographic distribution of *Lepus corsicanus*. Natural range (apricot); introduced (light apricot). (Based on Angelici and Spagnesi 2008a and new original data). (Map template: © Copyright Getty Images/iStockphoto)

Alternatively, the Apennine hare can be considered a post-glacial wreck that has differentiated from other populations, following geographic isolation (cf Angelici and Luiselli 1998, 2007). However, the two hypotheses of the paleobiogeographical scenario are not necessarily mutually exclusive, and both processes could actually take place.

Moreover, the morphological (Palacios 1996; Alves and Niethammer 2003; Schai-Braun and Hackländer 2018), genetic (Alves et al. 2008, Ferreira et al. 2021), and biogeographical (Acevedo et al. 2014) similarities between the *L. corsicanus* and *L. castroviejoi*, a Spanish endemic species relict with a single population in the Cantabrian Mountains (Palacios 1996; Alves and Niethammer 2003; Schai-Braun and Hackländer 2018) suggest that these species are most likely conspecific. So, a putative evolutionary scenario may involve two disjointed subspecies corresponding to relict portions of the original geographic range of a common ancestral. *L. castroviejoi* currently has a small range, but might be more widely distributed in the Iberian Peninsula, being limited by *L. europaeus* in the north and by the concomitant presence of *L. granatensis* in the southern part of its range (Perez-Suarez et al. 1994). In parallel, *L. corsicanus* appears confined to Central-Southern Italy and Sicily, following the natural expansion of *L. europaeus* from the North into the Italian peninsula. The particular morpho-geographical conformation of the 'boot', with the Alpine chain's geographic barrier, would have allowed a slightly slower expansion of *L. europaeus*, with a range of *L. corsicanus*, although in contraction, wider than that of *L. castroviejoi*.

Description

L. corsicanus has a general appearance similar to that of *L. europaeus*, but with distinctly slender forms. Comparing the two species, in *L. corsicanus* the lengths of head-body, tail, rear foot, and ears are proportionally longer, while, on the contrary, the average weight of adults is around 1 kg lighter than *L. europaeus*. They have three pairs of nipples, as in *L. europaeus*: a pectoral pair and two abdominal pairs.

Some adult body and skull measurements, recorded by Miller (1912) and Toschi (1965) are presented in Table 1.

More recent morphological measurements, both in mainland Italy and in Sicily, from specimens identified with certainty as *L. corsicanus*, are presented in Table 2.

The skull of the Apennine hare is smaller than the one from *L. europaeus* in all its dimensions. It also appears less massive and more rounded (Riga et al. 2003), and on the contrary, longer that those of *L. capensis* and *L. timidus* (Fig. 3).

Table 1 Body and skull measurements of *L. corsicanus* (Miller 1912; Toschi 1965)

Body:	
Rear foot length (mm)	113–127
Ear lenght (mm)	103–127
Skull:	
Occipito-nasal length (mm)	89.6–93.2
Cheekbone width (mm)	43.6–45.6
Post-orbital narrowing (mm)	12.8–14.8
Width of the skull (mm)	30.6–31.8
Length of nasals (mm)	38.4–40.4
Mandible length (mm)	65.4–70.0
Length of the upper dental row (mm)	16.8–17.6

Table 2 Body measurements of *L. corsicanus* (Lo Valvo et al. 1997; Angelici and Luiselli 1998; Riga et al. 2001, 2003; Angelici and Luiselli 2007)

Head-body length (n = 63) (mm)	441–612
Rear foot length (n = 67) (mm)	114–135
Ear lenght (n = 70) (mm)	90–126
Tail lenght (n = 63) (mm)	66–112
Body mass (n = 76) (kg)	1.85–3.8

In particular, the nasals of *L. europaeus* are longer and flattened, while those of *L. corsicanus* are shorter and curved (Palacios 1996).

The coat of the Apennine hare on the back and on the thighs has a characteristic color ranging from tawny-brown to dark yellowish ocher, with mixed dark brown hairs. The wad fur is generally more tending toward the reddish-fawn rather than the greyish-brown. The head, dorsally and in the upper part of the lateral regions, has the same color shade as the rest of the body, while in the lower part of the lateral sections the color is light gray-whitish. The nape and the dorsal portion of the neck have an area of a dark gray-anthracite color, almost black. The latter is a diagnostic characteristic of the species. The ears, longer than the head, are externally ash gray with a black apical part. The folded part of the auricle is on both sides black-brown-anthracite, while the inside of the pavilion is light brown-ocher. Another important feature in *L. corsicanus* is a clear separation, without intergradation, between the brown-reddish coloring of the back and the lateral portions and the belly, which is light white (Fig. 4). In *L. europaeus*, on the other hand, in the same sections of the body, there is a gradually faded transition zone that leads from the dorsal gray-brown color to the white of the abdomen, never as pure as *in L. corsicanus* (Fig. 4). The tail has a black dorsal portion, in the shape of a strip, while the lower parts are white. The hind legs are of a camel-fawn color in the external parts, while internally they are of the same color but lighter and less reddish, more tending to beige, and with mixed also light gray and whitish hairs.

The feet have the same color as the legs and the soles are tawny gray. In the adult, moreover, the basal portion of the hair in the dorsal area has a grayish color, at the height of the shoulder blades.

Rugge et al. (2009) conducted a comparative study of colorimetric determination using some indexes in coat sections of *L. corsicanus* comparing them with *L. europaeus*, and obtained statistical-numerical results that were consistent

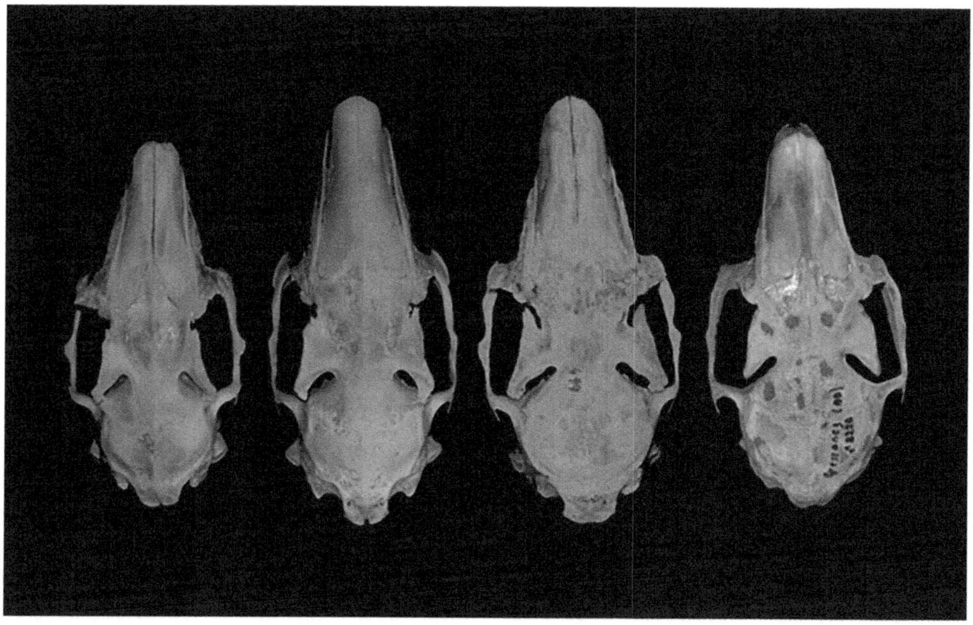

Fig. 3 Skulls in *norma dorsalis* (from left to right) of: *Lepus capensis*, *L. corsicanus*, *L. europaeus*, *L. timidus*. (From Angelici and Spagnesi 2008a). Copyright with the authors

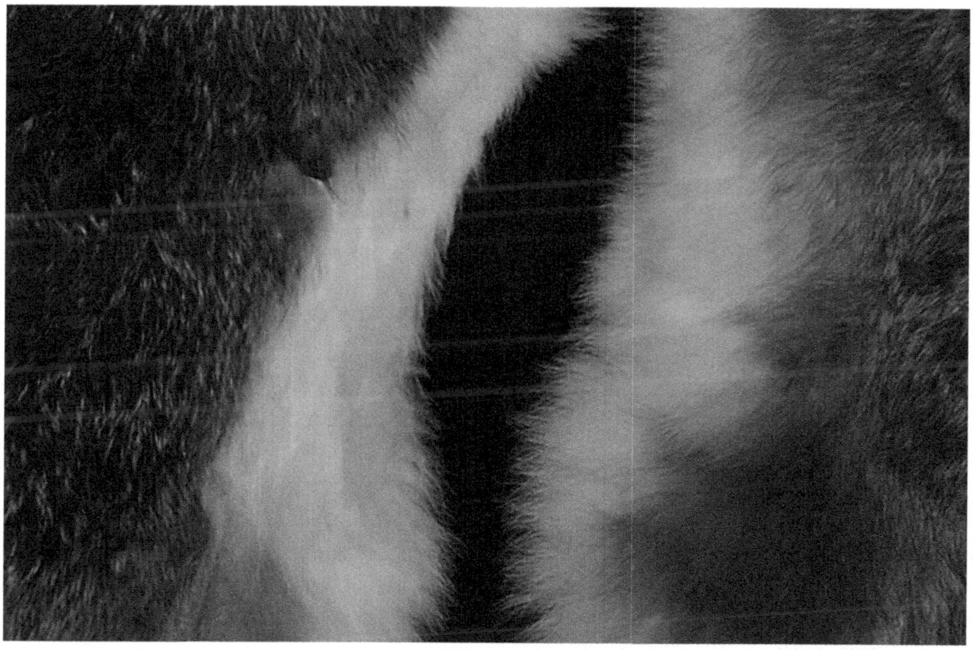

Fig. 4 Details of the ventro-lateral demarcation in *Lepus corsicanus* (left), and *L. europaeus* (right). (From Angelici and Spagnesi 2008a). Copyright with the authors

and assimilable with the qualitative descriptions known to date, and described here.

There is no clear sexual dimorphism and the differences in weight between the two sexes are not statistically significant (Angelici and Luiselli 2007). The winter fur, especially in the populations living in the Apennines and in Sicily at higher altitudes, is much thicker, longer and softer, than the summer fur.

Physiology

At the moment the data on the physiology of this species is very scarce and mostly absent. The only characteristic, highlighted in the section "Life History" and in accordance with the preliminary observations made by De Marinis et al. (2007), is that the Appennine hares might reproduce year-round.

Genetics

The Apennine hare, although there are no specific studies, should present, like all members of the genus *Lepus*, a chromosomal set of 2n = 48 (cf Robinson et al. 1983; Schröder et al. 1987).

The phylogenetic relationships between *L. corsicanus* and other species of the genus *Lepus* with European and African distribution have been reconstructed by means of molecular analysis (Pierpaoli et al. 1999; Alves et al. 2008, Melo-Ferreira et al. 2012, Ferreira et al. 2021). The results of these studies indicated that *L. corsicanus*, together with *L. timidus*, *L. castroviejoi*, and *L. granatensis*, belongs to an evolutionary lineage distinct from that to which *L. europaeus*, *L. capensis* and two other African species belong. These first data, based on a calibration of the molecular evolution rate (molecular clock), showed that these two lineages may have diversified about 2.5–3 million years ago. In a recent exhaustive work on the phylogenomics of Leporidae worldwide, highlighting the importance of hybridization and therefore of the introgression between different species in the radiation and diversification of the various species, the close phylogenetic relationship between *L. corsicanus* and *L. castroviejoi* is confirmed (Ferreira et al. 2021).

Genetic data indicate that *L. corsicanus* and *L. europaeus* are well distinct, suggesting a reproductive isolation; thus generally these species do not hybridize, although hybridization has been reported in Corsica, when often the two species are found in sympatry and syntopy (Pierpaoli et al. 1999; Riga et al. 2003, Pietri et al. 2011). Nevertheless, a compressive study conducted by Mengoni et al. (2015), where 458 samples belonging to the four species of *Lepus* distributed in Italy (*L. europaeus*, *L. timidus*, *L. capensis*, *L. corsicanus*) were analyzed using maternal and biparental markers (mtDNA control-region, 13 autosomal microsatellites, 9 autosomal SNPs), confirmed sharp interspecific genetic distinctions among the four species. These authors did not find interspecific hybrids in *L. corsicanus*. The Appenine hares from Sicily are genetically distinct (from continental Italy, Pierpaoli et al. 1999), as a result of long-lasting isolation, and it is therefore important to avoid any hare introduction to the island. In fact, the hares from Sicily have unique haplotypes exclusive to the island, different from the haplotypes found in the populations of *L. corsicanus* from mainland Italy (Pierpaoli et al. 1999).

Suchentrunk et al. (2007), Alves and Melo-Ferreira (2007), and Alves et al. (2008), using various molecular biology techniques, concluded that *L. corsicanus* and the broom hare *L. castroviejoi* most likely belong to the same species, as already hypothesized, biogeographically, by Angelici and Luiselli (2007). Ferreira et al. (2021), using most comprehensive genomic phylogenetic approach in the genus *Lepus*, confirm the very high genetic similarity between these two 'sister' species.

The genetic structure (i.e., by analyzing mitochondrial DNA and microsatellite DNA) of *L. corsicanus* validates the habitat suitability model and highlights the differences to *L. europaeus* (Fulgione et al. 2009; Buglione et al. 2020b). In fact, by calculating the correlation between genetic distance and ecological data intended as niche differentiation, these studies confirm the clear segregation between the two species and the lower ecological value of *L. europaeus* in the study area, due to the lesser adaptation to the environment following a recent introduction (see section "Habitat and Diet"). Recently, an accurate method for a rapid genetic identification of *L. corsicanus* has been developed by extracting the mitochondrial DNA from the feces (pellets) so that the species can be distinguished in areas where *L. europaeus* also occurs (Buglione et al. 2020a, b).

Life History

According to some observations made by Lo Valvo et al. (2012) in Sicily, on individuals of *L. corsicanus* captured and adapted to captivity, mating took place starting from the month of December, births starting from February to October. The average litter size was 1.63 (range between 1 and 3), and the average interpartum was 52.2 days (Lo Valvo et al. 2012). However, according to De Marinis et al. (2007), as well as Angelici and Spagnesi (2008b), *L. corsicanus* reproduces year-round as pregnant females and leverets have been found in autumn and winter, i.e., October–January). The number of litters per year is on average 2 with max 4 per year (Schai-Braun and Hackländer 2018). The number of leverets per birth varied from 1 to 4, with an average number of 1.86 (De Marinis et al. 2007).

The continuous breeding was also observed in *L. capensis* in Sardinia, while, on the contrary, there is a different situation to more 'continental' species, such as *L. europaeus*, which would concentrate the births only in the most favorable periods of the year, and with a greater number of leverets per litter (De Marinis et al. 2007). However, although these aspects are evident, they should be interpreted in an ecological and non-geographical view, bearing in mind that it would be appropriate to study a larger sample of *L. corsicanus*, namely, including the populations living in the mountains, where the climatic conditions are not strictly Mediterranean.

Habitat and Diet

General Ecology and Habitat

This species is adapted to live in different environments. In fact, if on the one hand it seems to tolerate Mediterranean climates and environments, but in the other hand it is well established on the Central-Southern Apennines up to about 2000 m a.s.l. and on the mountain ranges of Sicily up to 2400 m a.s.l. (Angelici and Luiselli 1998, 2007; Riga et al. 2003). The selected environments are represented by cultivated areas (more or less abandoned), bushy grasslands, and broad-leaved woods with large clearings (Angelici and Spagnesi 2008a). The latter aspect, that is, the presence of wooded areas in the environments of choice of the species, clearly differentiate it from *L. europaeus* which tends, much more decisively, to avoid woodlands and wooded areas (Fusco et al. 2007). In the purely Mediterranean environment, it occupies the maquis, even thick, including the coastal dune environments (cf. Riga et al. 2003). However, tree and shrub species are well represented and seem to be an essential component in many areas where the species is present (cf Rugge et al. 2007; Guglielmi et al. 2008).

Di Luzio and Barone (2012) have studied a small sample of Apennine hares in Latium (Central Italy) and have shown that these hares ($n = 3$) tend to select positively shrubs and oak forests, rather than cultivated areas, arid grasslands, and mixed deciduous forests. From this study, although preliminary, it emerges for the species the need to have available sufficiently covered areas, such as shrublands, but also open areas mixed with wooded areas, which can be suitable as foraging areas but also allow good anti-predatory coverage (Di Luzio and Barone 2012).

In Sicily, where it is currently the only hare species present, it inhabits many environmental typologies such as the hilly and mountain meadow-pastures, the clearings and the margins of deciduous woods, the uncultivated areas with bushes. Pastures, garrigues and cultivated areas are also well frequented, and among the latter the species seems rather recurrent in the large wheat extensions, but also in olive groves, vegetable gardens, vineyards, and citrus groves (Angelici and Spagnesi 2008a).

In comparative studies between *L. corsicanus* and *L. europaeus* in continental Italy, Angelici and Luiselli (1998, 2007) found for both species a clear preference, for the meadow-pastures and for the cultivated fields (cereals, olives, and vegetables) compared to the evergreen Mediterranean maquis and the broadleaf coppices. It is interesting to note that, according to the data of the latter authors, when *L. europaeus* is the only species present, it mainly occupies the cultivated fields, while if it occurs syntopic with *L. corsicanus*, it inhabits both the cultivated fields and the meadow-pastures. On the contrary, the Apeninne hare is most commonly found both in cultivated fields and in meadow-pastures when it is the only species present, but in the case of sympatry with *L. europaeus* it is found almost exclusively in meadow-pastures (Angelici and Luiselli 1998, 2007). So, in the event of syntopy, *L. europaeus* tends to be more generalist, while *L. corsicanus* proves to be more specialist in the choice of habitat.

Another mechanism of equal ripartition of space has also been observed, affecting the two species: in terms of altitude a.s.l. the two species do not seem to have different habits when they live in "allopatric" areas, while, when they are in sympatry (and potential syntopy), *L. corsicanus* tends to occupy areas at higher altitude compared to *L. europaeus* which instead is more represented at lower altitudes. The curious fact is that, in the absence of *L. corsicanus*, *L. europaeus* also occupies territories at higher elevation (Angelici and Luiselli 1998, 2007; cf. Guglielmi et al. 2008). All these results can be interpreted as a niche separation mechanism designed, which might reduce interspecific competition (Ricklefs 1980). Moreover, this seems to indicate a greater adaptability of *L. europaeus*, which apparently pushes *L. corsicanus*, under syntopic conditions, into mountainous areas, relatively less hospitable. A recent model of environmental suitability for *L. corsicanus* in Latium (Central Italy) confirms how, in this region, the Apennine hare prefers flat Mediterranean coastal areas or of modest altitude, and Pre-Apennine and Apennine internal areas at medium-high altitude (maximum 1700 m a.s.l.) (Angelici et al. 2010).

Fulgione et al. (2009) defined the ecological characteristics of *L. corsicanus* in comparison with *L. europaeus* in order to understand how landscape facilitates or impedes movement. Spatially explicit models were used to identify a species ecological niche and to build a landscape model of suitability. To validate ecological modeling of landscape, authors performed a population genetic analysis. Results suggest that the Apennine hare shows an ecological requirement close to average of available resources in the considered landscape. The genetic structure of this autochthonous species validates the habitat suitability model and highlights the differences with European hare.

Furthermore, the study by Acevedo et al. (2014) analyzes the ecological niches of *L. corsicanus* and *L. castroviejoi* biogeographically reaching the result that interestingly these two species, with allopatric geographic distribution, occupy a very similar ecological niche, apparently assimilable.

Foraging Activity and Diet

The twilight and nocturnal feeding behavior on pastures seems similar to that of *L. europaeus* (Riga et al. 2003). On the Mount-Etna volcano (Sicily), above timber line, on the pyroclastic substrates of ash and lapilli, characterized by pioneering groups and pulviniform vegetation *Rumici-Astragalion siculi, L. corsicanus* manages to survive, in the absence of snow, despite the potentially available plant species are very few (*Astralagus siculus*, *Juniperus haemispherica*, *Saponaria sicula*, *Rumex scutatus*, *Cerastium tomentosum*, *Tanacetum siculum*, *Anthemis aetnensis*, *Senecio aetnensis*, *Robertia taraxacoidea*, *Festuca stenantha*) (Riga et al. 2003).

Freschi et al. (2014a) examined the annual diet composition of *L. corsicanus* in two different sites within a southern Italy Regional Park. Microhistological analysis of fecal samples revealed that hares utilized 70 different species of plants during the year, indicating the capability of the Apennine hare of exploiting a wide variety of vegetation. Herbaceous plants (Hemicryptophytes, particularly graminoid grasses, and Geophytes) predominated in the diet. *Brachypodium sylvaticum* (9.44%) and *Allium subhirsutum* (8.28%) were the major contributors to the diet. Other taxa found most often in the diet were *Trifolium pretense* and *Prunus spinosa*. Significant differences were found between sites in terms of diet richness, diversity, and evenness. Nevertheless, both the used similarity indices showed that the food composition of the hare's diet was broadly the same in both sites. Some qualitative and quantitative differences between sites were due to the availability or consumption of some plant species and evidenced that the Apennine hare can modify its trophic niche in order to adapt its dietary requirements to the availability of food (Freschi et al. 2014a). Moreover, Freschi et al. (2014b) studied *L. corsicanus* diet by analyzing fecal pellets in a semi-natural habitat. The results showed that hares feed on 62 species of plants during the year, with a conspicuous presence of herbaceous ones, e.g., *Trifolium pratense*, *Brachypodium sylvaticum*, *Festuca arundinacea,* as these occurred at high frequencies in most of the fecal samples. In spring, diet composition was characterized by a high percentage of Graminaceae (>37% of fecal pellets). In the other seasons, hares also included fruits, e.g., *Prunus spinosa, Pyrus piraster, Malus sylvestris*), which, in autumn were detectable in >27% of the analyzed droppings. There were significant differences among seasons. The smallest values of richness and diversity were observed in spring. Dietary overlap was low between spring and the other seasons; conversely, there was substantial overlap (>70% of fecal pellets) in the diets during the other seasons with a more pronounced similarity between summer and autumn (Freschi et al. 2014b). Finally Freschi et al. (2015) analyzed in detail the seasonal variation of the Apennine hare diet. In this study, the seasonal pattern of diet composition of a population of *L. corsicanus* occupying a semi-natural landscape was estimated by using the micro-histological technique of fecal analysis. The results showed that hares had a diversified diet, consuming plant parts from over 70 species (Freschi et al. 2015). Like other *Lepus* species, the Apennine hare consumed a large amount of herbaceous plants, e.g., *Brachypodium sylvaticum*, *Trifolium pratense*, *Allium subhirsutum,* and *Festuca arundinacea*, although it complemented its diet seasonally with fruits of *Prunus spinosa*, *Pyrus piraster,* and *Malus sylvestris*. Analysis of similarities (ANOSIM) evidenced significant differences among seasons, as a consequence of the seasonal occurrence of the various food items. Spring and autumn, as well as spring and winter, showed low diet similarities; these results were supported by similarity percentage analysis (> 71% between spring and autumn; > 69% between spring and winter) with taxa like *P. spinosa*, *Cirsium strictus*, *T. pratense,* and *Rosa canina* making the greatest contributions to these differences. Higher similarities were instead found when comparing other seasons.

A study on feeding habits was carried out in Corsica (Rizzardini et al. 2019). This study is the first to consider the food selection of *L. corsicanus*. Grasses represented the basis of the diet, with frequencies around 50%, followed by non-leguminous forbes with an incidence of 29 to 31%. Leguminous shrubs complemented its diet. Poaceae were found to be the most preferred and selected family in the diet. A total of 79 species were detected in the diet, but only some of them had percentages higher than 5%. The most used species in the diet were *Brachypodium sylvaticum*, *Briza max, Trifolium angustifolium*, *Digitaria sanguinalis*, and *Daucus carota*. This study showed that in the areas considered, characterized also in the dry period by wide plant diversity, the Apennine hare used plants as a generalist (Rizzardini et al. 2019).

Buglione et al. (2015) showed, in a mountain environment in Campania (Southern Italy), that there is a 26% food overlap (competition) between *L. corsicanus* and *L. europaeus* on the used plant species. These authors also showed how this mechanism can be negative and dangerous for the survival of *L. corsicanus* following recent climate changes, due to the lower food adaptability. That is, with the gradual increase in temperature, some taxa of mountain plants, many of which are specifically preferred by the Apennine hare, could likely face extinction. This could be strongly negative for *L. corsicanus*, a species already stressed and relegated to marginal environments (Buglione et al. 2015).

Buglione et al. (2018) presented a non-invasive pilot study set to analyze the diet composition of the Apennine hare from Southern Italy, starting from fecal pellets, using, for the first time on this species, DNA metabarcoding and next generation sequencing to identify the plants. The findings of this study indicated that this genetic approach provides reliable qualitative and semi-quantitative information, allowing the characterization of the hare diet and its seasonal variation using 22 fecal samples. In a single experiment, through time and cost-effective screening of multiple DNA metabarcodes, Buglione et al. (2018) detected a broad diversity of plants (99 taxa). Unlike traditional methods, this approach can identify items that leave no solid remains or that simply are lacking in diagnostic taxonomic features.

Behavior

As for the use of space and territoriality, the species seems to be rather sedentary with home ranges of limited size; from sunset and for most of the night it frequents the same pasture areas, which are located close to the night shelters (M. Lo Valvo, pers. comm.). In a study in Sicily conducted by Lo Valvo (2007), it was able to ascertain how the annual home ranges detected on six individuals vary from 0.7 to 267.7 ha, using the method of the Convex Minimum Polygon (MCP). Some preliminary data collected on a small sample of Apennine hare in Latium (Central Italy), using the 95% Kernel method, have determined annual home range extension data in two females ranging from 8 to 28 ha, while in one male the home range varied from 59 to 73 ha (Di Luzio and Barone 2012). Some Apennine hares reintroduced on the Island of Elba (Scarselli et al. 2016; Lari 2019) and monitored with GPS technology have occupied an average home range varying between 87.33 and 97.41 ha.

Mori et al. (2020) in a multi-year study in Central Italy have shown that the home range (calculated both with the 95% MCP and with the 95% Kernel) is larger, both in males and females, during the warm months compared to winter. Furthermore, home ranges were larger during the night (hares more active for foraging) compared to daylight. In particular, in the warmer months the home ranges varied from about 40 to 70 ha, to decrease to a range of around 27 to 45 ha in the colder months; there were no significant differences between the two sexes (Mori et al. 2020).

The spatial behavior of seven male and five female Apennine hares was studied in Central Italy, through radiotracking data (Lovari et al. 2020). The median percentage of home range overlap was about five times lower in females (<15%) than in males (75%). Intrasexual aggression was recorded for females, but never for males. No species within the *Lepus* genus are known to be territorial, whereas authors suggest that interindividual female intolerance in this species may be compatible with territoriality (Lovari et al. 2020). The fact that *L. corsicanus* reproduces throughout the year, thus during food-poor months, might have triggered resource defense between females to maximize their reproductive fitness (Lovari et al. 2020).

Concerning the use and selection of the habitat, Apennine hares select Mediterranean scrubwood and cultivated areas, while they avoid mostly human settlements, and deciduous forests, in all seasons (Mori et al. 2020). According to these authors, at night, the crops were selected and the deciduous woods were underutilized; during the daylight hours the deciduous woods were selected during the cold months, and the Mediterranean shrubwoods during the summer, while the cultivations were always avoided. This is a consequence not only of the habits of foraging activities typical of the species and of the Leporidae in general, but it is also conditioned by the cover that the vegetation can offer for the antipredatory behavior. Greater consistency of bushes and trees is necessary during the daytime with hares largely inactive in the den, while during foraging and dispersal activities, it is important to have more open areas where herbaceous vegetation is more present as a primary food source, and visibility for antipredation and escape are favored (Mori et al. 2020).

From some observations collected by Riga et al. (2003) in the areas occupied by both *L. corsicanus* and *L. europaeus*, it has been shown that the Apennine hare, if discovered in thick vegetation tends to remain hidden by longer time; on the contrary, once discovered *L. europaeus* tends to run precipitously uncovered, and then re-occupy the den after long journeys. The Apennine hare runs very smoothly and makes long jumps over the vegetation, fully extending the limbs. From observations made by night with the aid of the headlights, there was a higher frequency, compared to *L. europaeus*, of the alert and observation behavior of the Apennine hare, in an upright position on the rear limbs (Riga et al. 2003).

According to Mori et al. (2020), during a 24 h period, the higher locomotory activity was between 5 pm and 5 am.

Parasites and Diseases

The first work on the endoparasites of the Apennine hare was conducted by Usai et al. (2012). In this study, in order to describe the helminths, gastro-intestinal tract of 29 Apennine hares, from mainland Italy (n = 14) and Sicily (n = 15), were collected between 1997 and 2009. Parasites were observed in 86% of the analyzed hares. Six parasite species were isolated: two cestodes., *Cittotaenia pectinata* with a prevalence of 3%, and *Paranoplocephala* sp. 3%; and four nematodes, *Trichostrongylus retortaeformis* 86%, *Graphidium strigosum* 14%, *Trichuris* sp. 10%, and *Teladorsagia circumcincta* 7%. Both morphotypes of *Teladorsagia circumcincta (T c. circumcincta* and *T. c. trifurcate)* were identified. *Teladorsagia circumcincta* and *Paranoplocephala* sp. have been reported in leporids for the first time.

The ticks that infest *L. corsicanus* found so far have been studied by Dantas-Torres et al. (2011) in a wildlife reserve in Southern Italy. Fifty-five ticks were collected by dragging. Ticks were identified as *Hyalomma marginatum*, *Dermacentor marginatus*, and *Rhipicephalus bursa*. In September 2009, ticks were collected from 17 Apennine hares and identified as *Ixodes ricinus* (2 larvae, 45 nymphs, 35 males, 37 females), *Rhipicephalus turanicus*, and *Hyalomma* sp. PCR amplification and sequencing of a partial region of the 12S rDNA gene of *Hyalomma* nymphs allowed their identification as *H. marginatum* (Dantas-Torres et al. (2011).

L. corsicanus can be affected by European Brown Hare Syndrome (EBHS), which is transmitted, at least originally, by *L. europaeus* (Guberti et al. 2000). On a sample of 24 Apennine hares from Sicily and mainland Italy, 19% of positive cases for this viral disease were found (Lavazza and Guberti 2007). This viral disease can have a severe impact on the population, with high mortality (Guberti et al. 2000).

Recently, the Rabbit Hemorrhagic Disease Virus type 2 (RHDV2) has also been reported in *L. corsicanus*, a viral pathology (*Lagovirus*) that does not only affects rabbits but also other leporids, such as hares (Camarda et al. 2014). This viral pathology produces a fatal hepatitis.

A sanitary surveillance for *L. corsicanus* was suggested by Lavazza and Guberti (2007). In summary, it should involve regular samplings on *L. europaeus* hunted during the hunting season, and on *L. corsicanus* sampled for scientific studies. This would allow to monitor the pathologies and viral diseases that mostly derive from imported European hares, and eventually starting some care and control actions. However, greater sanitary control of imported hares and a progressive decrease in the number of imported individuals, until their cessation, would be desirable. Although, there are no particular problems at the moment, it is important to continuously monitor the populations of this vulnerable species.

Population Ecology

Density

The density of the species is not well-known, although some estimates, done by countings with night lights in transects, were carried out both in mailand Italy and in Sicily: it was noted that the estimated density in mainland Italy in protected areas (5.54–7.06 individuals/100 ha) is significantly lower than that found in Sicily in comparable areas (11.73–11.98 individuals/100 ha) (Trocchi and Riga 2001). Lo Valvo et al. (1999) studied the dynamics of a small population of Apennine hares in a protected Sicilian area, through nocturnal counts made throughout the year. In this study, hare density increases from mid-February to June (average 17.5 individuals/100 ha) with a peak around mid-April, while in other seasons the number is fairly constant (average 12.0–14.5 individuals/100 ha). Lo Valvo (2007) reported density data in Sicily ranging from a minimum of 12.0 to a maximum of 103.0 individuals/100 ha.

Causes of Mortality and Predators

Apart from the possible pathologies, the species is often illegally killed in mainland Italy, during the hunting season, while being protected, due to the inevitable confusion that is done with the European hare, which can be hunted.

There are many predators of the *L. corsicanus*, namely, avian, such as Falconiformes and Strigiformes (see Contoli 1988), and mammalian carnivores (Angelici and Spagnesi 2008a). Even stray dogs and stray cats can prey on adults and young hares, while the wildboar, *Sus scrofa,* represents a danger for leverets (Angelici and Spagnesi 2008a). The grey wolf, *Canis lupus,* is certainly a predator of *L. corsicanus* (P. Ciucci, pers. comm.); while the red fox, *Vulpes vulpes,* is one of the most important predators in both mainland Italy and Sicily. In Sicily, the Apennine hare represents 9.5% of the prey captured (20.5% of the total biomass, n = 148) by the golden eagle, *Aquila chrysaetos* (M. Di Vittorio, pers. comm.). Lo Valvo (2007) also reports a case of pine marten, *Martes martes,* predation. In a very inaccessible and isolated area of Southern Latium (Central Italy), *L. corsicanus* was represented (12.9%) in the diet of the Eurasian eagle-owl, *Bubo bubo* (Angelici and Spagnesi 2008a).

Effects of Climate Change

As already mentioned in Habitat and diet, *L. corsicanus* partly consumes (26%) plant species avoided by *L. europaeus*, and these species, mostly mountain species, are seriously threatened by recent climate change. This could probably be a negative factor for the future survival of *L. corsicanus* (Buglione et al. 2015).

Conservation Status

L. corsicanus has been fully protected only in Sicily, from 2000 to 2004, because the Sicilian Regional Administration recognized it as a good species (see below). Furthermore, from 1999 to 2003, any form of hare hunting was prohibited on the island, in a general sense. It is also forbidden the introduction of *L. europaeus*, which up to 1995 has been introduced in large numbers but which, fortunately, does not seem to have formed naturalized populations in Sicily (Angelici and Spagnesi 2008a, b). Starting from the hunting calendar 2004–2005, in Sicily *L. corsicanus* has been included among the game species, albeit for a limited period of the year.

The situation in the continental Italy is different. The Apennine hare is not a game species, and it is protected. Nevertheless, a fundamental problem exists, as it is almost impossible for hunters to distinguish in the field *L. corsicanus* (protected species) from *L. europaeus* (a game species). As a consequence, many Apennine hares might be unintentionally killed annually during the hunting season, because they are confounded by *L. europaeus*. There is no quantitative data on these illegal kills, but it reasonable to assume that the number is not negligible. The only populations of *L. corsicanus* where this erroneous hunting does not occur are those present in protected areas (National Parks, Regional Parks, Natural Reserves, Restocking and capture areas, Oasis, etc.). Nevertheless, these populations may suffer also from poaching.

The Apennine hare in Italy was initially considered as Critically Endangered (CR), according to the categories of the International Union of Conservation of Nature (IUCN), since data were scarce (Amori et al. 1996, 1999; Angelici 1998). Subsequently, having available data, and following more strictly the 2000 IUCN categories (Hilton-Taylor 2000), it was decided to classify *L. corsicanus* as Endangered (EN) (Angelici and Luiselli 2001; Trocchi and Riga 2001). In the IUCN Red List *L. corsicanus* was included in 2008 (Angelici et al. 2008) as Vulnerable (VU), which was confirmed in the last assessment (Randi and Riga 2019).

Management

In order to effectively implement a species protection and conservation plan, a number of factors and components must be considered. First of all, since there is no comparison data with the past, it is difficult to understand the population trend in the last decades. It can, however, be reasonably assumed, given the continuous hunting of the European hare (which in a broad sense, might also include the Apennine hare, since hunters often cannot distinguish them), that several populations in mainland Italy have decline or even disappeared. However, some populations, mainly those present in protected areas have been preserved, or located in areas difficult to reach by hunters and hunting operators who introduce game species (namely, *L. europaeus*). As already have been mentioned (see section "Population Ecology") population densities (see section "Density"), at least on the Italian Peninsula, are low. Moreover, many populations seem to be isolated from each other. It is clear that range fragmentation is also a consequence of the alteration of the habitats suitable for the species, considering the increase in anthropogenization, both in the purely mountainous Apennine environments and in those that are more strictly Mediterranean. As a consequence of range fragmentation, there would also be an increase in inbreeding, which reduces individual fitness within single populations (see also Riga et al. 2003).

The introduction of allochthonous species contributes to the establishment of a competition for space and resources that surely harms an already remarkably stressed species. In fact, the introduction of *L. europaeus* in the whole Italian Peninsula has generated many situations of conflict between the two species. In situations of sympatry and syntopy, *L. europaeus* tends to "push" *L. corsicanus* toward higher elevated areas, reducing the effective extension of the range. It is also possible that other yet unknown competition phenomena exist. With the introduction of the European hare there is also the cross-interspecific dissemination of serious pathogens (see "Parasites and Diseases"). Fortunately, the hybridization between *L. corsicanus* and *L. europaeus* does not seem to be a danger for the conservation of the Apennine hare, at least in Italy and Sicily, also by studying it thoroughly from a genetic-molecular point of view (Mengoni et al. 2015). So far, hybridization has been only recorded in Corsica (Pietri et al. 2011).

Future Challenges for Research and Management

The actions to be taken as soon as possible (in part already implemented) will have to be articulated following these points:

(a) Increasing knowledge of the current range of *L. corsicanus*. This will allow to set up a network of protected areas, as much as possible in contact with each other, in order to promote a gene flow and avoid fragmentation.
(b) Creation of protected areas, with characteristics suitable for the species, where it is possible to breed pure Apennine hares and subsequently repopulate and/or reintroduce the species. These actions are ingoing at present (cf. De Filippo et al. 2018).

(c) Recognition in international biodiversity conventions as a threatened taxon (e.g., Bern, EU Habitats Directive).
(d) Prohibition of the introduction of *L. europaeus* in the historical and current range of *L. corsicanus*. This absolutely fundamental point is, however, difficult to implement, for economic and hunting reasons, but also because probably in Central-Southern Italy there were originally also some autochthonous populations of *L. europaeus* sensu *L. europaeus meridiei* (Hilzheimer 1906). The issue is being addressed by means of accurate genetic-molecular investigations.
(e) Implement a hunting management strategy of *L. europaeus* in Central and Southern Italy, and in Sicily (cf. FIDC 2017).
(f) Planning of a project of dissemination and knowledge of the species, at local, regional, national, and international level.

For a more in-depth and analytical analysis of the problems about protection and conservation of *L. corsicanus* in continental Italy and Sicily, see Trocchi and Riga (2001) and De Filippo et al. (2018).

References

Acevedo P, Melo-Ferreira J, Real R, Alves PC (2014) Evidence for niche similarities in the allopatric sister species *Lepus castroviejoi* and *Lepus corsicanus*. J Biogeogr 41:977–986

Alves PC, Melo-Ferreira J (2007) Are *Lepus castroviejoi* and *Lepus corsicanus* conspecific? Evidence from the analysis of nuclear markers. In: De Filippo G, De Riso L, Riga F, Trocchi V, Troisi SR (eds) Conservazione di *Lepus corsicanus* De Winton, 1898 e stato delle conoscenze. IGF, Napoli, pp 45–51

Alves PC, Niethammer J (2003) Iberische Hasen *Lepus granatensis* und *Lepus castroviejoi*. In: Krapp F (ed) Handbuch der Säugetiere Europas. Band 3/II: Hasentiere. Lagomorpha. Aula-Verlag, Wiebelsheim, pp 137–154

Alves PC, Melo-Ferreira J, Branco M, Suchentrunk F, Ferrand N, Harris DJ (2008) Evidence for genetic similarity of two allopatric European hares (*Lepus corsicanus* and *Lepus castroviejoi*) inferred from DNA sequences. Mol Phylogenet Evol 48:1191–1197

Amori G, Angelici FM, Prigioni C, Vigna Taglianti A (1996) Mammals of Italy: a review. Hystrix 8(ns):3–7

Amori G, Angelici FM, Boitani L (1999) Mammals of Italy: a revised list of species and subspecies. Senckenberg Biol 79:271–286

Angelici FM (1989) I Mammiferi dei Monti Cimini e Vicani: popolamento, confronto faunistico con altre aree del Lazio e considerazioni biogeografiche. Master Thesis, Università di Roma "La Sapienza"

Angelici FM (1998) Lepri: Lepre europea (autoctona) *Lepus europaeus meridiei*, Lepre appenninica *Lepus corsicanus*, Lepre sarda *Lepus capensis mediterraneus*. In: Bulgarini F, Calvario E, Fraticelli F, Petretti F, Sarrocco S (eds) Libro rosso degli animali d'Italia-Vertebrati. World Wildlife Fund, Roma, pp 116–117

Angelici FM (1999) *Lepus corsicanus* de Winton 1898. In: Mitchell-Jones AJ, Amori G, Bogdanowicz W, Krystufec B, Reijnders PJH, Spitzenberger F, Stubbe M, Thissen JBM, Vohralik V, Zima J (eds) Atlas of European mammals. Academic, London, pp 164–165

Angelici FM, Luiselli L (1998) Distribuzione, biogeografia dinamica e interazioni ecologiche tra Lepre appenninica *Lepus corsicanus* e Lepre europea *Lepus europaeus* in Italia. II Congresso Italiano di Teriologia, Varese, 28–30 Ottobre 1998. Riassunti: 14

Angelici FM, Luiselli L (2001) Conservation problems of the critically endangered Apennine hare *Lepus corsicanus* De Winton 1898, in Italy and Sicily. Oryx 35:245–249

Angelici FM, Luiselli L (2007) Body size, habitat choice, and altitude coexistence of two hare species (*Lepus europaeus* and *Lepus corsicanus*) in Italy. Wildl Biol 13:251–257

Angelici FM, Spagnesi M (2008a) *Lepus corsicanus* De Winton, 1898. In: Amori G, Contoli L, Nappi A (eds) Mammalia II. Erinaceomorpha, Soricomorpha, Lagomorpha, Rodentia. Fauna d'Italia, vol. XLIV. Calderini de Il Sole 24 ORE, Milano, pp 254–266

Angelici FM, Spagnesi M (2008b) *Lepus europaeus* Pallas, 1778. In: Amori G, Contoli L, Nappi A (eds) Mammalia II. Erinaceomorpha, Soricomorpha, Lagomorpha, Rodentia. Fauna d'Italia, vol. XLIV. Calderini de Il Sole 24 ORE, Milano, pp 267–282

Angelici FM, Randi E, Riga F, Trocchi V (2008) *Lepus corsicanus*. The IUCN Red List of Threatened Species 2008: e.T41305A10436746

Angelici FM, Petrozzi F, Galli A (2010) The Apennine hare *Lepus corsicanus* in Latium, Central Italy: a habitat suitability model and comparison with its current range. Hystrix (ns) 21:177–182

Azzaroli A (1980) Biogeografia dei Mammiferi della Sardegna. Lavori della Società Italiana di Biogeografia 8: 35–52

Buglione M, Fulgione D, Petrelli S, De Filippo G, Trapanese M (2015) Global climate changes affect food competition of two *Lepus* species. Poster presented at: Workshop 'The impact of climate changes on animal population', Erice, 9–14 novembre 2015 http://www.lepreitalica.org/doc/Buglione%20et%20al%202015.pdf

Buglione M, Maselli V, Rippa D, De Filippo G, Trapanese M, Fulgione D (2018) A pilot study on the application of DNA metabarcoding for non-invasive diet analysis in the Italian hare. Mamm Biol 88:31–42

Buglione M, Petrelli S, Notomista T, De Filippo G, Gregorio R, Fulgione D (2020a) Who is who? High resolution melting analysis to discern between hare species using non-invasive sampling. Conserv Genet Resour 12:727–732. https://doi.org/10.1007/s12686-020-01153-9

Buglione M, Petrelli S, De Filippo G, Troiano C, Rivieccio E, Notomista T, Maselli V, Di Martino L, Carafa M, Gregorio R, Latini R, Fortebraccio M, Romeo G, Biliotti C, Fulgione D (2020b) Contribution to the ecology of the Italian hare (*Lepus corsicanus*). Sci Rep 10:13071. https://doi.org/10.1038/s41598-020-70013-1

Calamai S (2019) Monitoraggio della lepre italica (*Lepus corsicanus* de Winton, 1898) nell'ambito del progetto di reintroduzione nel Parco Nazionale dell'Arcipelago Toscano. Thesis, Università di Pisa

Caloi L, Palombo MR (1989) La mammalofauna wurmiana di Grotta Barbara (Monte Circeo): implicazioni paleoeconomiche e paleoambientali. Hystrix (ns) 1:85–94

Camarda A, Pugliese N, Cavadini P, Circella E, Capucci L, Caroli A, Legretto M, Mallia E, Lavazza A (2014) Detection of the new emerging rabbit haemorrhagic disease type 2 virus (RHDV2) in Sicily from rabbit (*Oryctolagus cuniculus*) and Italian hare (*Lepus corsicanus*). Res Vet Sci 97:642–645

Contoli L (1988) La nicchia trofica di Allocco (*Strix aluco*) e Barbagianni (*Tyto alba*) in Italia: acquisizioni e problemi. Il Naturalista Siciliano S IV,12(Supp):129–143

Corbet GB (1978) A World List Of Mammalian Species. British Museum of Natural History, London, 256 pp

Dantas-Torres F, Testini G, Digeronimo PM, Lorusso V, Mallia E, Otranto D (2011) Ticks infesting the endangered Italian hare (*Lepus corsicanus*) and their habitat in an ecological park in southern Italy. Exp Appl Acarol 53:95–102

De Filippo G, Fulgione D, Riga F, Trocchi V (eds) (2018) 20 years of Italian hare. New researches. ISPRA, Ozzano dell'Emilia

De Marinis AM, Trocchi V, Mangiafico S, Fassò C, Mallia E (2007) Strategie riproduttive in tre specie di lepre (*Lepus* sp. pl.) in Italia. In: De Filippo G, De Riso L, Riga F, Trocchi V, Troisi SR (eds) Conservazione di *Lepus corsicanus* De Winton, 1898 e stato delle conoscenze. IGF, Napoli, pp 75–81

Di Luzio P, Barone V (2012) Risultati preliminari sul comportamento spaziale della Lepre italica (*Lepus corsicanus*) nella R.N.R. Monterano. In: Riga F, Scalisi M (eds) Atti del Workshop nazionale sulla conservazione della Lepre italica: azioni locali per la strategia nazionale. Roma, ISPRA, pp 111–130

Dubray D, Roux D (1984) Répartition des Lagomorphes (Lièvre commun et Lapin de garenne) en Corse. Bull Mensuel de l'Office National de la Chasse 85:36–38

Ellermann JR, Morrison-Scott TCS (1951) Checklist of Palaearctic and Indian mammals 1758 to 1946. British Museum (Natural History), London

Esu D, Kotsakis T (1980) Paleobiogeografia dei Vertebrati e dei Molluschi continentali del Terziario e del Quaternario della Sardegna. Lavori della Società Italiana di Biogeografia 8:53–82

Ferreira MS, Jones MR, Callahan CM, Farelo L, Tolesa Z, Suchentrunk F, Boursot P, Mills LS, Alves PC, Good JM, Melo-Ferreira J (2021) The legacy of recurrent introgression during the radiation of hares. Syst Biol 70:593–607

FIDC (Federazione Italiana della Caccia) (2017) Verso il prelievo venatorio sostenibile della Lepre italica (*Lepus corsicanus*) in Sicilia: buone pratiche e azioni di monitoraggio. Uffico Fauna Stanziale F.I.d.C, Roma

Frati F, Hartl GB, Lovari S, Delibes M, Markov G (1998) Quaternary radiation and genetic structure of the red fox *Vulpes vulpes* in the Mediterranean basin as revealed by allozymes and mitochondrial DNA. J Zool (Lond) 245:43–51

Freschi P, Fascetti S, Musto M, Mallia E, Blasi AC, Cosentino C, Paolino R (2014a) Diet of the Apennine hare in a southern Italy Regional Park. Eur J Wildl Res 60:423–430

Freschi P, Fascetti S, Musto M, Mallia E, Cosentino C, Paolino R (2014b) Diet of Italian hare (*Lepus corsicanus*) in a semi-natural landscape of southern Italy. Mammalia 79:51–59

Freschi P, Fascetti S, Musto M, Cosentino C, Paolino R, Valentini V (2015) Seasonal variation in food habits of the Italian hare in a south Apennine semi-natural landscape. Ethol Ecol Evol 28:148–162

Fulgione D, Maselli V, Pavarese G, Rippa D, Rastogi RK (2009) Landscape fragmentation and habitat suitability in endangered Italian hare (*Lepus corsicanus*) and European hare (*Lepus europaeus*) populations. Eur J Wildl Res 55:385–396

Fusco L, Troisi SR, Accardo Y, Vaccaro L, Caliendo MF, De Filippo G (2007) Selezione d'habitat interspecifica tra popolazioni simpatiche di *Lepus corsicanus* e *L. europaeus*. In: De Filippo G, De Riso L, Riga F, Trocchi V, Troisi SR (eds) Conservazione di *Lepus corsicanus* De Winton, 1898 e stato delle conoscenze. IGF, Napoli, pp 119–122

Guberti V, De Marco MA, Riga F, Lavazza A, Trocchi V, Capucci L (2000) Virology and species conservation: the case of EBHSV and the Italian hare (*Lepus corsicanus* De Winton, 1898). In: Lavazza A, Brocchi E (eds) Proceedings of V International Congress of European Society for Veterinary Virology. Brescia, 27–30 Agosto 2000, pp 198–199

Guglielmi S, Properzi S, Riga F, Sorace A, Trocchi V, Scalisi M (2008) Primi dati sull'uso dell'habitat di *Lepus corsicanus* e *L. europaeus* (Mammalia, Lagomorpha) nel Lazio. Atti VI Congresso Italiano di Teriologia. Cles (TN), 16-18 aprile 2008. Hystrix (ns) 29(Suppl):93

Hilton-Taylor C. (ed.) 2000 IUCN red list of threatened species. Gland, xvii+61 pp
Hilzheimer M (1906) Die europäischen Hasen. Zool Anz 30:510–513
Hoffmann RS (1993) Order Lagomorpha. In: Wilson DE, Reeder DAM (eds) Mammals species of the world. A taxonomic and geographic reference. Smithsonian Institution Press, Washington/London, pp 807–827
Iacoangeli O (1997) - I Lagomorfi pleistocenici di Melpignano (LE). Tesi di laurea A.A. 1996/'97, Universitá degli Studi di Roma "La Sapienza", pp 75
Kotsakis T, Pandolfi L, Maiorino L, Petronio C, Sansalone G, Salari L (2020) Small vertebrates from the late Pleistocene of Avetrana (Apulia, southern Italy) karst filling. Estud Geol 76(1):e122. https://doi.org/10.3989/egeol.43378.508
Lari A (2019) Reintroduzione della Lepre italica (*Lepus corsicanus*) all'Isola d'Elba. Monitoraggio mediante Radiotracking (VHF) e impiego di radiocollari GPS. Due tecniche a confronto. Thesis Università di Pisa
Lavazza A, Guberti V (2007) Malattie di *Lepus corsicanus* e programmi di monitoraggio e gestione sanitaria. In: De Filippo G, De Riso L, Riga F, Trocchi V, Troisi SR (eds) Conservazione di *Lepus corsicanus* De Winton, 1898 e stato delle conoscenze. IGF, Napoli, pp 83–88
Lo Valvo M (2007) Status di *Lepus corsicanus* in Sicilia. In: De Filippo G, De Riso L, Riga F, Trocchi V, Troisi SR (eds) Conservazione di *Lepus corsicanus* De Winton, 1898 e stato delle conoscenze. IGF, Napoli, pp 89–95
Lo Valvo M, Barera A, Seminara S (1997) Biometria e status della Lepre appenninica (*Lepus corsicanus*, De Winton 1898) in Sicilia. Il Naturalista Siciliano IV 21: 67–74
Lo Valvo M, Seminara S, Di Vittorio M (1999) Censimento di Lepre appenninica (*Lepus corsicanus* De Winton, 1898) in alcune aree campione del Parco delle Madonie (Sicilia) (p 129). In: IV Convegno Nazionale dei Biologi della Selvaggina. Bologna, 28–30 Ottobre 1999
Lo Valvo M, Mallia E, Galante G (2012) Nuovi dati sull'allevamento della Lepre italica in Sicilia. In: Riga F, Scalisi M (eds) Atti del Workshop nazionale sulla conservazione della Lepre italica: azioni locali per la strategia nazionale. ISPRA, Roma, pp 139–148
Lorenzini R, Lovari S, Masseti M (2002) The rediscovery of the Italian roe deer: genetic differentiation and management implications. Ital J Zool 69:367–379
Lovari S, Fattorini N, Calosi M, Cozzi F, Gabbrielli C, Giari C, Torniai L, Romeo G, Mori E (2020) Spatial behaviour of the Apennine hare: are females territorial? Ethology 126:839–843
Masseti M (2003) Fauna toscana. Galliformi non migratori, Lagomorfi e Artiodattili. Agenzia Regionale per lo Sviluppo e l'Innovazione nel settore Agricolo-Forestale (ARSIA). Regionde Toscana, Firenze
Masseti M, De Marinis AM (2007) Prehistoric and historic artificial Disperal of lagomorphs on the Mediterranean islands. In: Alves PC, Ferrand N, Hackländer K (eds) Lagomorph biology: evolution, ecology, and conservation. Springer-Verlag, Berlin/Heidelberg, pp 13–24
Melo-Ferreira J, Boursot P, Carneiro M, Esteves PJ, Farelo L , Alves PC (2012) Recurrent Introgression of Mitochondrial DNA Among Hares (Lepus spp.) revealed by Species-tree Inference and Coalescent Simulations. Systematics Biology 61, 367–381.
Mengoni C, Mucci N, Randi E (2015) Genetic diversity and no evidences of recent hybridization in the endemic Italian hare (*Lepus corsicanus*). Conserv Genet 16: 477–489
Miller GS (1912) Catalogue of the mammals of Western Europe. British Museum (Natural History), London
Mori E, Lovari S, Cozzi F, Gabbrielli C, Giari C, Torniai L, Romeo G, Ferretti F, Fattorini N (2020) Safety or satiety? Spatiotemporal behaviour of a threatened herbivore. Mamm Biol 100:49–61
Palacios F (1996) Systematics of the indigenous hares of Italy traditionally identified as *Lepus europaeus* Pallas 1778 (Mammalia: Leporidae). Bonn Zool Beitr 46: 59–91
Palacios F, Orueta JF, Tapia GG (1989) Taxonomic review of the *Lepus europaeus* group in Italy and Corsica. In: Boitani L and Capanna E (eds) Fifth international theriological congress, Roma, 22–29 Agosto 1989. Abstracts of papers and posters, vol I, 189–190
Perez-Suarez G, Palacios F, Boursot P (1994) Speciation and paraphyly in Western Mediterranean hares (*Lepus castroviejoi*, *L. europaeus*, *L. granatensis*, and *L. capensis*) revealed by mitocondrial DNA phylogeny. Biochem Genet 32:423–437
Pierpaoli M, Riga F, Trocchi V, Randi E (1999) Species distinction and evolutionary relationships of the Italian hare (*Lepus corsicanus*) as described by mitochondrial DNA sequencing. Mol Ecol 8:1805–1817
Pietri C (2007) Caccia e protezione delle popolazioni di lepri (*Lepus* sp.) in Corsica. In: De Filippo G, De Riso L, Riga F, Trocchi V, Troisi SR (eds) Conservazione di *Lepus corsicanus* De Winton, 1898 e stato delle conoscenze. IGF, Napoli, 53–62
Pietri C (2015) Range and status of the Italian hare *Lepus corsicanus* in Corsica. Hystrix 26:166–168
Pietri C, Alves PC, Melo-Ferreira J (2011) Hares in Corsica: high prevalence of *Lepus corsicanus* and hybridization with introduced *L. europaeus* and *L. granatensis*. Eur J Wildl Res 57:313–321
Randi E, Riga F (2019) *Lepus corsicanus*.The IUCN Red List of Threatened Species2019: e.T41305A2952954
Ricklefs RE (1980) Ecology, 2nd edn. Nelson, Sunbury-on-Thames
Riga F, Trocchi V, Toso S (2001) Morphometric differentiation between the Italian hare (*Lepus corsicanus* De Winton, 1898) and European brown hare (*Lepus europaeus* Pallas, 1778). J Zool (Lond) 253:241–252
Riga F, Trocchi V, Angelici FM, Randi E, Pierpaoli M (2003) *Lepus corsicanus* De Winton 1898 – Apenninenhase. In: Krapp F (ed) Handbuch der Säugetiere Europas. Band 3/II: Hasentiere. Lagomorpha. Aula-Verlag, Wiebelsheim, pp 117–135

Rizzardini G, Fascetti S, Pietri C, Riga F, Cosentino C, Freschi P (2019) Feeding preferences in dry season of the Italian hare (*Lepus corsicanus*) in two sites of Corsica. Eur J Wildl Res 65:43

Robinson TJ, Elder FFB, Chapman JA (1983) Karyotypic conservatism in the genus *Lepus* (order Lagomorpha). Can J Genet Cytol 25:540–544

Rugge C, Mallia E, Delorenzo M (2007) Analisi vegetazionale degli habitat maggiormente frequentati da *Lepus corsicanus*, all'interno dell'area Sic-Zps "Foresta di Gallipoli Cognato", nel Parco Regionale Gallipoli Cognato. In: De Filippo G, De Riso L, Riga F, Trocchi V, Troisi SR (eds) Conservazione di *Lepus corsicanus* De Winton, 1898 e stato delle conoscenze. IGF, Napoli, pp 137–148

Rugge C, Mallia E, Perna A, Trocchi V, Freschi P (2009) First contribute to the characterization of coat in *Lepus corsicanus* and in *Lepus europaeus* by colorimetric determinations. Ital J Anim Sci 8:802–804

Sarà M (1998) I mammiferi delle isole del Mediterraneo. L'Epos, Palermo

Sarà M (2005) Resti faunistici dal castro normanno di Calathamet (XIII sec. Sicilia nord-occidentale). In: Fiore I, Malerba G, Chilardi S (eds) Atti del 3° Convegno Nazionale di Archeozoologia, Siracusa, 3–5 novembre 2000. Istituto Poligrafico dello Stato, Roma, pp 493–499

Scalera R, Angelici FM (2003) Rediscovery of the Apennine hare *Lepus corsicanus* in Corsica. Bollettino del Museo Regionale di Scienze Naturali, Torino 20: 161–166

Scarselli D, Vecchio G, Gasperini M, Oliviero F, Riccetti A, Petrini R, Mencarelli C, Giannini F, Riga F (2016) Analisi dell'home range e della sopravvivenza di due gruppi di Lepre italica (*Lepus corsicanus*) rilasciate in un progetto di reintroduzione all'isola d'Elba. In: De Filippo G, Fulgione D, Riga F, Trocchi V (eds) International Congress: '20 years of Italian hare', Vallo della Lucania, 13 December 2016 pp 27–32

Schai-Braun S, Hackländer K (2018) *Lepus corsicanus* de Winton, 1898 Corsican hare. In: Smith AT, Johnston CH, Alves PC, Hackländer K (eds) Lagomorphs: Pikas, rabbits, and hares of the world. Princeton University Press, Baltimore

Schröder J, Soveri T, Soumalainen HA, Lindberg LA, van der Loo W (1987) Hybrids between *Lepus timidus* and *Lepus europaeus* are rare although fertile. Hereditas 107:185–189

Shuckburgh ES (1889) The history of Polybius. Translated from the text of F. Hultsch. Macmillan and Co., London/New York

Suchentrunk F, Ben Slimen H, Scandura M, Apollonio M (2007) Molecular genetic aspects of *Lepus corsicanus*. In: De Filippo G, De Riso L, Riga F, Trocchi V, Troisi SR (eds) Conservazione di *Lepus corsicanus* De Winton, 1898 e stato delle conoscenze. IGF, Napoli, pp 23–35

Tagliacozzo A (1993) Archeozoologia della Grotta dell'Uzzo, Sicilia. Supplemento al Bollettino di Paletnologia Italiana, ns II 84:1–278

Toschi A (1965) Mammalia. Lagomorpha, Rodentia, Carnivora, Ungulata, Cetacea. Fauna d'Italia, vol VII. Calderini, Bologna

Trocchi V, Riga F (eds) (2001) Piano d'azione nazionale per la Lepre italica (*Lepus corsicanus*). Quaderni di Conservazione della Natura 9. Ministero dell'Ambiente e della Tutela del Territorio. Roma – Istituto Nazionale per la Fauna Selvatica, Ozzano dell'Emilia

Usai F, Rinnovati R, Trocchi V, Stancampiano L (2012) *Lepus corsicanus* gastro-intestinal helminths: first report. Helmintologia 49:71–77

Vigne J-D (1988) Les mammifères post-glaciaires de Corse. Etude archéozoologique, XXVIe supplement à Gallia Préhistoire. Editions du Centre National de la Recherche Scientifique, Paris

Vigne J-D (1990) Biogeographical history of the mammals on Corsica (and Sardinia) since the final Pleistocene. Atti dei Convegni Lincei 85:369–392

Vigne J-D (1992) Zooarchaeology and the biogeographical history of the mammals of Corsica and Sardinia since the last ice age. Mammal Rev 22:87–96

Winton WE (De) (1898) - On the hares of Western Europe and North Africa. Ann Mag Nat Hist London. 1: 149–158

European Hare *Lepus europaeus* Pallas, 1778

7

Klaus Hackländer

Contents

K. Hackländer (✉)
University of Natural Resources and Life Sciences, Vienna, Department of Integrative Biology and Biodiversity Research, Institute of Wildlife Biology and Game Management, Vienna, Austria

Deutsche Wildtier Stiftung (German Wildlife Foundation), Hamburg, Germany
e-mail: klaus.hacklaender@boku.ac.at

K. Hackländer, P. C. Alves (eds.), *Primates and Lagomorpha*, Handbook of the Mammals of Europe,
https://doi.org/10.1007/978-3-030-34043-8_9

Common Names

English	European hare, brown hare, European brown hare, and common hare
German	Feldhase
French	Lièvre d'Europe, Lièvre européen, Lièvre commun, Lièvre brun, hase
Spanish	Liebre europea, Liebre común
Italian	Lepre commune, Lepre europea
Russian	За́яц-руса́к

Taxonomy and Systematics

As it is the case with many *Lepus* species (see Introduction to this volume), taxonomy and systematics are a matter of an ongoing debate also for *L. europaeus* (Fig. 1). The species has no type locality, but according to Trouessart (1910), Pallas had his specimen from S Poland and Pannonia. The true distribution range and the taxonomic status remain still unclear due to smooth transitions between potential species, hybridization within the genus *Lepus,* and vast ecological plasticity. In the past, *L. europaeus* was included in *L. capensis* (see species chapter in this volume). Today, there are contradictory results from genetic studies (e.g., Ben Slimen et al. 2008; Ferreira et al. 2021; Lado et al. 2019). Mitochondrial DNA (mtDNA) data suggest that *L. europaeus* is closer than expected to *L. capensis* from Near East and East Africa regions, despite demonstrating divergence at nuclear level, which might indicate ancestral admixture and mtDNA introgression (Lado et al. 2019). A recent phylogenomic study confirms that the European hare is closer to African hare species, but shows levels of genomic differentiation similar to the ones observed among other pairs of hare species (Ferreira et al. 2021). As long as these uncertainties are dominating the scientific debate, *L. europaeus* remains a distinct species.

Out of the sixteen recognized subspecies of *L. europaeus,* twelve have been described for Europe. However, these are all based on differences in size and coat color and thus are doubtful in the light of the ongoing debate on the species status itself:

L. e. europaeus Pallas, 1778, in W Europe; *L. e. caspicus* Hemprich and Ehrenberg, 1832, in Lower Volga and Kalmykia (Russia); *L. e. creticus* Barrett-Hamilton, 1903, in Crete (Greece); *L. e. cypricus* Barrett-Hamilton, 1903, in Cyprus; *L. e. hybridus* Desmarest, 1822, in the Baltic states, Belarus, Ukraine, Finland, and W Russia; *L. e. karpathorum* Hilzheimer, 1906, in the Carpathian mountains; *L. e. medius* Nilsson, 1820, in Denmark; *L. e. occidentalis* de Winton, 1898, in Great Britain; *L. e. parnassius* Miller, 1903, in C Greece; *L. e. ponticus* Ognev, 1929, at the Black sea coast; *L. e. rhodius* Festa, 1914, in Rhodes (Greece); *L. e. transsylvanicus* Matschie, 1901, in E and SE Europe.

Fig. 1 Two adult European hares. © Rolf Giger with kind permission

Paleontology

The place of origin of *L. europaeus* is most likely the Middle East. The oldest reliable fossil remains of *L. europaeus* (*euxinicus*) within Europe are from the Late Pleistocene and were found in Crimea and Moldova (Averianov et al. 2003) as well as in C Italy (Angelone et al. 2020). From there on, *L. europaeus* was quite common in S Europe, with fossils reported in more than 60 occurrences from Spain, France, Italy, former Yugoslavia, Bulgaria, Hungary, Moldova, the Ukraine, Crimea, the European Russia, Kazakhstan, and the Caucasus region (Averianov et al. 2003). The Holocene fossil findings of *L. europaeus* come from more than 120 locations in Switzerland, Austria, Poland, former Yugoslavia, Hungary, Israel, Moldova, Ukraine, and Crimea, as well as the Caucasus and the Urals regions. Both the Pleistocene and the Holocene fossil records of *L. europaeus* lie within the current distribution range of the species. The most reliable morphological characters for identifying *L. europaeus* fossil material are dental features. The size of skulls or other bones play only a minor role in the determination of fossil *L. europaeus* as some European Pleistocene populations of *L. timidus* show a similar size (Averianov et al. 2003). In most cases, the determination of fossil remains of *L. europaeus* cannot be verified by descriptions and illustrations of the underlying material; they often only appear in fauna lists. All of these determinations should be taken with great care and classified as indetermined Leporidae (Angelone et al. 2020). Pleistocene hare remains from Spain and Italy, which were ascribed to either *L. europaeus* or *L. capensis*, may also belong to *L. granatensis*, *L. castroviejoi*, or *L. corsicanus* (Kotsakis et al. 2003).

Current Distribution

Originally, *L. europaeus* was distributed over steppes, steppe forests, and open areas in woodlands in SE and C Europe, the Middle East, the Caucasus, the Russian lowlands, and W Kazakhstan. After colonization of continental Europe following human agriculture (Lado et al. 2018), hares increased locally their densities but were still uncommon in the late medieval time (Huber 1973). The rise of European hare population started in the eighteenth and nineteenth century with increasing agricultural efficiency, e.g., through drainage of swamps and field melioration (Huber 1973). Currently, the European hare can be found throughout Europe except the very SW (C and S Iberian Peninsula, south of Cantabria, and the river Ebro) as well as far N (C and N Fennoscandia, N Russia as well as the islands in the Arctic Sea) (Fig. 2). European hares have been introduced to many European islands including several British isles and numerous ones in the Mediterranean Sea (e.g., Crete, Rhodes, Cyprus, and Corsica) (Masseti and De Marinis 2008).

The distribution range in Europe is expanding. Already Ognev (1940 in Angermann 1972) described a shift to N and E in Russia and Kazakhstan beginning 200 years ago. Whereas land use change, namely deforestation and increase of arable land, was the main driver since the beginning of the twentieth century (Angermann 1972), global warming is an additional factor fostering the range expansion of the European hare, as the example in Fennoscandia shows (Jansson and Pehrson 2007; Thulin 2003). In line with that, European hare range expansion toward higher altitude has been reported for the Alps (Schai-Braun et al. 2021a).

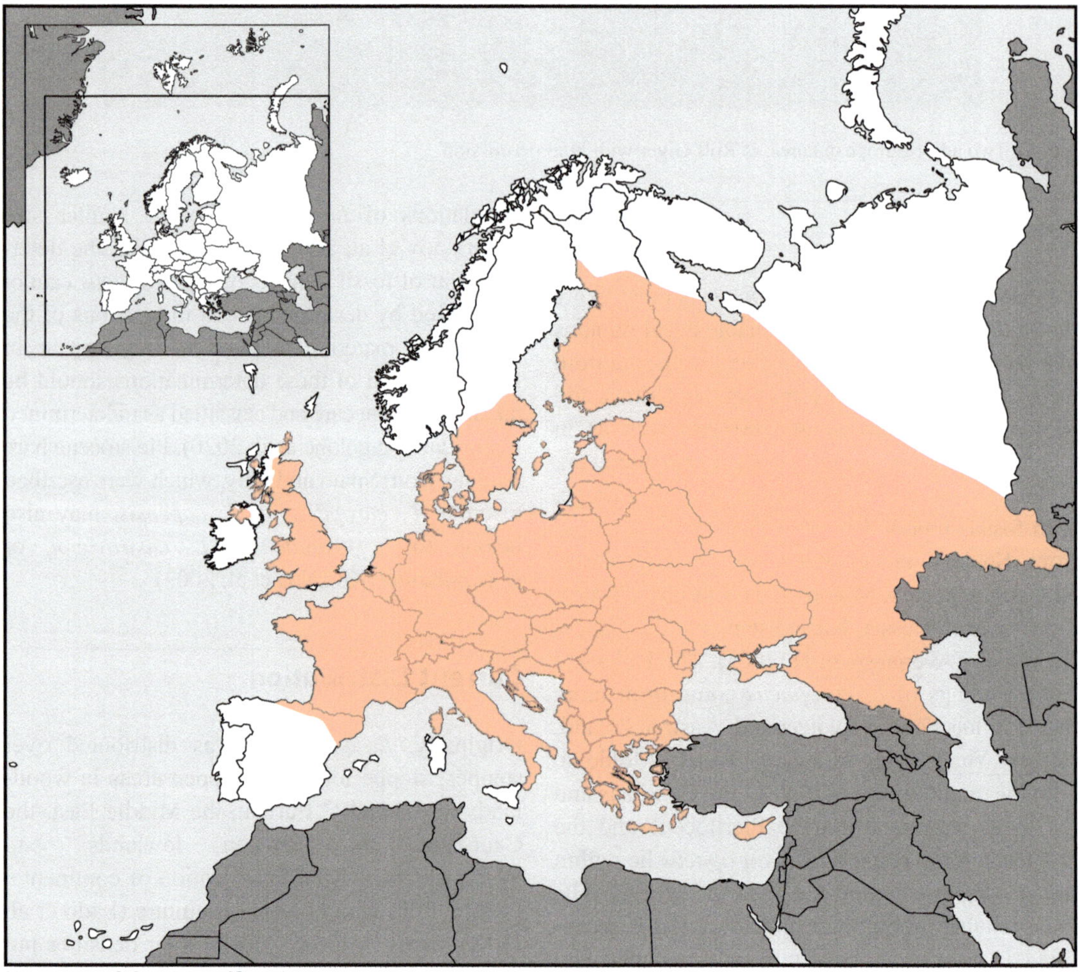

Fig. 2 Current distribution of *Lepus europaeus* following Schai-Braun and Hackländer (2016). (Map template: © Copyright Getty Images/iStockphoto)

Description

Size and Morphology

The European hare is a typical leporid with longer hind limbs than forelimbs. It is the largest hare in W Europe, however with huge variation across the distribution range: head-body length 530–760 mm, tail length 70–140 mm, hind-foot length 120–185 mm, ear length 90–150 mm, and body weight 2.5–8 kg (Angermann 1972; Grimmberger et al. 2009; Schai-Braun and Hackländer 2016; Zörner 1981). An increase in size following Bergmann's rule from SW to NE can be observed. Moreover, Pielowski (1969) described an increase in body weight from W to E, i.e., with increasing continentality. In sum, morphological variation within the species can be explained mainly by geographical variation, differences in habitat and climate. However, some morphological features might be forged due to the strong anthropogenic intermixing of hare populations in the last centuries, since hares have been translocated across Europe, and even back from S America (e.g., Uruguay and Argentina, Suchentrunk et al. 2006) where they were previously introduced, mainly for hunting purposes.

There is no pronounced sexual dimorphism, and a distinction between the sexes in the field is quite impossible from the distance, if single hares are observed. However, females are slightly larger (Bujalska 1964) and heavier than males (Caboń-Raczyńska 1974; Flux 1967a; Kröning 1963). Sexing according to the primary sexual characteristics can lead to errors, because the penis and the clitoris look similar at first glance. The penis is pointed, tubular, and curved slightly backward. The clitoris, on the other hand, has a groove with a similar shape like the penis. In addition, during times of sexual inactivity or in younger individuals, males do not have their testicles in the scrotum, i.e., not visible from outside.

The eyes of European hares are large with rounded pupils. The iris color is highly variable ranging from dark brown to bright yellow. Hares can close their nostrils with two skin folds. The upper lip is split in the middle ("harelip"). A hare's forefeet have 5 and the hind feet 4 clawed toes. The sole is densely haired and no pads are visible. Hares have three pairs of teats (but see Massányi et al. 1997) and thus follow Aristotle's "one-half rule" (Diamond 1987), saying that the number of teats in a mammal is twice as high as the average litter size (see "Reproduction").

The skull of European hares is relatively large and elongated, with a broad, dorsally distended nasal part (Fig. 3). The anterior and posterior wings of the supraorbital processes are well developed and mostly free. The maxillaria have a sieve-like appearance. The fossa mesopterygoidea is broad, the bony palate short (ca. 8% of the Cbl), and the choans wide (ca. 130% of the Cbl; Corbet 1983).

Characteristic for all leporids is the relatively large caecum, which exceeds the total body length of a hare. As in other hindgut fermenter, this part of the digestive tract is used to increase the digestibility of fibers with the help of microbial activity (see "Diet"). According to data provided by Caboń-Raczyńska (1964a) and Slamečka et al. (1997), the heart is large, weighing more than 1% of the body weight, which exceeds the general pattern for mammals (0.55%, Lindstedt and Schaeffer 2002). This can be interpreted as an adaption to persistent and high-speed running (see "Locomotion").

Pelage

The pelage is soft and smooth. Hair lengths range between 4 and 10 mm (Borowski 1964). The fur is dominated by dense down hair, which are about 12 μm thick and roughened on the surface (Kayser s.a. in Koenen 1956). Awn hairs are sparse, and only a few guard hairs can be found mainly on the lateral parts of the body. Awn hair are about 84–90 μm thick. According to Toldt (1935), ventral hairs are thinner than dorsal ones. Hairs in the auricles are short, otherwise alternating between about 5 and 10 mm in length on the body.

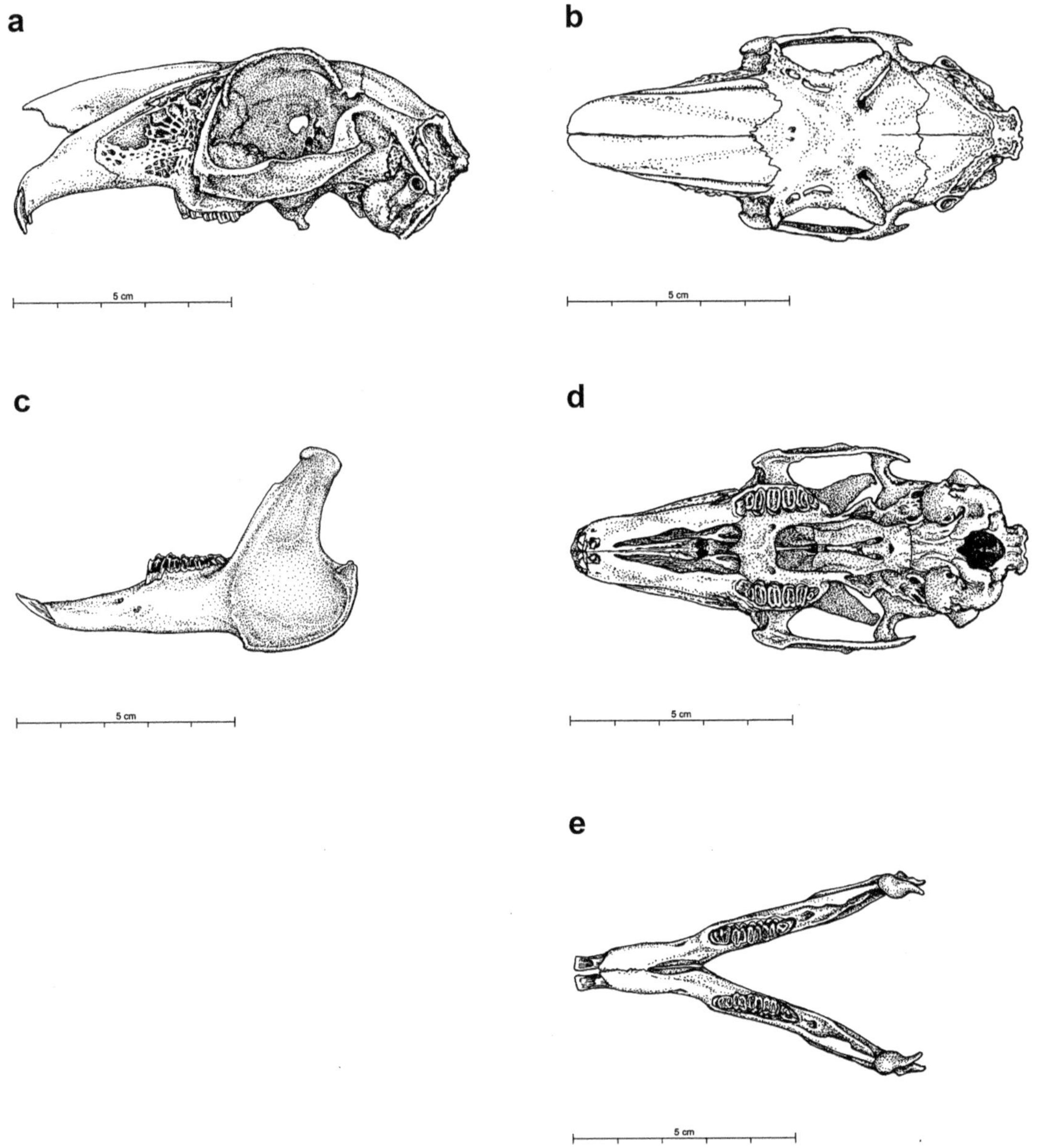

Fig. 3 Skull of *Lepus europaeus*. (**a, b, d**) lateral, dorsal, and ventral view of skull, (**c**) lateral view of left mandible (view from outside), and (**e**) dorsal view of mandible (© Franz Müller with kind permission)

Particularly long hair was found in the anal region, 72 mm in winter coat. Fur density is up to 175 hairs/mm^2 (Waldeyer-Petri s.a. in Koenen 1956). Whisker hairs on the lids and lips are up to 110 mm long, the longer ones are predominantly white.

Hair molt of *L. europaeus* was described in all months with the exception of December and January, but especially in the months of March to June and September. This is interpreted as changing hair twice a year (autumn and spring). Molt is triggered by changing photoperiod and might be accelerated by temperature (Sokolov et al. 2009). According to Borowski (1964), the course of the molt both in spring and autumn begins on the back, then moves to the chest, and later includes the head, ears and neck, and finally the ventral parts. In contrast, according to Hewson (1963), the molting process in autumn is the opposite of that in spring.

In general, the fur is tawny or rusty with a whitish ventral part. The color of the summer fur varies considerably between individuals, from ocher-gray or brownish-yellow or brown to yellowish-gray with a distinct blackish tinge on the back, which is caused by the black or black-brown tips of the awn and down hairs. The latter is silver gray or white with black or brownish black tips. The back of the ears is white or grayish-white with a black or black-brown spot on the tip. The tips of the ears have a large triangular patch on the back that contrasts with the paler gray of the rest of the ear color. The cheeks are ocher-brown or ocher-gray. The chest and flanks are tan or rust-colored. The outer sides of the limbs are ocher-brown, and their inner surfaces and the belly side are white. The tail has a black spot on top but is pure white below. For more information on coat color and inter-individual variation, see Rieck (1953), Koenen (1956), and Borowski (1964).

The winter fur is thicker and woollier. All three hair types are also longer in winter, and the hair density is higher. The winter coat is lighter and grayer compared to the summer coat. All winter coat characteristics (hair length, hair density, and coat color change) get more pronounced along a gradient from S to NE (Rieck 1953). In the S, the winter coat is colored almost like in summer (Sokolov et al. 2009), but in winter there is some white on the sides of the head and at the base of the ears, some gray on the thighs. In individuals from the N (from Belarus eastward), the winter fur is mostly white except for the front back, head, and ears that are darker (Sokolov et al. 2009). Note that the white hairs of *L. europaeus* are blackish at their base in contrast to pure white hairs in *L. timidus* winter coat (Grimmberger et al. 2009).

Dentition

As all European leporids, *L. europaeus* has 28 teeth with I^1, I^2, P^1, P^2, and P^3 and M^1, M^2, and M^3 as well as I_1, P_1, and P_2 and M_1, M_2, and M_3. The incisivi are surrounded by enamel all around. The I^1 have a mesially located longitudinal furrow that, in rare cases, may contain some cement. Typical of the species' I^1 compared to *L. timidus* is the rectangular, nonsquare cross-section (Hauser 1921) with a width-to-thickness ratio of around 1.6. Despite the premolar milk teeth, all teeth are rootless and grow continuously. Problems might arise when a tooth breaks in asymmetric skulls. In this case, the teeth have no counterparts and grow spiraled (Müller-Using 1974).

Age Determination

Due to the size variation within the species, age determination by skull size, body length, or weight is hazardous, both in adults (Suchentrunk et al. 1991) and in leverets (Bray et al. 2002).

In the field, age class (subadult vs. adult) can be assessed by Stroh's sign (Stroh 1931), an epiphyseal protrusion that can be palpated as a lateral ulnar knob close to the carpal joint. In European hares, this protrusion can no longer be palpated at an age of approximately 6–8 months. However, Suchentrunk et al. (1991) reported that some young of the year lost Stroh's sign already at an age of 4 months, and some individuals might still have this sign at an age older than 12 months. However, on average young hares born very early in the year are no longer recognized as subadults during autumn using Stroh's sign. Therefore, the reliability of this age determination method decreases rapidly toward the end of the year (Rieck 1966; Zörner 1977). Thus, the precision of age structure analysis using Stroh's sign in autumnal hunting bag depends on the hunting date: The percentage of undetected young hares ranges between 10 and 20% for hunting dates in November; in December it can be up to 50% (Rieck 1967). Bujalska et al. (1965) mentioned an error rate of up to 70%. Despite these restrictions, Stroh's sign is still the most applicable age determination method in the field if season is considered. This age determination method may be valid in other hare species, too.

A more appropriate age determination is done by determining the dried eye-lens weight (DLW) as described by Andersen and Jensen (1972) and further elaborated by Broekhuizen and Maaskamp (1979) as well as Suchentrunk et al. (1991). This method allows distinguishing between young of the year and adults (hares from the previous year) as well as estimation of date of birth within young of the year to cluster them into birth cohorts. In populations with reproductive pause (see "Reproduction"), young of the year and adults can easily be distinguished when frequencies of DLW are plotted as the distribution has two peaks then. In areas where hares reproduce all year round, the DLW method is less useful, since such frequency peaks might not appear.

Physiology

In optimal conditions, body weight of adults seems to be stable throughout the year (Caboń-Raczyńska 1964a), while body composition varies. For example, body fat reserves fluctuate seasonally. Fat depots are built up during autumn until winter, with males having higher body fat indices (approximately 2.7% of body weight) than females (approximately 2%, Zörner 1981). The body fat reserves are depleted gradually from winter onward (Flux 1967a; Packer 1989; Rieck 1953; Schneider 1978) until no retroperitoneal fat is left in summer. Hence, in terms of reproduction, female hares rely on body fat resources early in the year (capital breeder) and on energy sources from their diet in summer (income breeder) (Valencak et al. 2010). Peak sustained energy assimilation rates in lactating females are highest in autumn when food intake increases to cover the energetic costs of milk production for the precocial young (Valencak et al. 2010).

At rest (in the thermoneutral zone), body temperature is slightly above 38 °C (Paci et al. 2006), heart rate between 95 (Paci et al. 2006) and 140 beats min^{-1} (Noszczyk-Nowak et al. 2009), and oxygen consumption about 0.54 ml g^{-1} h^{-1} (Kronfeld and Shkolnik 1996). The upper critical ambient temperature is about 32 °C (Kronfeld and Shkolnik 1996). In winter, for every 1 °C decrease, adult hares increase food intake by 3.6 kcal kg^{-1} day^{-1} (Myrcha 1968), whereas leverets respond by 7.4 kcal kg^{-1} day^{-1} (Pilarska 1969). All these values have to be considered in the light of allometric effects, but body weights of examined hares were not always given in the studies mentioned above.

The precocial leverets are able to maintain normothermy from day 1 of life when exposed to temperatures down to −8 °C (Hackländer et al. 2002a). Resting metabolic rates in leverets are comparably low during cold, and heat loss might be reduced by peripheral vasoconstriction. This is believed to be an adaption to precociality and low maternal care (see "Behavior"). Heat loss is reduced in the first weeks of life with increasing body weight and decreasing thermal conductance. The slopes of regression of resting metabolic rates (RMR, i.e., in the thermoneutral zone) versus ambient temperatures for each week of life decreased from 0.54 W kg^{-1} C^{-1} in week 1 to 0.21 W kg^{-1} C^{-1} in weeks 3–5 of life (Hackländer et al. 2002a).

Muscles are dark and long-fibered. Polyunsaturated fatty acids (PUFAs) are enriched in muscle phospholipids compared to white adipose tissue lipids and liver phospholipids (Valencak et al. 2003). Especially, the content of n-6 PUFAs in muscles from *L. europaeus* is high, characterizing the hare as a typical persistent runner (Ruf et al. 2006).

In areas with reproductive pause, the reproductive cycle of hares is triggered by photoperiod (Martinet 1976). Reproductive hormones are following this seasonal cycle (Massányi et al. 2000; Slamečka et al. 2001; Stavy et al. 1978). Studies on female sexual hormones concentrated mainly on the phenomenon of superfetation (see "Reproduction"). Progesterone levels dropped after day 35 of gestation indicating the onset of an estrus during pregnancy (Caillol and Martinet 1976). Prolactin increased before that drop prepartum and then increased stronger after the first suckling stimulus (Caillol et al. 1990). Luteinizing hormone (LH) and follicle-stimulating hormone (FSH) are also increasing after day 34 of gestation, indicating estrus and readiness for the induced ovulation while pregnant (Caillol and Martinet 1983; Caillol et al. 1991a).

Genetics

Chromosomes: 2n = 48, NP = 88 (Gustavsson 1971). Genetics studies using mtDNA markers have revealed the existence of two major phylogeographic lineages, the European and the Anatolian (e.g., Ashrafzadeh et al. 2018; Djan et al. 2017; Minoudi et al. 2018). Due to the postglacial spread of *L. europaeus* across the continent and translocations, the species has a low genetic differentiation in Europe, suggesting a high gene flow among populations (Fickel et al. 2005; Hartl et al. 1990, 1993; Mamuris et al. 2001, 2002; Suchentrunk et al. 2000). Nevertheless, rare alleles can be found in the glacial refugal areas (Amoutzias et al. 2016; Suchentrunk et al. 2003).

Hybridization within the genus *Lepus* has been described, especially between *L. europaeus* and *L. timidus* (Conrado zu Baldenstein 1863; Gustavsson 1971; Notini 1941; Thulin et al. 2006). Hybrids are fertile (Schröder et al. 1987) and, in natural populations, are predominantly the result of a unidirectional mating between male *L. europaeus* and female *L. timidus* (Levänen et al. 2018a; Thulin and Tegelström 2002). Moreover, limited genetic admixture was found in the contact zone between *L. granatensis* and *L. europaeus* in N Spain, despite existing evidences of hybridization being found (Melo-Ferreira et al. 2014). It should be highlighted, however, that the great majority of European hare populations in Spain have massive *L. timidus* mitochondrial introgression (Alves et al. 2003; Melo-Ferreira et al. 2009, 2014). In fact, ancient and recent introgression between hare species has been detected in many populations, and in several cases it involves the European hare (Alves et al. 2008; Ferreira et al. 2021; Lado et al. 2019; Levänen et al. 2018b; Melo-Ferreira et al. 2012; Suchentrunk et al. 1999; Thulin et al. 2006; Zachos et al. 2010). The report on *L. timidus* transcriptome allows to identify 4.672 putative diagnostic sites between *L. europaeus*, and a subset of 85 random independent single nucleotid polymorphisms (SNPs) which can be valuable to assess population status and monitor hybridization between these two species (Marques et al. 2017). Nevertheless, microsatellites have been shown to be a useful molecular tool for assessing hybridization between *L. europaeus* and *L. timidus* in the Alps (Rehnus and Bollmann 2016; Schenker et al. 2020).

Life History

Reproduction

It remains unclear whether *L. europaeus* has an estrous cycle (Caillol and Martinet 1983; Caillol et al. 1991b). As in other leporids, ovulation is triggered by copulation and is thus induced and not spontaneous (Martinet 1980). Gestation length is usually 42 days but might last up to 45 days (Sokolov et al. 2009). However, interbirth intervals might be shorter due to superfetation. This phenomenon (see Roellig et al. 2011 for a review) was already described by the Greek philosopher Aristotle in the fourth century BCE (Jennings 2017) and allows females to get fertilized again while pregnant, beginning with day 34 of pregnancy (Caillol and Martinet 1983; Caillol et al. 1991a), and thus to bear two litters of different age in the uterus at the same time. Consequently, interbirth intervals are shorter than gestation length itself. Interbirth intervals can even be as short as 24 (Slamečka and Šebová 1991) or 25 days (Sackmann 1977). The two litters of superfetation were never observed sharing the same of the two uterine horns (Roellig et al. 2011). Although superfetation enables females to produce more young per time (Roellig et al. 2010b), this phenomenon seems to be rather rare in the wild (Broekhuizen and Maaskamp 1981; Flux 1967a; Horáček and Uher 1965; Lloyd 1968; Möller 1971; Raczyński 1964).

Higher probability of superfetation in captivity might be due to improved body condition of female hares, permanent presence of a male during pregnancy, and unlimited availability of high-quality food (Hediger 1948; Martinet et al. 1970). While discussing superfetation, it is worth notifying that males produce a substance that reduces the sperm motility, thus enabling them to survive for a longer time (at least 36 days according to Martinet and Raynaud 1972).

Reproductive pattern of *L. europaeus* is quite variable across the distribution range, whereas reproduction is restricted to spring and summer months in C, E, and N Europe (e.g., Lincoln 1974; Möller 1971; Stieve 1952; Raczyński 1964). Irrespective of this general pattern, sporadic winter reproduction has been found all over the distribution range, e.g., E Austria (Schai-Braun et al. 2020), but is more likely in oceanic climates (e.g., Hackländer et al. 2011 for Belgium; Hewson and Taylor 1975 for Scotland) and in the Mediterranean range (Antoniou et al. 2008 for Greece). In line with this, the pattern of annual testicular activity in males varies between study sites (e.g., Blottner et al. 2000; Eskens et al. 1999; Štrbenc et al. 2003).

Where leverets can be found in all months, number of litters is larger but mean litter size is smaller. In areas without reproduction in autumn/winter, mean litter size is larger but number of litters small (Flux 1981a; Sokolov et al. 2009). Therefore, number of litters per female and year varies from 2 (Kolosov and Bakeev 1947 in Sokolov et al. 2009) to 5 in C Europe (Möller 1976; Petrusewicz 1970; Pielowski 1976a; Rieck 1956a) in the wild, and up to 6 in captivity with unnatural photoperiod (Martinet 1976) or with superfetation (Slamečka and Šebová 1991). The number of litters per female in S or NW Europe, where reproductive pause is less pronounced or even absent (see above), remains unclear, however might be larger than 5.

In areas with reproductive pause (and thus climatic differences between seasons), litter size at the beginning and at the end of the breeding period is smaller than in the middle (Lincoln 1974; Raczyński 1964). Litter size has been described between 1 and 6 individuals in the wild (Pielowski 1976a). However, although reproductive pattern varies across the distribution range, yearly reproductive output per female is similar across the distribution range with 10 leverets per female and year (Flux 1981a). Despite that, the average reproductive output of adult females in a population might reach 13 leverets per year (Schai-Braun et al. 2020; Velek and Semizorova 1976 in Pépin 1989). Maximum number of placental scars per female and year is 22 (Bray 1998; Schai-Braun et al. 2020). The vast amount of studies on litter size and number of litters per year have to be interpreted in the light of female age and body condition as well as reproductive pattern. As a general characteristic, Schai-Braun et al. (2021b) showed that: i) Heavier females have heavier litters at birth, ii) in summer and spring, total litter mass is larger than in winter, and iii) at the end of lactation, the litters of multiparous females are heavier than those of primiparous females.

One has to bear in mind that differences between results on reproductive performance in *L. europaeus* might be due to methodological differences as nearly all studies are based on counting corpora lutea, placental scars or embryos/fetuses in hares that were killed or found dead (but see Göritz et al. 2001 for ultrasound analysis in hares caught in the wild). Differences between litter size at birth and prepartum might be due to prenatal or early postnatal mortality. In fact, a litter size larger than the maximum of 6 (Pielowski 1976a) has been reported, but only prepartum (up to 10 in Broekhuizen and Maaskamp 1981 and even 13 young in von Thüngen 1878 in Zörner 1981).

Prenatal mortality might reach up to 90% depending on climate, season, and female age or body condition (Lincoln 1974; Raczyński 1964). Prenatal mortality includes failed implantation of fertilized eggs, resorption of embryos, and death of fetuses. This late stage of prenatal mortality leads to stillbirths of some or all leverets of a litter. In rare cases, mummified leverets can be found in a female body cavity, i.e., outside the uterus (Zörner 1981).

The proportion of reproducing females in a population depends on age, body condition, and health status of females (Broekhuizen and Maaskamp 1981; Ciberej 1993; Schai-Braun et al. 2021b). In adult females, proportion of reproducing individuals is increasing from winter to summer (Broekhuizen and Maaskamp 1981; Möller 1976). Within 1 year, between 85 and 100% of adult females reproduce (Schai-Braun et al. 2019). Reproductive senescence is reached between 4 and 6.3 years (Schai-Braun et al. 2021b) and is accompanied by pathological alterations in the uterus (Bensinger et al. 2000; Hackländer et al. 2001; Šebová 1991).

Subadults are usually fertile with an age of 4–6 months, with some variation due to time of birth (Caillol et al. 1992; Lincoln and MacKinnon 1976; Möller 1976; Pépin 1977; Slamečka and Šebová 1991): Hares born later in the year and dwelling in areas with reproductive pause in autumn have a delayed puberty for several months, and those females reach sexual maturity at an age of 9–15 months (Sokolov et al. 2009). In moderate to warm climates, hares might take part in reproduction already in the year of birth, although with lower leveret survival (Schai-Braun et al. 2021b). The youngest female to become pregnant was 67 days old, however in captivity (Schai-Braun et al. 2021b). Bray (1998) reported that 14% of the reproducing females in S France were less than 12 months. Toward the N and NE ranges, the proportion of reproductive young of the year diminishes (Averianov et al. 2003). Belova (1987 in Sokolov et al. 2009) did not find any female younger than 12 months reproducing in the Baltic region.

Growth

Prenatal growth is sigmoidal with a maximum growth rate between week 3 and 4 prepartum (Hackländer et al. 2003; Roellig et al. 2010a). Hares produce precocial young, i.e., leverets are relatively mature at birth, with open eyes and ears and fur (for camouflage). Note that hares give birth in forms above ground. Birth weight of European hare leverets depends on litter size and ranges between 40 (Martinet et al. 1970) and 192 g (Zörner 1981), with larger individuals in smaller litters. Crown-rump lengths of newborn European hares range between 90 and 140 mm, averaging 115 mm (Broekhuizen and Martinet 1979; Štěrba 1981). Young are suckled for a period of 11 (Notini 1941) to 60 days (Sokolov et al. 2009), depending on litter size, season (ambient temperature, food availability), and gestational status of the mother. There is an obvious trend that lactation length is longer in harsher environments as highest values are reported from the N and E of Europe.

Females provide a milk rich in fat (at least 20%, Broekhuizen and Maaskamp 1980; Hackländer et al. 2002b; Kučera 1991), with a comparably high content of long-chain monounsaturated fatty acids (Demarne et al. 1978; Lhuillery et al. 1984). Fatty acid composition depends on time of breeding (Valencak et al. 2009). Fat content might be lower (down to about 15%) in suboptimal habitats (Hackländer et al. 2002b) or when fed a low-fat diet in captivity (Lhuillery et al. 1984; Martinet and Demarne 1984). Milk intake has three phases with an increase until day 12, a plateau until day 22, and a decrease until day 30 (Lhuillery et al. 1984). The change of milk teeth starts in week 2 of life (Arkhipchuk 1983 in Sokolov et al. 2009) and is terminated in the fourth week of life. Between an age of 7 and 15 days, solid food is consumed by leverets (Broekhuizen and Maaskamp 1976; Cooper 1970; Hackländer et al. 2002a; Lhuillery et al. 1984; Sackmann 1977). Daily weight gain postpartum is highest between day 20 and 80 and ranges between 22.5 and 32.5 g d^{-1}. Leverets increase their body weight 8.5-fold within 35 days (Hackländer et al. 2002a) in a sigmoidal growth curve (Pielowski 1971a; Pilarska 1969). Usually, leverets are no longer provided with milk when body weight approaches 1000 g (Flux 1967a).

European hares reach adult body size at about 8 months, with some variability due to environmental factors mentioned above. Body weights still increase until the age of 4 years (Pielowski 1971a). The skull increases in some parameters throughout life, at least in hunted populations (Caboń-Raczyńska 1964b) where maximum life span is shorter than in nonhunted or captive populations.

Sex ratio at birth is close to 1:1 and independent of female body mass at mating, season of birth, female age, diet quality or a female's individual reproductive history (Schai-Braun et al. 2021b). The balanced sex ratio remains in the adult population (Frylestam 1979; Jezierski 1965; Misiorowska et al. 2014; Rieck 1953; Semizorová 1982; Zörner 1981), i.e., mortality rate is only slightly affected by sex (Hušek et al. 2015; Marboutin and Hansen 1998; Marboutin and Péroux 1995). Deviations from a sex ratio of 1:1 in adults are discussed in the light of potential effects of sampling (hunting) method and seasonal effects (Möller 1975; Pielowski and Raczyński 1976; Szederei 1959; Zörner 1981).

Survival

In adults, survival rate is depending on study site and year and averages to about 50% (Broekhuizen 1979; Kovács 1983; Pépin 1987; Petrusewicz 1970; Pielowski 1971b). Adult survival is slightly higher in males than in females (Marboutin and Hansen 1998; Marboutin and Péroux 1995), but independent of age (Broekhuizen 1979; Kovács 1983; Petrusewicz 1970). Maximum age reported is 12.5 years in the wild (Pielowski 1971b).

Most leverets die within the early stage of life (Karp and Gehr 2020) leading to very low survival rates of young of the year ranging between 5% (Schai-Braun et al. 2020) and 56% (Abildgård et al. 1972). Averianov et al. (2003) reviewed that juvenile survival rate of 25% is the average. In line with this, the proportion of subadults in the autumnal hunting bag varies predominantly according to juvenile survival. In long-term examinations of game bags, the average proportion of hares with Stroh's sign (Stroh 1931) ranges between 10 and 70% (e.g., Ahrens et al. 1993; Eskens et al. 1999).

Reasons for low survival rates in leverets are predominantly adverse weather, predation, and agricultural activities. Unfavorable weather conditions, especially low temperatures and high precipitation in late spring and early summer (Beuković et al. 2013; Hackländer et al. 2001; Karp and Gehr 2020), are hard on leverets as they rely on their own thermoregulatory capacities (Hackländer et al. 2002a) and are not protected by their mother when there is rain, cold wind, or wet snow (see "Parental Care"). Apart from that, leveret survival will decrease in years with summer droughts (Bresiński and Chlewski 1976), as females might be unable to provide enough milk through food processing. In line with that, adverse weather is associated with lower hare abundances and hunting bags (Eiberle and Matter 1982; Rödel and Dekker 2012; Siivonen 1956). Unfavorable weather decreases also adult body condition (Caboń-Raczyńska 1974) and might lead to reduced reproductive performance, too.

Leveret survival might be negatively affected in areas of high predator abundance and/or a lack of protecting vegetation (Knauer et al. 2010; Reynolds and Tapper 1995; Schmidt et al. 2004). Predators range from small mustelids, like the least weasel (*Mustela nivalis*), to wolf (*Canis lupus*), and from small birds of prey and Corvidae to eagle owls (*Bubo bubo*) (Hell and Soviš 1997; Pielowski 1993). Smaller predators prey on leverets while larger ones can kill adults. Survival rate in adult hares is less affected by predation, except for individuals in bad condition (Severtsov et al. 2017).

Mortality causes include also agricultural activities like mowing of meadows, use of weeder in cereal crop fields, etc. (e.g., Durdík 1981; Kałuziński and Pielowski 1976; Kittler 1979), and road kills (e.g., Haerer et al. 2001; Heigl et al. 2016; Reichholf 1981; Roedenbeck and Voser 2008). Pépin (1986) demonstrated that losses by agricultural activities might have more negative impact on leveret survival than adverse weather conditions. Last, but not least, in all age classes, diseases can significantly contribute to mortality rates (see "Parasites and Diseases").

Habitat and Diet

Habitat

The European hare lives above ground year-round. Originally dwelling in steppe habitats, forest steppes, and open areas in the forest zone, *L. europaeus* followed humans on their range expansion during the Neolithic revolution from SE Europe north- and westward, where the settlers opened the landscape for agriculture (Huber 1973; Lado et al. 2018). European hares can now be found predominantly in arable and pastural landscapes, but also in open woodland, moorland, heathland, or salt marshes. Apart from that, they inhabit also parks and airfields. *L. europaeus* live from sea level up to 2800 m above sea (Huber 1973). Deep and loose snow cover (Sokolov et al. 2009), pastures with livestock (Lundström-Gilliéron and Schlaepfer 2003; Lush et al. 2014), and the proximity of roads (Roedenbeck and Voser 2008) are avoided. However, hares tolerate anthropogenic structures and can be found in cities, too (e.g., Köhler 2008; Mayer and Sunde 2020b).

Hares prefer areas rich in structure providing food and shelter all year round (Pavliska et al. 2018). The distribution of food and cover is of particular importance. At night, open land with low and growing vegetation is preferred for food intake; during the day, protected areas are necessary as resting places (Neumann et al. 2011; Schai-Braun and Hackländer 2014). The day rest areas can be integrated in the nocturnal grazing area but can also be outside the latter.

Arable land is preferred over pastures (McLaren et al. 1997; Vaughan et al. 2003). However, if cereal fields are very compact making it nearly impossible for hares to penetrate, Hewson (1977) and Rühe (1999) showed that hares use only the field edges. Most important for hares is good visibility, at least for the active period during the night (Bresiński 1976, 1983). In line with this, the expanding maize cultivation in Europe leads to an increasing proportion of arable land unsuitable for hares (Mayer and Sunde 2020a; Sliwinski et al. 2019). The same is true for other biomass energy crops like *Miscanthus* (Petrovan et al. 2017) or other tall vegetation (Schrama et al. 2015). Dense thickets of trees and shrubs are avoided at night but are important features of a hare's home range during the day, especially in arable land in winter (Bresiński and Chlewski 1976; Matuszewski 1981; Pielowski 1966). In line with that, in arable land nonfarmed features such as hedges (Cardarelli et al. 2011; Pépin and Angibault 2007; Tapper and Barnes 1986) and fallow land (Cardarelli et al. 2011; Schai-Braun et al. 2013; Smith et al. 2004) are positively selected for shelter during the day. Moreover, field edges are preferred, especially during the inactive phase (day), both for leverets (Voigt and Siebert 2019) and adults (Lewandowski and Nowakowski 1993; Schai-Braun and Hackländer 2014). In the absence of higher vegetation, hares prefer ploughed fields in winter for the resting period (Pépin 1986).

As in other leporids, home range size has a positive allometry with body size (Swihart 1986). Besides that, home range size is affected by habitat heterogeneity, average field size, hare density, season, sex, and, of course, the used method for its estimation (e.g., VHF vs. GPS telemetry, MCP vs. Kernel, length of the observation period, and sample size; for discussion, see Marboutin 1997; Schai-Braun and Hackländer 2014). Usually, males have larger home ranges than females (Belova 1987 in Sokolov et al. 2009; Homolka 1985; Reitz and Leonard 1994; Schneider 1977; Zörner 1981), and males are more active during the night (Zaccaroni et al. 2013). These differences can be explained by reproductive strategies and reproductive behavior: During lactation, females stay near their litter and thus have smaller home ranges. Males, on the other hand, can maximize their lifetime reproductive success by mating with as many as possible females resulting in a higher activity and home range.

Occasionally, the daytime resting place and the nocturnal feeding grounds can be several hundred meters away, e.g., when feeding sites are not providing shelter and a nearby forest is used as a retreat during the day, leading to large home ranges (Homolka 1985). Home range sizes thus depend on the distances between shelter (during the day) and food patches (during the night). Hares travel up to 15 km while feeding in one night (Flux 1967b). In areas with large fields and low variety of crops, annual home range sizes can exceed 330 ha (Pielowski 1972). As hares are nocturnal, the daytime home ranges are much smaller than those at night (Schai-Braun and Hackländer 2014). In contrast, if heterogeneity of habitat is high, home ranges do not differ much between day and night and are rather small (Kunst et al. 2001;

Lewandowski and Nowakowski 1993; Schai-Braun and Hackländer 2014).

The home range of an individual is neither a rigid spatial unit nor a defended territory. As hares do not use self-dug burrow systems like rabbits, their space use is quite flexible and is affected by habitat quality (availability of food, cover, mating partners, and predation pressure). Consequently, home range sizes are larger when observing hares for a longer period and when habitats are changing over the year (as they do especially in arable land). Hares move the center of their home range rapidly during cereal harvest periods (Marboutin and Aebischer 1996; Ullmann et al. 2020; but see Reitz and Leonard 1994), but only in landscapes of low heterogeneity (Schai-Braun et al. 2014). Cereal harvest seems to not affect the stress level of hares if landscape heterogeneity is high (Cybulska et al. 2020). Fields might be used after harvesting as they might provide food and short vegetation (Ullmann et al. 2020). The hare's plasticity in space use leads to comparably low adult mortality rates associated to harvest (Reitz and Leonard 1993).

Some studies showed a high site fidelity of hares (e.g., Bray et al. 2007; Broekhuizen and Maaskamp 1982; Hewson and Taylor 1968), but this can be found only in optimal habitats with high heterogeneity and low levels of disturbance (Avril et al. 2012). Interestingly, *L. europaeus* has an impressive "home instinct," i.e., translocated hares might come back to their original home range, occasionally even up to distances of 464 km (Jezierski 1967).

Disturbances through agricultural activities (Ullmann et al. 2020), hunting (Avril et al. 2014), natural events like floodings (Schrama et al. 2015), or harsh winter conditions might lead to migrations (Sokolov et al. 2009). Food shortages in winter might even lead to mass migrations of several thousand individuals (Fortunatow s.a. and Neschenzew s.a. in Angermann 1972).

While some studies suggested that hares are generally philopatric (Broekhuizen and Maaskamp 1982; Pielowski 1972) and show limited dispersal behavior, Bray et al. (2007) reported a dispersal rate of 43% and a median dispersal distance of 209 m for philopatric individuals and 1615 m for dispersers (maximum 17.35 km). Natal dispersal starts with sexual maturity (between 4–6 months), and males disperse more frequently than females (Bray et al. 2007). In sum, juvenile dispersion is more pronounced in habitats with low quality and low hare density.

Diet

L. europaeus is an herbivore using cultivated crops, weeds, and grasses (predominantly Poaceae). Among weeds in particular species of Fabaceae, Asteraceae, Brassicaceae, and Plantaginaceae are consumed (Brüll 1976; Chapuis 1990; Homolka 1982, 1987; Jennings et al. 2006; Reichlin et al. 2006). In winter, European hares feed also on seeds, buds, twigs, bark, and other woody parts of plants (Frylestam 1986; Rödel et al. 2004; Sokolov et al. 2009), especially when the ground vegetation is covered by snow. Much has been published on plant species used by hares. Information on plant use and preference of plants is of course depending on availability, which is depending on study site. Hence, it is not surprising that the numerous studies differ in their results. Sampling size is quite important, too, as there is a huge inter-individual variation within study sites (Katona and Altbäcker 2002). However, while the food spectrum can include several 100 plant species in one area, the preferred plants are very few. Hence, the European hare has been described as a picky herbivore (Schai-Braun et al. 2015): Of the 349 plant taxa identified in a study site in E Austria, hares used 47 taxa and only 10 were positively selected. Food selection is partly driven by energy content, i.e., crude fat and crude protein, whereas crude fiber is avoided (Schai-Braun et al. 2015; Smith et al. 2005b). While the average crude fat content of food plants for herbivorous mammals is approximately 3.5%, a hare's gut can contain more than 30% fat (Popescu et al. 2011). Dietary fat is important for hares in two ways: first, because reproductive performance is higher in females with access to food rich in fat (Hackländer et al. 2002b) as leverets will have access to a fattier milk. Second, fat serves as source of water when this nutrient is metabolized. Therefore, a positive selection for fatty plant parts is to be expected for a mammal deriving from steppe habitats (Kronfeld and Shkolnik 1996). In line with that, water requirements are usually covered by food metabolization (Kummer 1970), but direct water uptake by hares or feeding of snow is possible (Sokolov et al. 2009) and has been often described for hares living in captivity and fed with dry food (e.g., Landois 1885).

As all lagomorphs, *L. europaeus* uses feces to increase nutrient uptake (Hirakawa 2001), both hard pellets (coprophagy) in times of food shortage (Flux 1981b) and – more importantly – soft pellets (cecotrophy). The cecum is designed as a large fermentation chamber in which bacterial digestive processes of nondigestible fiber and an enrichment of the intestinal contents with proteins and vitamins take place. Leporids incapable of consuming cecotrophs show a protein deficiency (Hirakawa 2001).

Behavior

Activity Pattern

The European hare is mostly active at night (Rühe and Hohmann 2004; Tapper and Barnes 1986). However, in spring and summer hares are more active during the day due to mating activities and shorter nights, respectively (Holley 2001; Homolka 1986; Schai-Braun et al. 2012; Sokolov et al. 2009). In general, the activity pattern is affected by habitat quality, hare density, and disturbances, e.g., by agricultural activities or predation (Pépin and Cargnelutti 1994; Schneider 1978). When active, hares spend most of the time for foraging, between 11 (Flux 1981b) and 14 hours (Homolka 1986), usually in an open area with proper visibility and a vegetation height between 1 and 25 cm (Mayer et al. 2018; Naldi et al. 2020).

During the resting phase, hares use forms, i.e., self-scratched, flat hollows in the ground or simply depressions in higher vegetation (Schneider 1978). The dimensions of these forms are 35–50 cm length and 13–27 cm width (De Vos and Dean 1967; Sokolov et al. 2009) and thus resemble the size of a resting hare. They are deeper (up to 29 cm) in open areas with stronger winds (Sokolov et al. 2009). Usually, hares have their face directed to the open area (Sokolov et al. 2009) or against the main wind direction (De Vos and Dean 1967). In times of heavy snow fall, hares might even dig holes or burrows into the snow (De Vos and Dean 1967; Sokolov et al. 2009) or allow snow to cover them (De Vos and Dean 1967). Formozov (1929 in Angermann 1972) reported that hares dig burrows into sandy soil. However, digging burrows remains an exception.

Forms are reused occasionally (De Vos and Dean 1967). Hares enter these forms usually by approaching them using "backtracking," i.e., at dawn they move to their daylight range, sit and groom for some minutes at a spot, and then use the same track back to enter the form (Flux 1981b; Koenen 1956). This behavior is interpreted to confuse terrestrial predators. In their forms, they mostly stay immobile relying on their camouflage. The eyes are fully closed when sleeping. Even in deep rest hares are alert and flush full speed out of their forms when in danger. Flight distance is shorter, when protection is given (e.g., in high vegetation); temperatures are higher (danger of overheating while running) and when the soil is muddy (Jerzierski 1973) as mud will stick to the furred paws and thus impede full speed running. While at rest in their forms, *L. europaeus* takes up cecotrophs, starting around 2–3 hours before noon (Kummer 1970; Schneider 1978; Watson and Taylor 1955) directly from the anus (this behavior looks like grooming the anogenital region). Hence, a hare killed before noon will contain a large proportion of fine food particles (less than 0.25 mm length, Naumova et al. 2015) in the stomach.

Locomotion

European hares have different locomotor patterns. The "hobbling" is the calm, normal, and typical gait. The "escape" is a hopping accelerated to a gallop with jumps creating a floating phase without contact to the ground. When a predator comes close, hares speed up by 15.84 km h^{-1} (Taylor

2017). Maximum speed has been reported by Rieck (1953) to be up to 80 km h^{-1}. Hares can jump distances up to 2.7 m long, and 3 m high (Kummer 1970; Rieck 1953). Hence, *L. europaeus* can jump over fences of considerable height or even climb over them. In addition, they can do their typical zigzag escape trick (Kuznetsov et al. 2017). Several other locomotor patterns have been described for *L. europaeus*: When "slipping," individual steps are only carried out with the front legs until the body is more or less stretched out. The hind legs are then jerked forward. "Tip-toeing" is a hopping or walking around, where hare straightens up on all four legs and with largely stretched hind legs. This is probably an impressing position when meeting conspecifics or predators. The European hare can swim distances of more than 1 km (Schneider 1978).

Communication and Signals

Numerous ways of intra- and interspecific communication have been described (Schneider 1981). Among all senses, hearing and smell are most important for hares. As eyes are positioned laterally, hares cannot see clearly but are enabled to detect moving objects around them.

Various fragrances produced in special scent glands are used for individual recognition, for individual orientation in its home range, and as a sign of presence for conspecifics. Apart from that, scents play an important role in mating behavior. Scent glands are in the chin (submandibular gland), the cheek, close to the eyes (Harderian gland), inguinal area (on both sides of the genital opening), and anal region. There are no glands on the paws. Age and hormonal status influence the size of all scent glands (Zörner 1981).

By rubbing against branches, poles, or stones, hares bring out the secretions of the chin gland. The secretions of all head glands are spread over the body with the feet when cleaning the fur. The secretion of the inguinal glands accumulates in hairless skin pockets that are located in the groins. It can also spread over the wider genital area with yellowish crust formation. Inguinal glands are larger in female hares than in males (Mykytowycz 1966).

Schneider (1978) interprets a special type of urination (when the back of the body is lifted and the tail is folded up) as marking behavior. In addition, urine spraying aimed at conspecifics is also observed in the field hare.

Acoustic communication plays a subordinate role in the European hare. A "plaintive" is expressed when hares are in a desperate situation, e.g., when caught by a predator. This call, similar to a bloodcurdling human baby cry, might serve as a warning signal or call for help to conspecific. A "grumble" has been described in various situations, such as fighting or copulating. "Grinding teeth" is interpreted as a threatening or fearful sound, often produced by leverets when disturbed (Hediger 1948). Schneider (1978) also describes a gentle contact sound of leverets before suckling and a warning knocking with the hind legs.

Social Behavior

The European hare lives predominantly solitary, especially at low densities, otherwise in pairs (during mating or mate guarding), or in groups. Grouping is common in spring (Flux 2009), when peak mating starts. At food shortage, hares might gather at food patches, where unusual hierarchies might arise (Lindlöf 1978; Monaghan and Metcalfe 1985; Sokolov et al. 2009). In addition, hares are observed in loose groups at high densities, especially at night during feeding (Broekhuizen and Maaskamp 1982; Marboutin and Péroux 1999). In general, hares are not territorial, and thus there is a high degree of tolerance toward conspecifics, even from the same sex, when food is not limited (Broekhuizen and Maaskamp 1982). When two hares meet, they sniff each other's body and nose. Real fights are rare, even during the breeding season (Schneider 1977), but might get severe and even lethal in captivity where the opponents cannot separate from each other (Landois 1885). The "boxing" behavior, when two individuals stand upright on their hind legs and beat the conspecific's head and fore feet, is usually a fight between a female and a male interpreted as a test of mating partners or preventing males from mating (Holley and Greenwood 1984; Schneider 1978). However, male-male boxing has been reported, too (Flux 2009; Schneider 1977).

Mating Behavior

In spring, hares show a lively mating behavior in groups, even during the day ("March madness," Lincoln 1974). These games include chasing, impressing positions, immediate stoppings, sniffing, urinating, and boxing. Fragrances that convey sexual attraction and individual identity play a major role in this, but also visual stimuli through the ears and tail, both rich in contrast, are important (Schneider 1976). During fights, tufts of hair might fly through the air (Landois 1885). Especially younger females in intensive chases might suffer from severe injuries on their back from male scratches. Copulations take place in segregating couples only after extensive chases and fights. Females ready for copulations sit on their fore legs with erased back (called "lordosis" posture). Belova (1987 in Sokolov et al. 2009) described that lordosis lasts for 6 seconds, mounting by the male 2 seconds and the actual copulation between 5 and 30 seconds. The female terminates the copulation by lifting its hind feet and throwing away the male (Schneider 1977). Usually several matings follow in continued mating runs or in the same place (Schneider 1977; Sokolov et al. 2009). Males show mate guarding and accompany a female after copulation for some days (Flux 1981b; Schneider 1976). Therefore, one can observe a female in her form and a male sitting next to her in alert mode. Competitors are expelled when approaching the couple closer than 3–4 m (Bock 2020).

Parental Care

Leverets are born in the morning (Belova 1987 in Sokolov et al. 2009). The birth takes place in a form, often in a protected and dry place. Usually females do not provide any special preparation of the site, but at cold weather females cover their young with hay (Sokolov et al. 2009). Young-borns have open eyes upon birth and can move immediately after birth, although just in a crawling mode (Hediger 1948). Soon after birth, or after 3–5 days at the latest, leverets separate from each other and hide near the suckling area (Broekhuizen and Maaskamp 1976, 1980; Pielowski 1976a; Zörner 1981). Usually females visit the young only once a day, shortly after sunset, for two to six minutes for suckling (Broekhuizen and Maaskamp 1976, 1980; Landois 1885; Martinet and Demarne 1984). The female and the suckled young are sitting opposite each other, with the female being vigilant (Broekhuizen and Maaskamp 1976; Kummer 1970). Leverets accept several feeding events per day, if provided (Cooper 1970). According to Broekhuizen and Maaskamp (1976), the young get active before the female arrives and gather at the suckling spot which is the birth place. The female approaches the suckling spot always from the same direction, and leverets come toward approaching females with increasing age. Thus, the suckling spot therefore moves from the birthplace toward the direction from which the female reaches the suckling spot every night (Broekhuizen and Maaskamp 1976). Young hares do not recognize their mother by olfactory cues (Broekhuizen and Maaskamp 1980; Martinet and Demarne 1984; Stavy et al. 1985), explaining that they might suckle on other lactating females by chance (Landois 1885; Sokolov et al. 2009). After suckling, a female licks up the leveret urine and finishes the nursing by jumping away (Broekhuizen and Maaskamp 1980). Thereafter, leverets separate themselves again waiting to be suckled 24 h later.

Antipredator Behavior

To avoid being killed by predators, juvenile hares mainly remain on their camouflage and press themselves to the ground staying immobile until the danger has gone (Voigt and Siebert 2019). When disturbed, leverets might grumble or grind their teeth, sometimes even attack the opponent with their teeth (Hediger 1948) or their claws of the forefeet. The adult hare relies both on its camouflage as well as on its locomotor capacities. Hares remain motionless when resting in their forms.

When flushed, they use full speed (with ears pressed to the neck), zig-zag running, jumping, swimming through rivers, or even running through a herd of grazing cattle to escape predation (Sokolov et al. 2009). This flight mode is shown when chased by terrestrial as well as aerial predators. In severe danger, hares might use burrows from other species (Angermann 1972; Sokolov et al. 2009) or holes (even in low trees, Zörner 1981). To confuse terrestrial predators, hares use backtracking (Flux 1981b; Koenen 1956) or large jumps to the side (Sokolov et al. 2009). When flushed, hares often run in loops with variable diameter and come back to their form.

Although red foxes (*Vulpes vulpes*) are the main predator for European hares (see "Population Ecology"), this is not generally true for adults. Red foxes might kill adults only by stalking (when vegetation is high). In habitats with good visibility, hares approached by a red fox will show alertness by standing upright on the hind feet. Note that a hare's nails can be dangerous to predators. Therefore, red foxes do not come as close as 30 m to hares when they are in alert mode (Holley 1992). However, red fox presence affects hare space use. Hares use edge habitat more when red foxes are present (Weterings et al. 2019), tend to feed on patches of lower food quality (Weterings et al. 2018), and are more vigilant and less feeding (Mayer et al. 2020b). Living in groups reduces predation risk in hares, and the frequency of individual vigilant behavior is lower in groups of hares than in individually grazing hares (Broekhuizen and Maaskamp 1982; Marboutin and Aebischer 1996), explaining why hares feed in groups, if possible. Single hares reduce activity in bright moon nights, when red foxes are more active (Viviano et al. 2021). Hence, red foxes might have not only direct effects on hares by predation but also indirect effects by creating a landscape of fear.

Parasites and Diseases

Numerous diseases have been described for European hares (reviewed by, e.g., Boch and Schneidawind 1988; Frölich et al. 2001; Sokolov et al. 2009), many of them with severe effects on population dynamics. Parasite infestations can reach epidemic status and in some years can cause large losses in European hare populations. Among ectoparasites, several species of ticks (Ixodida, most importantly *Ixodes ricinus* and *I. reticulatus*) and fleas (Neoptera, e.g., *Chaetopsylla trichosa*) have been described. The ectoparasite load usually increases from N to S. Among the infectious diseases (often associated with ectoparasites), hares might suffer from pseudotuberculosis (*Yersinia pseudotuberculosis*, e.g., Bartling et al. 2004), pasteurellosis (*Pasteurella multocida*, e.g., Haerer et al. 2001), staphylomycosis (*Staphylococcus* sp., e.g., Tomaso et al. 2018), brucellosis (*Brucella* sp., e.g., Gyuranecz et al. 2011), tularemia (*Pasteurella tularensis*, e.g., Rijks et al. 2013), toxoplasmosis (*Toxoplasma gondii,* e.g., Sedlák et al. 2000), and leishmaniasis (*Leishmania infantum*, e.g., Ruiz-Fons et al. 2013; Tsokana et al. 2016). Note that some of these infectious diseases are serious zoonoses (Tsokana et al. 2020), making hares' sanitary surveillance a key factor in the One Health approach.

Among endoparasites, various nematode (e.g., *Nematodirus aspinosus*), cestode (e.g., *Dicrocoelium dendriticum*), and trematode species have been reported. Endoparasitic diseases include lung strongylosis (*Protostrongylus commutatus*, e.g., Pajerský et al. 1992), stomach strongylosis (*Trichostrongylus retortaeformis,* e.g., Soveri and Valtonern 1983), and sarcosporidiosis; however, coccidiosis (*Eimeria* sp., e.g., Posautz et al. 2015) is the most abundant and important one. Disease prevalence fluctuates between years and differs between study sites, age classes, and sex (Lamarque et al. 1996; Rieck 1956b), partly explained by density and weather effects.

In the recent decades, emerging viral diseases through a calicivirus from the genus *Lagovirus* caused severe declines in hare populations (Salvioli et al. 2017). The first calicivirus infection in *L. europaeus* was described in the early 1980s (Gavier-Widén and Mörner 1991) associated with a disease called European Brown Hare Syndrome (EBHS). This calicivirus (EBHSV) is closely related to the virus causing Rabbit Hemorrhagic Disease (RHD), a fatal hepatitis in European rabbits (*Oryctolagus cuniculus*, Frölich and Lavazza 2008). Hare populations dramatically decline after an EBHS outbreak, especially in low-density populations, where the virus persists longer and the proportion of young individuals (which are less affected) is rather low (Paci et al. 2011). The critical density according to these authors is 8–15 hares 100 ha^{-1}. In the early 2010s, a new calicivirus (RHDV2) spilled over from European rabbits to European hares (Velarde et al. 2017). The rapid mutation of the caliciviruses will be a challenging problem for many hare populations (Le Gall-Reculé et al. 2017), not only in Europe.

Besides this, also the myxoma virus can jump from European rabbits to *L. europaeus* (reviewed in Barlow et al. 2014). Myxomatosis in European hares is usually associated with high prevalence in sympatrically living European rabbit populations (Wibbelt and Frölich 2005).

The European hare is a suitable bioindicator for the local or regional distribution of some agrochemicals and other environmental chemicals (Kleinmann and Wang 2017; Topping et al. 2016). A hare's pesticide uptake has been described through feeding or grooming after pesticide overspray (Mayer et al. 2020a). There are numerous studies on hunted hares regarding residues of heavy metals, pesticides, polychlorinated biphenyls, or microplastics (e.g., Bukovjan et al. 1997; Hornek-Gausterer et al. 2021; Massányi et al. 2003; Nösel and Ahrens 1996; Petrović et al. 2014; Škrivanko et al. 2008; Wajdzik et al. 2017), and even experimental applications in captive individuals (Črep and Švický 1993). However, no clear connections between residue findings and population dynamics have been demonstrated yet (e.g., Edwards et al. 2000; Stubbe and Stubbe 1997), although these contaminants might be detrimental for hares (Linšak et al. 2013; Nováková et al. 1976; Orthwein 1984a, b) and reach values that might be dangerous for humans consuming the affected individuals (Slamečka et al. 2012). Agrochemicals had also no effects on the reproductive performance of a hare population, both for females (Hackländer et al. 2001) or males (Blottner et al. 2001).

Population Ecology

Recruitment and Density

The density of hares in steppe environments is low, approximately 2 hares 100 ha^{-1}. Agricultural activity, especially on fertile, but dry, soil types, lead to higher densities (Sokolov et al. 2009). Population density might reach up to 339 hares 100 ha^{-1} in optimal conditions, as it was observed on a Danish island without agriculture, hunting, and terrestrial predators (Abildgård et al. 1972). Habitat quality influences fecundity, (juvenile) survival, thus yearly recruitment and population density. Across the distribution range, spring densities of 1 (Kilias and Ackermann 2001; Rühe et al. 2000) to 156 hares 100 ha^{-1} (Klansek 1996) are described (reviewed in Averianov et al. 2003). As recruitment might double spring numbers, autumn densities can reach up to 275 hares 100 ha^{-1} (Klansek 1996).

Population growth depends predominantly on juvenile survival rate (see "Survival"), which varies between study sites (habitat quality, predation pressure, and climate) and year (e.g., Frylestam 1979; Haerer et al. 2001). Leveret mortality is high and exceeds nearly 65% within the first month (Voigt and Siebert 2020). Until autumn, the overall postpartum mortality reaches up to 81% (Ciberej and Kačúr 1991) or even 95% (Schai-Braun et al. 2020). Therefore, the proportion of young of the year in relation to older females in the hunting bag varies between 10% and 70% (e.g., Ahrens et al. 1993; Eskens et al. 1999).

In general, a complex interaction of numerous factors is responsible for hare recruitment and abundance, namely landscape heterogeneity, land use (practices), soil type, climate (or weather), diseases, and predation (Petrov 1976; Pielowski 1981; Smith et al. 2005a). Studies trying to figure out the relevance of each mortality factor reached different results. However, in many studies, land use practices were more important than weather and diseases (Eiberle and Matter 1982; Frölich et al. 2003; Lundström-Gilliéron and Schlaepfer 2003; Reynolds and

Tapper 1995; Ristić et al. 2021; Schmidt et al. 2004; Smith et al. 2005a; Weber et al. 2019).

Unfavorable weather associated with high humidity and cold temperatures might increase the proportion of infected hares and thus will allow predators to find more easy prey in hares or make use of carcasses. Impact of predation on yearly recruitment strongly depends on the availability of alternative prey (Goszczyniski et al. 1976; Sokolov et al. 2009). Hence, in the peak of small mammal cycles, young hares are less likely predated (Pegel 1986).

The main predator for European hares in C Europe is the red fox. In Poland, hares formed 12–46% of red fox diet (Goszczyński and Wasilewski 1992; Pielowski 1976b), mainly depending on the yearly variation of alternative prey availability (e.g., voles). Note that impact of red foxes increases after large-scale rabies vaccination (Martini 1983; Nösel and Ahrens 1996; Ponjiger et al. 2019; Spittler 1972). Juveniles are the predominant age class that red foxes prey on. However, up to 50% of adult hare mortality might be attributed to the red fox (Goszczyński and Wasilewski 1992). With respect to this report, it should also be considered that red foxes also use carcasses, e.g., from road kills, and thus (adult) hare mortality and red fox predation are not always linked to each other. According to Goszczyński and Wasilewski (1992), hare biomass in red fox diet was highest in spring (43%), lower in summer (12–15%), and intermediate in winter (28%). This pattern is explained by both leveret availability and vegetation cover. In general, red fox density is negatively correlated with population increase or density of hares (Ahrens 1996; Frölich et al. 2003; Goszczyński and Wasilewski 1992; Pegel 1986; Späth 1989).

Interspecific Competition

Where European hares live in sympatry with other species of the genus *Lepus*, they usually outcompete them (Sokolov et al. 2009). With climate change (less snow, increase of arable land use), European hares tend to expand into the range of mountain hares, where they replace them by direct competition and hybridization (Acevedo et al. 2012; Jansson and Pehrson 2007; Thulin 2003). In addition, introduced *L. europaeus* also outcompete *L. corsicanus* (Angelici and Luiselli 2001); however, they do not interbreed (Mengoni et al. 2015).

Where *L. europaeus* lives in sympatry with rabbits (*O. cuniculus* or *Sylvilagus floridanus*), competition for food and aggressive behavior have been reported (Bertolino et al. 2013; Homolka 1987; Kuijper et al. 2004) but are low when both species are at low densities (Flux 2008; Vidus-Rosin et al. 2012). Rabbits depend on their burrow systems and might be dominant over hares within their territories, but this seems to be a rather local phenomenon.

Conservation Status

The European hare's Red List status has recently been confirmed as "Least Concern" on a global scale (Hackländer and Schai-Braun 2019). It is widespread and abundant across its geographic range. However, especially European populations of *L. europaeus* declined in the twentieth century (Hackländer and Schai-Braun 2018). Note, that despite numerous studies claiming that the decline started in the 1960s (e.g., Bock 2020; Farkas et al. 2020; Smith et al. 2005a), there is evidence from long-term hunting bag statistics (e.g., Schwenk 1985), that the decline started some decades earlier, namely with the beginning of the agricultural industrialization (early twentieth century). The European hare is listed under Appendix III of the Bern Convention in Europe (as "*Lepus capensis* (*europaeus*)," Council of Europe 1979). Although the Red List status in Europe and the European Union is "Least Concern" (Temple and Terry 2007), several European countries (e.g., Austria, Germany, Norway, and Switzerland) have classified *L. europaeus* as "Near Threatened" or "Threatened" in their national Red Lists (Reichlin et al. 2006). Despite that, in most European countries, densities of European hares are above the density known from steppe habitats where *L. europaeus* came from prior to their range expansion (see "Current Distribution"). In line with this, lowered densities might not lead to an extinction of the species. Apart from that, there is increasing concern regarding the status of native local populations of this species (Flux and Angermann 1990; Schai-Braun and Hackländer 2016) as restockings for hunting purposes took and take place in several countries including Denmark, France, Greece, Italy, and Spain (Suchentrunk et al. 2006).

Management

The European hare is the most important small game species across Europe (Flux and Angermann 1990), despite the recent decline (see "Conservation Status"). Usually, hunting is open in autumn (October–December), occasionally extended to January or February in vineyards or orchards to prevent damage (Suchomel et al. 2019). As the European hare is reproductively active already in December in C Europe, already Kutzer et al. (1976) called for a shorter hunting season. In England and Wales, where European hares are allochthonous, a closed season is absent (Jennings 2017) though claimed for animal welfare reasons (Butterworth et al. 2017). At higher densities, hunters harvest hares with shotguns using drive hunts, often with the help of dogs and stalkers. In high-density areas, falconers might hunt European hares with goshawks (*Accipiter gentilis*) or golden eagle (*Aquila chrysaetos*). In addition, single hunt with a small caliber rifle is common. Traditional hunting types in the UK like beagling or hare coursing (hounds tracking hares by sight in arenas) are no longer allowed (Jennings 2017), irrespective of the fact that their impact on hare numbers was low compared to shooting (Stoate and Tapper 1993). However, coursing is still practiced in the Republic of Ireland and beagling in France (Jennings 2017). As hares often run loops when flushed (see "Antipredator Behavior"), hunters with hounds use this behavior by just waiting for the hare at the flushing site while listening to the barking hound chasing the hare on its way back to the form (Hackländer 2011).

Hares are hunted for their meat, which is treated as a delicacy in some areas. The hare meat is low in fat and rich in essential fatty acids (Frunza et al. 2019; Slamečka et al. 1998; Škrivanko et al. 2008; Valencak et al. 2015). Slow stewing or roasting was the most common way of preparing hare meat. A modern trend is toward simple seasoning and grilling (Jennings 2017). In former times, hunted hares were used not only for their meat, but also for their fur to produce felts or fishing flies (Jennings 2017).

A sustainable use is based on monitoring hare abundance or density in spring and autumn (Andrzejewski and Jezierski 1966). Langbein et al. (1999) reviewed appropriate monitoring methods. For counting at night (when hares are active), spotlights (Ahrens et al. 1995; Frylestam 1981; Strauß et al. 2008) or night vision devices (Focardi et al. 2001) are in use. Spring densities and population increase are then used to calculate sustainable harvest rates (Marboutin et al. 2003; Schai-Braun et al. 2019). In areas where night counting is compromised (e.g., high proportion of forest cover), hunting bags have to be checked for age composition before continuing hunting on the same day or later. Evaluating the proportion of young of the year can be done either by Stroh's sign (Stroh 1931) or by determining the dried eye-lens weight (Ciberej et al. 1997; Kovács and Heltay 1981): If proportion of young of the year in the hunting bag is lower than, e.g., 50% (and thus recruitment was weak), hunting should be stopped for the year in that specific area.

The superfactor leading to declining European hare densities is the intensification of agriculture (Lundström-Gilliéron and Schlaepfer 2003; Mayer et al. 2019; Panek 2018; Panek and Kamieniarz 1999; Petrak 1990; Schäfers 1996; Schröpfer and Nyenhuis 1982; Smith et al. 2005a). While the European hare benefitted from extensive agriculture in the past, it is now negatively affected by the intensive use of the cultivated land. One has to distinguish between the reasons for the decline and the measure considered to increase hare populations. To counteract the decline of hare populations in Europe, protected areas (Canova et al. 2020), food supplementation in winter (Matuszewski 1966;

Reichlin et al. 2006), habitat improvements (Genghini and Capizzi 2005; Kamieniarz et al. 2013; Meichtry-Stier et al. 2014; Petrovan et al. 2013; Santilli et al. 2014; Schai-Braun et al. 2020; Sliwinski et al. 2019), and predator control (Panek 2013; Reynolds et al. 2010) have been shown to be beneficial, especially by increasing juvenile survival rate. It is clear that habitat improvement will have long-lasting effects (not only for hares), whereas other measures are just fighting the symptoms, though very effectively. Note that habitat improvements have to be planned in a proper way in order to avoid that they act as ecological traps (Hummel et al. 2017). In addition, organic farming is not a beneficial to population growth per se (Santilli and Galardi 2016).

Hare restocking is still a management measure in several hunting sites across Europe (e.g., Cukor et al. 2018; Ferretti et al. 2010; Fischer and Tagand 2012; Misiorowska 2013; Modesto et al. 2011; Spyrou et al. 2013), although mortality rate of translocated individuals is usually very high in the first month and lies between 40% (Misiorowska and Wasilewski 2012) and 79% (Angelici et al. 2000), mainly due to red fox predation (see also Marboutin et al. 1990). Other adverse effects of hare translocations are disease transmissions and genetic introgression (Pierpaoli et al. 1999; Stamatis et al. 2007; Suchentrunk et al. 2006). Translocated hares derive from captive breeding or have an origin from sites with higher hare densities, sometimes several hundred kilometers away from the release site (i.e., from potentially allochthonous populations). They might even come from areas where they have been introduced, like from Argentina and Uruguay (Suchentrunk et al. 2006). Transport and release are debated also in the light of animal welfare (Paci et al. 2006).

Future Challenges for Research and Management

Clearly, the European hare is the best-studied *Lepus* species in the world with far more than 1000 entries in the modern literature databases. Much of the sources, including leading monographies, are written in former scientific languages like German, French, or Russian. Other milestone publications are conference proceedings (Myers and McInnes 1981; Pielowski and Pucek 1976). All these literature sources are rarely used by modern researchers as most of the relevant sources from the last century have not been digitalized, yet. Consequently, there are plenty of study repetitions during the last century, making it difficult to keep the overview on the literature regarding European hares. Despite the ongoing repetition of knowledge, there are still some major gaps to fill (see Hackländer et al. 2008). We need basic research both in the field as well as in captive European hare populations. In addition, comprehensive meta-analyses on available literature will help to understand the species' evolution and ecology.

Acknowledgments This chapter is dedicated to my wife Sandra who shares my enthusiasm and love for hares. My dear friend Paulo C. Alves provided numerous valuable comments and suggestions to an earlier draft of the manuscript.

References

Abildgård F, Andersen J, Barndorff-Nielsen O (1972) The hare population (*Lepus europaeus* Pallas) of Illumø Island, Denmark. A report on the analysis of data from 1957 to 1970. Dan Rev Game Biol 6:1–32

Acevedo P, Jiménez-Valverde A, Melo-Ferreira J, Real R, Alves PC (2012) Parapatric species and the implications for climate change studies: a case study on hares in Europe. Glob Change Biol 18:1509–1519

Ahrens M (1996) Untersuchungen zu Einflussfaktoren auf die Besatzentwicklung bei Feldhasen in verschiedenen Gebieten Sachsen-Anhalts [in German]. Beitr Jagd Wildforsch 21:229–235

Ahrens M, Tottewitz F, Gleich E (1993) Zur Altersstruktur von Feldhasen (*Lepus europaeus* Pallas, 1778) aus verschiedenen Gebieten Ostdeutschlands [in German]. Beitr Jagd Wildforsch 18:129–133

Ahrens M, Goretzkj J, Stubbe C, Tottewitz F, Gleich E, Sparing H (1995) Untersuchungen zur Entwicklung des Hasenbesatzes auf Wittow/Rügen [in German]. Beitr Jagd Wildforsch 20:191–200

Alves PC, Ferrand N, Suchentrunk F, Harris DJ (2003) Ancient introgression of *Lepus timidus* mtDNA into *L. granatensis* and *L. europaeus* in the Iberian Peninsula. Mol Phylogen Evol 27:70–80

Alves PC, Melo-Ferreira J, Freitas H, Boursot P (2008) The ubiquitous mountain hare mitochondria: multiple introgressive hybridization in hares, genus *Lepus*. Phil Trans R Soc Lond B: Biol Sci 363:2831–2839

Amoutzias GD, Giannoulis T, Moutou KA, Psarra AM, Stamatis C, Tsipourlianos A, Mamuris Z (2016) SNP identification through transcriptome analysis of the European brown hare (*Lepus europaeus*): cellular energetics and mother's curse. PLoS One 11:e0159939

Andersen J, Jensen D (1972) Studies on the European hare XXVIII. The weight of the eye in European hares of known age. Acta Theriol 17:87–92

Andrzejewski R, Jezierski W (1966) Studies on the European hare. XI. Estimation of population density and attempt to plan the yearly take of hares. Acta Theriol 11:433–448

Angelici FM, Luiselli L (2001) Conservation problems of the critically endangered Apennine hare *Lepus corsicanus* De Winton 1898, in Italy and Sicily. Oryx 35:245–249

Angelici FM, Riga F, Boitani L, Luiselli L (2000) Fate of captive-reared brown hares *Lepus europaeus* released at a mountain site in central Italy. Wildl Biol 6:173–178

Angelone C, Moncunill-Solé B, Kotsakis T (2020) Fossil Lagomorpha (Mammalia) of Italy: systematics and biochronology. Riv It Paleontol Strat 126:157–187

Angermann R (1972) Die Hasentiere [in German]. In: Altevogt R, Angermann R, Dathe H, Grzimek B, Herter K, Müller-Using D, Rahm U, Thenius E (eds) Grzimeks Tierleben. Enzyklopädie des Tierreichs, Band 12: Säugetiere 3. Kindler, Zürich, pp 419–465

Antoniou A, Kotoulas G, Magoulas A, Alves PC (2008) Evidence of autumn reproduction in female European hares (*Lepus europaeus*) from southern Europe. Eur J Wildl Res 54:581–587

Ashrafzadeh MR, Djan M, Szendrei L, Paulauskas A, Scandura M, Bagi Z, Ilie DE, Kerdikoshvili N, Marek P, Soós N, Kusza S (2018) Large-scale mitochondrial DNA analysis reveals new light on the phylogeography of Central and Eastern-European brown hare (*Lepus europaeus* Pallas, 1778). PLoS One 13:e020465

Averianov A, Niethammer J, Pegel M (2003) *Lepus europaeus* Pallas, 1778—Feldhase [in German]. In: Niethammer J, Krapp F (eds) Handbuch der Säugetiere Europas, Band 3/II Hasentiere. AULA, Wiesbaden, pp 35–104

Avril A, Letty J, Pradel R, Léonard Y, Santin-Janin H, Pontier D (2012) A multi-event model to study stage-dependent dispersal in radio-collared hares: when hunting promotes costly transience. Ecology 93:1305–1316

Avril A, Letty J, Léonard Y, Pontier D (2014) Exploration forays in juvenile European hares (*Lepus europaeus*): dispersal preludes or hunting-induced troubles? BMC Ecol 14:6

Barlow A, Lawrence K, Everest D, Dastjerdi A, Finnegan C, Steinbach F (2014) Confirmation of myxomatosis in a European brown hare in Great Britain. Vet Rec 175:75–76

Bartling C, Wölfel R, Nikolaou K, Petry T, Thiede S, Hildebrandt T, Fassbender M, Göritz F, Blottner S, Spittler H, Neubauer H (2004) Prävalenz von Anti-Yersinia Antikörpern bei Feldhasen in Nordrhein-Westfalen [in German with English summary]. Dtsch Tierärztl Wochenschr 111:259–264

Ben Slimen H, Suchentrunk F, Stamatis C, Mamuris Z, Sert H, Alves PC, Kryger U, Basyouni Shahin A, Ben Ammar Elgaaied A (2008) Population genetics of cape and brown hares (*Lepus capensis* and *L. europaeus*): a test of Petter's hypothesis of conspecificity. Biochem Syst Ecol 36:22–39

Bensinger S, Kugelschafter K, Eskens U, Sobiraj A (2000) Untersuchungen zur jährlichen Reproduktionsleistung von weiblichen Feldhasen (*Lepus europaeus* Pallas, 1778) in Deutschland [in German with English summary]. Z Jagdwiss 46:73–83

Bertolino S, Cordero di Montezemolo N, Perrone A (2013) Habitat use of coexisting introduced eastern cottontail and native European hare. Mamm Biol 78(4):235–240

Beuković M, Beuković D, Popović Z, Đorđević N, Đorđević M (2013) Impact of climatic factors to the percentage of young in the population of brown hare (*Lepus europaeus* P) in the Bačka district. Acta Vet Brno 63:111–122

Blottner S, Faber D, Roelants H (2000) Studies on European hare 56. Seasonal variation of testicular activity in European brown hare *Lepus europaeus*. Acta Theriol 45:385–394

Blottner S, Lange A, Göritz F, Faßbender M, Broich A, Quest M, Gilles M, Lengwinat T, Hildebrandt TB (2001) Untersuchungen zur reproduktiven Fitness an lebenden männlichen Feldhasen aus unterschiedlichen Habitaten [in German with English summary]. Z Jagdwiss 47:84–91

Boch J, Schneidawind H (1988) Krankheiten des jagdbaren Wildes [in German]. Paul Parey, Hamburg

Bock A (2020) *Lepus europaeus* (Lagomorpha: Leporidae). Mamm Spec 52:125–142

Borowski S (1964) Studies on the European hare I. Moulting and coloration. Acta Theriol 9:217–231

Bray Y (1998) Vers une meilleure connaissance des flux demographiqu es chez le lièvre d'Europe (*Lepus europaeus*) [in French]. Diplome superieur d'Etudes et de recherche, Universite de Bourgogne, Dijon, France

Bray Y, Champely S, Soyez D (2002) Age determination in leverets of European hare *Lepus europaeus* based on body measurements. Wildl Biol 8:31–39

Bray Y, Devillard S, Marboutin E, Mauvy B, Péroux R (2007) Natal dispersal of European hare in France. J Zool 273:426–434

Bresiński W (1976) Agrarian structure vs. the European hare population density. In: Pielowski Z, Pucek Z (eds) Ecology and management of European hare populations. The proceedings of an international symposium held in Poznań on December 23–24, 1974. Państwowe Wydawnictwo Rolnicze I Leśne, Warszawa, pp 195–197

Bresiński W (1983) The effect of some habitat factors on the spatial distribution of a hare population during winter. Acta Theriol 28:435–441

Bresiński W, Chlewski A (1976) Tree stands in fields and spatial distribution of hare populations. In: Pielowski Z, Pucek Z (eds) Ecology and management of European hare populations. The proceedings of an international symposium held in Poznań on December 23–24, 1974. Państwowe Wydawnictwo Rolnicze I Leśne, Warszawa, pp 185–193

Broekhuizen S (1979) Studies on the European hare. XXXVI. Survival in adult European hares. Acta Theriol 24:465–473

Broekhuizen S, Maaskamp F (1976) Behaviour and maternal relations of young European hares during the nursing period. In: Pielowski Z, Pucek Z (eds) Ecology and management of European hare populations. The proceedings of an international symposium held in Poznań on December 23–24, 1974. Państwowe Wydawnictwo Rolnicze I Leśne, Warszawa, pp 59–67

Broekhuizen S, Maaskamp F (1979) Age determination in the European hare (*Lepus europaeus* Pallas) in The Netherlands. Z Säugetierkd 44:162–175

Broekhuizen S, Maaskamp F (1980) Behaviour of does and leverets of the Euroepean hare (*Lepus europaeus*) whilst nursing. J Zool 191:487–501

Broekhuizen S, Maaskamp F (1981) Annual production of young in European hares (*Lepus europaeus*) in The Netherlands. J Zool 193:499–516

Broekhuizen S, Maaskamp F (1982) Movement, home range and clustering in the European hare (*Lepus europaeus* Pallas) in The Netherlands. Z Säugetierkd 47:22–32

Broekhuizen S, Martinet L (1979) Growth of embryos of the European hare (*Lepus europaeus* Pallas). Z Säugetierkd 44:175–179

Brüll U (1976) Nahrungsbiologische Studien am Feldhasen in Schleswig-Holstein. Ein Beitrag zur Äsungsverbesserung [in German]. In: Pielowski Z, Pucek Z (eds) Ecology and management of European hare populations. The proceedings of an international symposium held in Poznań on December 23–24, 1974. Państwowe Wydawnictwo Rolnicze I Leśne, Warszawa, pp 93–99

Bujalska G (1964) Studies on European hare. IV. Variations in the pelvis and sacrum. Acta Theriol 11:287–304

Bujalska G, Caboń-Raczyńska K, Raczyński J (1965) Studies on the European hare. VI. Comparison of different criteria of age. Acta Theriol 10:1–9

Bukovjan K, Slamečka J, Jurčík R, Hell P (1997) Contamination with polychlorated biphenyls. In: Slamečka J, Hell P, Jurcík R (eds) Brown hare in the Westslovak lowland. Acta Sc Nat Brno 31(3–4):49

Butterworth A, Turner KME, Jennings N (2017) Minimising orphaning in the brown hare *Lepus europaeus* in England and Wales: should a close season be introduced? Wildl Biol 2017:wlb.00279

Caboń-Raczyńska K (1964a) Studies on the European hare. II. Variations in the weight and dimensions of the body and weight of certain internal organs. Acta Theriol 9:233–248

Caboń-Raczyńska K (1964b) Studies on the European hare. III. Morphologcial variation of the skull. Acta Theriol 9:249–285

Caboń-Raczyńska K (1974) Studies on the European hare. XXX. Variability of the body weight of European hares. Acta Theriol 19:69–80

Caillol M, Martinet L (1976) Preliminary results on plasma progesterone levels during pregnancy and superfetation in the hare, *Lepus europaeus*. J Reprod Fert 46:61–64

Caillol M, Martinet L (1983) Mating periods and fertility in the doe hare (*Lepus europaeus*) bred in captivity. Acta Zool Fennica 174:65–68

Caillol M, Mondain-Monval M, McNeilly AS (1990) Pattern of serum concentrations of prolactin and progesterone during pregnancy and lactation in the brown hare (*Lepus europaeus*). J Endocrinol 124:11–17

Caillol M, Mondain-Monval M, Meunier M, Rossano B (1991a) Pituitary and ovarian responses to luteinizing-hormone-releasing hormone during pregnancy and after parturition in brown hares (*Lepus europaeus*). J Reprod Fert 92:89–97

Caillol M, Mondain-Monval M, Rossano B (1991b) Gonadotrophins and sex steroids during pregnancy and natural superfoetation in captive brown hares (*Lepus europaeus*). J Reprod Fert 92:299–306

Caillol M, Mondain-Monval M, Meunier M, Rossano B (1992) Influence of season of birth on onset of gonadotrophic and ovarian functions in young doe hares (*Lepus europaeus*). J Reprod Fert 96:747–753

Canova L, Gazzola A, Pollini L, Balestrieri A (2020) Surveillance and habitat diversity affect European brown hare (*Lepus europaeus*) density in protected breeding areas. Eur J Wildl Res 66:66

Cardarelli E, Meriggi A, Brangi A, Vidus-Rosin A (2011) Effects of arboriculture stands on European hare *Lepus europaeus* spring habitat use in an agricultural area of northern Italy. Acta Theriol 56:229–238

Chapuis JL (1990) Comparison of the diets of two sympatric lagomorphs, *Lepus europaeus* (Pallas) and *Oryctolagus cuniculus* (L.) in an agroecosystem of the Ile-de-France. Z Säugetierkd 55:176–185

Ciberej J (1993) Growth of embryos and losses during the prenatal development in European hare [in Slovak with English summary]. Folia Venatoria 23:99–112

Ciberej J, Kačúr M (1991) Coefficient of survival and losses during the reproduction cycle in hare (*Lepus europaeus*) in the East Slovakian lowland [in Slovak with English summary]. Folia Venatoria 21:123–130

Ciberej J, Halász J, Kačúr M (1997) The study of reproduction coefficients of brown hare population in the years 1980–1996 in eastern Slovakia [in Slovak with English summary]. Folia Venatoria 26-27:65–71

Conrado zu Baldenstein T (1863) Der Bastardhase [in German]. Jahresber d Naturforsch Ges Graubündens 8:98–101

Cooper J (1970) A note on hand-rearing European hares *Lepus europaeus* at London Zoo. Int Zoo Yearbook 10: 138–139

Corbet GB (1983) A review of classification in the family Leporidae. Acta Zool Fennica 174:11–15

Council of Europe (1979) Convention on the conservation of European wildlife and natural habitats. European Treaty Series - No.104. Council of Europe, Bern

Črep K, Švický E (1993) Experimental study of the influence of herbicides on European hare [in Slovak with English summary]. Folia Venatoria 23:169–181

Cukor J, Havránek F, Linda R, Bukovjan K, Painter MS, Hart V (2018) First findings of brown hare (*Lepus europaeus*) reintroduction in relation to seasonal impact. PLoS One 13:e0205078

Cybulska N, Hackländer K, Palme R, Frey-Roos A, Schai-Braun SC (2020) Influence of cereal harvest on adrenocortical activity in European hares (*Lepus europaeus*). Eur J Wildl Res 66:95

De Vos A, Dean PB (1967) The distribution of and the use of forms by European hares, *Lepus europaeus hybridus* (Desmarest, 1822), in southern Ontario. Säugetierkd Mitt 15:57–61

Demarne Y, Lhuillery C, Pihet J, Martinet L, Flanzy J (1978) Comparative study of triacylglycerol fatty acids in mild fat from two leporid species: rabbit (*Oryctolagus cuniculus*) and hare (*Lepus europaeus*). Comp Biochem Physiol 61B:223–226

Diamond JM (1987) Aristotle's theory of mammalian teat number is confirmed. Nature 325:200

Djan M, Stefanović M, Veličković N, Lavadinović V, Alves PC, Suchentrunk F (2017) Brown hares (*Lepus europaeus* Pallas, 1778) from the Balkans: a refined phylogeographic model. Hystrix 28:186–193

Durdík M (1981) Losses of small field game caused by mechanization of crop production [in Slovak with English summary]. Folia Venatoria 10-11:95–107

Edwards PJ, Fletcher MR, Berny P (2000) Review of the factors affecting the decline of the European brown hare, *Lepus europaeus* (Pallas, 1778) and the use of wildlife incident data to evaluate the significance of paraquat. Agric Ecosyst Environ 79:95–103

Eiberle K, Matter J-F (1982) Ergebnisse einer Streckenanalyse beim Feldhasen [in German with English summary]. Z Jagdwiss 28:178–193

Eskens U, Kugel B, Bensinger S, Bitsch N (1999) Untersuchungen über mögliche Einflussfaktoren auf die Populationsdichte des Feldhasen [in German with English summary]. Z Jagdwiss 45:60–65

Farkas P, Kusza S, Balogh P, Majzinger I (2020) Examination of fertility indicators of European hares (*Lepus europaeus*) in Eastern Hungary. J Anim Plant Sci 30: 634–641

Ferreira MS, Jones MR, Callahan CM, Farelo L, Tolesa Z, Suchentrunk F, Boursot P, Mills LS, Alves PC, Good JM, Melo-Ferreira J (2021) The legacy of recurrent introgression during the radiation of hares. Syst Biol 70:593–607

Ferretti M, Paci G, Porrini S, Galardi L, Bagliacca M (2010) Habitat use and home range traits of resident and relocated hares (*Lepus europaeus*, Pallas). Ital J Anim Sci 9:3

Fickel J, Schmidt A, Putze M, Spittler H, Ludwig A, Streich W, Pitra C (2005) Genetic structure of populations of European brown hare: implications for management. J Wildl Manag 69:760–770

Fischer C, Tagand R (2012) Spatial behaviour and survival of translocated wild brown hares. Anim Biodivers Conserv 35:189–196

Flux JEC (1967a) Reproduction and body weights of the hare *Lepus europaeus* Pallas in New Zealand. NZ J Sci B 10:357–401

Flux JEC (1967b) Hare numbers and diet in an alpine basin in New Zealand. Proc NZ Ecol Soc 14:27–33

Flux JEC (1981a) Reproductive strategies in the genus *Lepus*. In: Myers K, McInnes CD (eds) Proceedings of the world lagomorph conference. University of Guelph, Guelph, pp 155–174

Flux JEC (1981b) Field observations of behaviour in the genus *Lepus*. In: Myers K, McInnes CD (eds) Proceedings of the world lagomorph conference. University of Guelph, Guelph, pp 377–394

Flux JEC (2008) A review of competition between rabbits (*Oryctolagus cuniculus*) and hares (*Lepus europaeus*). In: Alves PC, Ferrand N, Hackländer K (eds) Lagomorph biology: evolution, ecology, and conservation. Springer, Heidelberg, pp 241–249

Flux JEC (2009) Notes on New Zealand mammals 10. Effect of observer position on group size and level of aggression of mating brown hares. NZ J Zool 36:81–82

Flux JEC, Angermann R (1990) The hares and jackrabbits. In: Chapman JA, Flux JEC (eds) Rabbits, hares and pikas. Status survey and conservation action plan. International Union for the Conservation of Nature and Natural Resources, Gland, pp 61–94

Focardi S, De Marinis AM, Rizzotto M, Ucci A (2001) Comparative evaluation of thermal infrared imaging and spotlighting to survey wildlife. Wildl Soc Bull 29: 133–139

Frölich K, Lavazza A (2008) European Brown Hare syndrome. In: Alves PC, Ferrand N, Hackländer K (eds) Lagomorph biology: evolution, ecology, and conservation. Springer, Heidelberg, pp 253–261

Frölich K, Thiede S, Wisser J (2001) Infektionskrankheiten des Feldhasen [in German]. NUA-Seminarber 7:34–46

Frölich K, Wisser J, Schmüser H, Fehlberg U, Neubauer H, Grunow R, Nikolaou K, Priemer J, Thiede S, Streich WJ, Speck S (2003) Epizootiologic and ecologic investigations of European brown hare (*Lepus europaeus*) in selected populations from Schleswig-Holstein, Germany. J Wildl Dis 39:751–761

Frunza G, Simeanu D, Pop C, Lazar R, Boisteanu PC, Stefan M (2019) Contributions on the sensorial, physico-chemical and nutritional characterization of hare meat (*Lepus europaeus* Pallas). Rev Chim 70: 174–180

Frylestam B (1979) Studies on the European hare. XXXV. Structure, size, and dynamics of three European hare populations in Southern Sweden. Acta Theriol 24: 449–464

Frylestam B (1981) Studies on the European hare. XXXVII. Estimating by spotlight the population density of the European hare. Acta Theriol 26:419–427

Frylestam B (1986) Agricultural land use effects on the winter diet of brown hare (*Lepus europaeus* Pallas) in southern Sweden. Mamm Rev 16:157–161

Gavier-Widén D, Mörner T (1991) Epidemiology and diagnosis of the European brown hare syndrome in Scandinavian countries. Rev Sci Tech OIE 10:453–458

Genghini M, Capizzi D (2005) Habitat improvement and effects on brown hare *Lepus europaeus* and roe deer *Capreolus capreolus*: a case study in northern Italy. Wildl Biol 11:319–329

Göritz F, Faßbender M, Broich A, Quest M, Lange A, Blottner S, Gilles M, Lengwinat T, Spittler H, Hildebrandt TB (2001) Untersuchungen zur reproduktiven Fitness lebender weiblicher Feldhasen [in German with English summary]. Z Jagdwiss 47: 92–99

Goszczyniski J, Ryszkowski L, Truszkowski J (1976) The role of the European hare in the diet of predators in cultivated field systems. In: Pielowski Z, Pucek Z (eds) Ecology and management of European hare populations. The proceedings of an international symposium held in Poznań on December 23–24, 1974. Państwowe Wydawnictwo Rolnicze I Leśne, Warszawa, pp 269–286

Goszczyński J, Wasilewski M (1992) Studies on the European hare. 46. Predation of foxes on a hare population in central Poland. Acta Theriol 37:329–338

Grimmberger E, Rudloff K, Kern C (2009) Atlas der Säugetiere Europas, Nordafrikas und Vorderasiens [in German]. NTV Natur und Tier-Verlag, Münster

Gustavsson I (1971) Mitotic and meiotic chromosomes of the variable hare (*Lepus timidus* L.), the common hare (*Lepus europaeus* Pall.) and their hybrids. Hereditas 67:27–34

Gyuranecz M, Erdélyi K, Makrai L, Fodor L, Szépe B, Ráczné Mészáros Á, Dán A, Dencső L, Fassang E, Szeredi L (2011) Brucellosis of the European brown hare (*Lepus europaeus*). J Comp Pathol 145:1–5

Hackländer K (2011) Der Hase im Bergrevier. Der Anblick 2/2011:6–9

Hackländer K, Schai-Braun SC (2018) *Lepus europaeus* Pallas, 1778. European hare. In: Smith AT, Johnston CH, Alves PC, Hackländer K (eds) Lagomorphs: pikas, rabbits, and hares of the world. Johns Hopkins University Press, Baltimore, pp 187–190

Hackländer K, Schai-Braun SC (2019) *Lepus europaeus*. The IUCN red list of threatened species 2019:e. T41280A45187424

Hackländer K, Frisch C, Klansek E, Steineck T, Ruf T (2001) Die Fruchtbarkeit weiblicher Feldhasen (*Lepus europaeus*) aus Revieren mit unterschiedlicher Populationsdichte [in German with English summary]. Z Jagdwiss 47:100–110

Hackländer K, Arnold W, Ruf T (2002a) Postnatal development and thermoregulation in the precocial European hare (*Lepus europaeus*). J Comp Physiol B 172:183–190

Hackländer K, Tataruch F, Ruf T (2002b) The effect of dietary fat content on lactation energetics in the European hare (*Lepus europaeus*). Physiol Biochem Zool 75:19–28

Hackländer K, Miedler S, Beiglböck C, Zenker W (2003) The assessment of female reproduction and gestational age in European hares (*Lepus europaeus*) using ultrasonography. Mamm Biol 68:187–191

Hackländer K, Ferrand N, Alves PC (2008) Overview of lagomorph research: what we have learned and what we still need to do. In: Alves PC, Ferrand N, Hackländer K (eds) Lagomorph biology: evolution, ecology, and conservation. Springer, Heidelberg, pp 381–391

Hackländer K, Zeitlhofer C, Ceulemans T, Suchentrunk F (2011) Continentality affects body condition and size but not yearly reproductive output in female European hares (*Lepus europaeus*). Mamm Biol 76:662–664

Haerer G, Nicolet J, Bacciarini L, Gottstein B, Giacometti M (2001) Causes of mortality, zoonoses and reproductive performance in European brown hares in Switzerland [in German with English summary]. Schweiz Arch Tierheilkd 143:193–201

Hartl GB, Markowski J, Kovács G, Grillitsch M, Willing R (1990) Biochemical variation and differentiation in the brown hare (*Lepus europaeus*) of Central Europe. Z Säugetierkd 55:186–193

Hartl GB, Suchentrunk F, Nadlinger K, Willing R (1993) An integrative analysis of genetic differentiation in the brown hare *Lepus europaeus* based on morphology, allozymes, and mitochondrial DNA. Acta Theriol 38: 33–57

Hauser WM (1921) Osteologische Unterscheidungsmerkmale der schweizerischen Feld- und Alpenhasen [in German]. Vierteljahrschr Naturforsch Ges Zürich 66:213–220

Hediger H (1948) Die Zucht des Feldhasen (*Lepus europaeus* Pallas) in Gefangenschaft [in German]. Physiol Comp Oecol 1:46–62

Heigl F, Stretz CR, Steiner W, Suppan F, Bauer T, Laaha G, Zaller JG (2016) Comparing road-kill datasets from hunters and citizen scientists in a landscape context. Remote Sens 8:832

Hell P, Soviš B (1997) Influence of predators on hare population. In: Slamečka J, Hell P, Jurcík R (eds) Brown hare in the Westslovak lowland. Acta Sc Nat Brno 31(3–4):51–57

Hewson R (1963) Moults and pelages in the brown hare *Lepus europaeus occidentalis* De Winton. Proc Zool Soc Lond 141:677–688

Hewson R (1977) Food selection by brown hares (*Lepus capensis*) on cereal and turnip crops in north-east Scotland. J Appl Ecol 14:779–785

Hewson R, Taylor M (1968) Studies on the European hare. XVI. Movements of European hares in an upland area of Scotland. Acta Theriol 13:31–34

Hewson R, Taylor M (1975) Studies on the European hare. XXXIII. Embryo counts and length of the breeding season in European hares in Scotland from 1960–1972. Acta Theriol 20:247–254

Hirakawa H (2001) Coprophagy in leporids and other mammalian herbivores. Mamm Rev 31:61–80

Holley AJF (1992) Studies on the biology of the brown hare (*Lepus europaeus*) with particular reference to behaviour. Dissertation. University of Durham, UK

Holley AJF (2001) The daily activity period of the brown hare (*Lepus europaeus*). Mamm Biol 66:357–364

Holley AJF, Greenwood PJ (1984) The myth of the mad March hare. Nature 309:549–550

Homolka M (1982) The food of *Lepus europaeus* Pallas in a meadow and woodland complex. Folia Zool 31: 243–253

Homolka M (1985) Spatial activity of hares (*Lepus europaeus*). Folia Zool 34:217–226

Homolka M (1986) Daily activity patterns of the European hare (*Lepus europaeus*). Folia Zool 35:33–42

Homolka M (1987) The diet of brown hare (*Lepus europaeus*) in Central Bohemia. Folia Zool 36:103–110

Horáček V, Uher J (1965) Zur Kenntnis des Geschlechtszyklus und der Superfötation bei dem Feldhasen *Lepus europaeus* Pall [in German]. Zool Listy 14:107–116

Hornek-Gausterer R, Oberacher H, Reinstadler V, Hartmann C, Liebmann B, Lomako I, Scharf S, Posautz A, Kübber-Heiss A (2021) A preliminary study on the detection of potential contaminants in the European brown hare (*Lepus europaeus*) by suspect and microplastics screening. Environ Adv 4:100045

Huber W (1973) Biologie und Ökologie des Feldhasen [in German with English summary]. Beih Z Schweiz Forstver 52:223–237

Hummel S, Meyer L, Hackländer K, Weber D (2017) Activity of potential predators of European hare (*Lepus europaeus*) leverets and ground-nesting birds in wildflower strips. Eur J Wildl Res 63:102

Hušek J, Panek M, Tryjanowski P (2015) Predation risk drives habitat-specific sex ratio in a monomorphic species, the brown hare (*Lepus europaeus*). Ethology 121: 593–600

Jansson G, Pehrson Å (2007) The recent expansion of the brown hare (*Lepus europaeus*) in Sweden with possible implications to the mountain hare (*L. timidus*). Eur J Wildl Res 53:125–130

Jennings N (2017) Hares. RSPB spotlight. Bloomsbury, London

Jennings N, Smith RK, Hackländer K, Harris S, White PCL (2006) Variation in demography, condition and dietary quality of hares *Lepus europaeus* from high-density and low-density populations. Wildl Biol 12: 179–189

Jerzierski W (1973) Environmental conditioning of the space structure and shyness in hares (*Lepus europaeus* Pallas, 1778). Ekol Polska 21:1–12

Jezierski W (1965) Studies on the European hare. VII. Changes in some elements of the structure and size of population. Acta Theriol 10:11–25

Jezierski W (1967) Four cases of 'home instinct' in the European hare. Acta Theriol 12:173–174

Kałuziński J, Pielowski Z (1976) The effect of technical agriculture operations on the hare population. In: Pielowski Z, Pucek Z (eds) Ecology and management of European hare populations. The proceedings of an international symposium held in Poznań on December 23–24, 1974. Państwowe Wydawnictwo Rolnicze I Leśne, Warszawa, pp 205–211

Kamieniarz R, Voigt U, Panek M, Strauss E, Niewęgłowski H (2013) The effect of landscape structure on the distribution of brown hare *Lepus europaeus* in farmlands of Germany and Poland. Acta Theriol 58:39–46

Karp D, Gehr B (2020) Bad hare day: very low survival rate in brown hare leverets. Wildl Biol 2020:wlb.00645

Katona K, Altbäcker V (2002) Diet estimation by faeces analysis: sampling optimization for the European hare. Folia Zool 51:11–15

Kilias H, Ackermann W (2001) Zur Bestandssituation des Feldhasen (*Lepus europaeus* Pallas) in Bayern [in German with English summary]. Z Jagdwiss 47: 111–124

Kittler L (1979) Wildverluste durch den Einsatz landwirtschaftlicher Maschinen nach einer Erhebung aus dem Jagdjahr 1976/77 in Nordrhein-Westfalen [in German with English summary]. Z Jagdwiss 25: 22–32

Klansek E (1996) Biotoprevitalisierung und neue Überlegungen zur jagdlichen Nutzung des Feldhasen [in German]. Schriftenreihe Landesjagdverbandes Bayern eV 2:31–36

Kleinmann JU, Wang M (2017) Modeling individual movement decisions of brown hare (*Lepus europaeus*) as a key concept for realistic spatial behavior and exposure: A population model for landscape-level risk assessment. Environ Toxicol Chem 36:2299–2307

Knauer F, Küchenhoff H, Pilz S (2010) A statistical analysis of the relationship between red fox *Vulpes vulpes* and its prey species (grey partridge *Perdix perdix*, brown hare *Lepus europaeus* and rabbit *Oryctolagus cuniculus*) in Western Germany from 1958 to 1998. Wildl Biol 16:56–65

Koenen F (1956) Der Feldhase. Neue Brehm Bücherei 169 [in German]. A. Ziemsen, Wittenberg Lutherstadt, GDR

Köhler D (2008) Beobachtungen zur Urbanisierung des Feldhasen (*Lepus europaeus*) in Berlin – ein weiteres Beispiel für seine Anpassungsfähigkeit [in German]. Säugetierkd Inf 6:233–255

Kotsakis T, Abbazzi L, Angelone C, Argenti P, Barisone G, Fanfani F, Marcolini F, Masini F (2003) Plio-Pleistocene biogeography of Italian mainland micromammals. Deinsea 10:313–342

Kovács G (1983) Survival pattern in adult European hares. Acta Zool Fennica 174:69–70

Kovács G, Heltay I (1981) Study of a European hare population mosaic in the Hungarian lowland. In: Myers K, McInnes CD (eds) Proceedings of the world

lagomorph conference. University of Guelph, Guelph, pp 508–528
Kronfeld N, Shkolnik A (1996) Adaptation to life in the desert in the brown hare (*Lepus capensis*). J Mammal 77:171–178
Kröning F (1963) Was wiegt ein Göttinger Hase? [in German]. Z Jagdwiss 9:107–109
Kučera O (1991) Contents of heavy metals and milk fat in hare in artificial breeding [in Slovak with English summary]. Folia Venatoria 21:131–146
Kuijper DPJ, van Wieren SE, Bakker JP (2004) Digestive strategies in two sympatrically occurring lagomorphs. J Zool 264:1–8
Kummer J (1970) Beobachtungen bei der Aufzucht und Haltung von Feldhasen (*Lepus europaeus* Pallas) [in German]. Zool Garten 38:138–140
Kunst PJG, van der Wal R, van Wieren S (2001) Studies on the European hare. 57. Home ranges of brown hares in a natural salt marsh: comparisons with agricultural systems. Acta Theriol 43:287–294
Kutzer E, Thiemann G, Grünberg W, Frey H (1976) Beiträge zu einer Feldhasenstudie. 1. "Gesundenuntersuchung" an erlegten Feldhasen aus österreichischen Revieren [in German with English summary]. Z Jagdwiss 22:50–61
Kuznetsov AN, Luchkina OS, Panyutina AA, Kryukova NV (2017) Observations on escape runs in wild European hare as a basis for the mechanical concept of extreme cornering with special inference of a role of the peculiar subclavian muscle. Mamm Biol 84:61–72
Lado S, Farelo L, Forest V, Acevedo P, Dalén L, Melo-Ferreira J (2018) Post-glacial range revolutions in South European hares (*Lepus* spp.): insights from ancient DNA and ecological niche modelling. J Biogeogr 45:2609–2618
Lado S, Alves PC, Islam MZ, Brito JC, Melo-Ferreira J (2019) The evolutionary history of the Cape hare (*Lepus capensis* sensu lato): insights for systematics and biogeography. Heredity 123:634–646
Lamarque F, Barrat J, Moutou F (1996) Principal diagnosis for determining causes of mortality in the European brown hare (*Lepus europaeus*) found dead in France between 1986 and 1994. Gibier Faune Sauvage, Game Wildl 13:53–72
Landois H (1885) Hasenzucht in enger Gefangenschaft [in German]. Zool Garten 26:359–361
Langbein J, Hutchings MR, Harris S, Stoate C, Tapper SC, Wray S (1999) Techniques for assessing the abundance of brown hares *Lepus europaeus*. Mamm Rev 29: 93–116
Le Gall-Reculé G, Lemaitre E, Bertagnoli S, Hubert C, Top S, Decors A, Marchandeau S, Guitton J-S (2017) Large-scale lagovirus disease outbreaks in European brown hares (*Lepus europaeus*) in France caused by RHDV2 strains spatially shared with rabbits (*Oryctolagus cuniculus*). Vet Res 48:70
Levänen F, Kunnasranta M, Pohjoismäki J (2018a) Mitochondrial DNA introgression at the northern edge of the brown hare (*Lepus europaeus*) range. Ann Zool Fennici 55:15–24
Levänen R, Thulin C-G, Spong G, Pohjoismäki JLO (2018b) Widespread introgression of mountain hare genes into Fennoscandian brown hare populations. PLoS One 13:e0191790
Lewandowski K, Nowakowski JJ (1993) Studies on the European hare. 49. Spatial distribution of brown hare *Lepus europaeus* populations in habitats of various types of agriculture. Acta Theriol 38:435–442
Lhuillery C, Martinet L, Demarne Y, Lecourtier MJ (1984) Food intake in captive leverets before weaning and the composition of the milk in the brown hare (*Lepus europaeus*). Comp Biochem Physiol 78A:73–76
Lincoln GA (1974) Reproduction and "March madness" in the brown hare, *Lepus europaeus*. J Zool 174:1–14
Lincoln GA, MacKinnon PCB (1976) A study of seasonally delayed puberty in the male hare, *Lepus europaeus*. J Reprod Fert 46:123–128
Lindlöf B (1978) Aggressive dominance rank in relation to feeding by European hare. Viltrevy 10:143–158
Lindstedt SL, Schaeffer PJ (2002) Use of allometry in predicting anatomical and physiological parameters of mammals. Lab Anim 36:1–19
Linšak Ž, Tomić Linšak D, Špirić Z, Srebočan E, Glad M, Milin Č (2013) Effects of mercury on glutathione and glutathione-dependent enzymes in hares (*Lepus europaeus* Pallas). J Environ Sci Health A 48: 1325–1332
Lloyd HG (1968) Observations on breeding in the brown hare (*Lepus europaeus*) during the first pregnancy of the season. J Zool 156:521–528
Lundström-Gilliéron C, Schlaepfer R (2003) Hare abundance as an indicator for urbanization and intensification of agriculture in Western Europe. Ecol Model 168: 283–301
Lush L, Ward AI, Wheeler P (2014) Opposing effects of agricultural intensification on two ecologically similar species. Agri Ecosyst Environ 192:61–66
Mamuris Z, Sfougaris AI, Stamatis C (2001) Genetic structure of Greek brown hare (*Lepus europaeus*) populations as revealed by mtDNA RFLP-PCR analysis: implications for conserving genetic diversity. Biol Conserv 101:187–196
Mamuris Z, Sfougaris AI, Stamatis C, Suchentrunk F (2002) Assessment of genetic structure of Greek brown hare (*Lepus europaeus*) populations based on variation in random amplified polymorphic DNA (RAPD). Biochem Gen 40:323–338
Marboutin E (1997) A note on home range size in the European hare (*Lepus europaeus*). Gibier Faune Sauvage, Game Wildl 14:349–357
Marboutin E, Aebischer NJ (1996) Does harvesting arable crops influence the behaviour of the European hare *Lepus europaeus*. Wildl Biol 2:83–91
Marboutin E, Hansen K (1998) Survival rates in a non harvested brown hare population. J Wildl Manag 62: 772–779

Marboutin E, Péroux R (1995) Survival pattern of European hare in a decreasing population. J Appl Ecol 32:809–816

Marboutin E, Péroux R (1999) Some aspects of the spatial distribution of hares (*Lepus europaeus*) at night. Gibier Faune Sauvage, Game Wildl 16:143–156

Marboutin E, Benmergui M, Pradel R, Fiechter A (1990) Survival patterns in wild and captive-reared leverets (*Lepus europaeus* Pallas) determined by telemetry. Gibier Faune Sauvage 7:325–342

Marboutin E, Bray Y, Péroux R, Mauvy B, Lartiges A (2003) Population dynamics in European hare: breeding parameters and sustainable harvest rates. J Appl Ecol 40:580–591

Marques JP, Ferreira MS, Farelo L, Callahan CM, Hackländer K, Jenny H, Montgomery WI, Reid N, Good JM, Alves PC, Melo-Ferreira J (2017) Mountain hare transcriptome and diagnostic markers as resources to monitor hybridization with European hares. Sci Data 4:170178

Martinet L (1976) Seasonal reproduction cycles in the European hare, *Lepus europaeus*, raised in captivity. Role of photoperiodicity. In: Pielowski Z, Pucek Z (eds) Ecology and management of European hare populations. The proceedings of an international symposium held in Poznań on December 23–24, 1974. Państwowe Wydawnictwo Rolnicze I Leśne, Warszawa, pp 55–57

Martinet L (1980) Oestrous behaviour, follicular growth and ovulation during pregnancy in the hare (*Lepus europaeus*). J Reprod Fert 59:441–445

Martinet L, Demarne Y (1984) Nursing behaviour and lactation in the brown hare (*Lepus europaeus*) raised in captivity. Acta Zool Fennica 171:187–190

Martinet L, Raynaud F (1972) Méchanisme possible de la superfetation chez la hase [in French]. C R Acad Sci Paris 274:2683–2686

Martinet L, Legouis J-J, Moret B (1970) Quelques observations sur la reproduction du lièvre européen (*Lepus europaeus* Pallas) en captivité [in French with English summary]. Ann Biol Anim Bioch Biophys 10:195–202

Martini H (1983) Die Ursachen auffallend hoher Hasenstrecken des Saarlandes in den Jagdjahren mit sehr niedrigen tollwutbedingten Fuchsdichten 1968/69 bis 1974/75 [in German with English summary]. Z Jagdwiss 29:1–12

Massányi P, Lukač N, Hluchý N, Slamečka J, Jurčik R (1997) Histological and morphometric study of reproductive organs. In: Slamečka J, Hell P, Jurcík R (eds) Brown hare in the Westslovak lowland. Acta Sc Nat Brno 31(3–4):60–66

Massányi P, Slamečka J, Lukác N, Jurčik R (2000) Seasonal variations in the morphometric analysis of the ovary and uterus and in progesterone and 17β-oestradiol production in the brown hare (*Lepus europaeus*). J Anim Feed Sci 9:697–708

Massányi P, Tataruch F, Slamecka J, Toman R, Jurik R (2003) Accumulation of lead, cadmium, and mercury in liver and kidney of the brown hare (*Lepus europaeus*) in relation to the season, age, and sex in the West Slovakian Lowland. J Environ Sci Health A: Toxic Hazard Subst Environ Eng 38:1299–1309

Masseti M, De Marinis AM (2008) Prehistoric and historic artificial dispersal of lagomorphs on the Mediterranean islands. In: Alves PC, Ferrand N, Hackländer K (eds) Lagomorph biology: evolution, ecology, and conservation. Springer, Heidelberg, pp 13–25

Matuszewski G (1966) Studies on the European hare. XIII. Food preference in relation to trees' branches experimentally placed on the ground. Acta Theriol 11: 485–496

Matuszewski G (1981) Circadian activity of European hares in spring, on the Kampinos forest border. In: Myers K, McInnes CD (eds) Proceedings of the world lagomorph conference. University of Guelph, Guelph, pp 357–365

Mayer M, Sunde P (2020a) The role of maize cultivation on European hare abundance. Agric Ecosyst Environ 295:106909

Mayer M, Sunde P (2020b) Colonization and habitat selection of a declining farmland species in urban areas. Urban Ecosyst 23:543–554

Mayer M, Ullmann W, Sunde P, Fischer C, Blaum N (2018) Habitat selection by the European hare in arable landscapes: the importance of small-scale habitat structure for conservation. Ecol Evol 8:11619–11633

Mayer M, Ullmann W, Heinrich R, Fischer C, Blaum N, Sunde P (2019) Seasonal effects of habitat structure and weather on the habitat selection and home range size of a mammal in agricultural landscapes. Landsc Ecol 34: 2279–2294

Mayer M, Duan X, Sunde P, Topping CJ (2020a) European hares do not avoid newly pesticide-sprayed fields: overspray as unnoticed pathway of pesticide exposure. Sci Total Environ 715:136977

Mayer M, Fog Bjerre DH, Sunde P (2020b) Better safe than sorry: the response to a simulated predator and unfamiliar scent by the European hare. Ethology 126: 704–715

McLaren GW, Hutchings MR, Harris S (1997) Why are brown hares (*Lepus europaeus*) rare in pastoral landscapes in Great Britain? Gibier Faune Sauvage, Game Wildl 14:335–348

Meichtry-Stier KS, Jenny M, Zellweger-Fischer J, Birrer S (2014) Impact of landscape improvement by agri-environment scheme options on densities of characteristic farmland bird species and brown hare (*Lepus europaeus*). Agric Ecosyst Environ 189:101–109

Melo-Ferreira J, Alves PC, Freitas H, Ferrand N, Boursot P (2009) The genomic legacy from the extinct *Lepus timidus* to the three hare species of Iberia: contrast between mtDNA, sex chromosomes and autosomes. Mol Ecol 18:2643–2658

Melo-Ferreira J, Boursot P, Carneiro M, Esteves PJ, Farelo L, Alves PC (2012) Recurrent introgression of mitochondrial DNA among hares (*Lepus* spp.) revealed by species-tree inference and coalescent simulations. Syst Biol 61:367–381

Melo-Ferreira J, Farelo L, Freitas H, Suchentrunk F, Boursot P, Alves PC (2014) Home-loving boreal hare mitochondria survived several invasions in Iberia: the relative roles of recurrent hybridisation and allele surfing. Heredity 112:265–273

Mengoni C, Mucci N, Randi E (2015) Genetic diversity and no evidences of recent hybridization in the endemic Italian hare (*Lepus corsicanus*). Cons Gen 16:477–489

Minoudi S, Papapetridis I, Karaiskou N, Chatzinikos E, Triantaphyllidis C, Abatzopoulos TJ (2018) Genetic analyses of brown hare (*Lepus europaeus*) support limited migration and translocation of Greek populations. PLoS One 13:e0206327

Misiorowska M (2013) Annual and seasonal home range and distances of movements of released hares (*Lepus europaeus* Pallas, 1778) in Central Poland. Folia Zool 62:133–142

Misiorowska M, Wasilewski M (2012) Survival and causes of death among released brown hares (*Lepus europaeus* Pallas, 1778) in Central Poland. Acta Theriol 57: 305–312

Misiorowska M, Ludwisiak Ł, Nasiadka P (2014) Population parameters of brown hare (*Lepus europaeus* L.) in regions of the species highest density in Poland [in Polish with English summary]. Sylwan 158: 901–910

Modesto P, Cava PL, Peletto S, Colussi S, Vidus Rosin A, Meriggi A, Acutis PL (2011) Evaluation of genetic variability in brown hare populations from two protected areas in northern Italy. Hystrix 22:257–268

Möller D (1971) Beitrag zur Reproduktion des Feldhasen (*Lepus europaeus* Pall.) in der Deutschen Demokratischen Republik [in German with English summary]. Tag-Ber dt Akad Landwirtsch-Wiss Berlin 113:191–202

Möller D (1975) Zum Altersaufbau der Hasenpopulationen in der Deutschen Demokratischen Republik [in German with English summary]. Beitr Jagd Wildforsch 9:315–325

Möller D (1976) Die Fertilität der Feldhasenpopulationen [in German]. In: Pielowski Z, Pucek Z (eds) Ecology and management of European hare populations. The proceedings of an international symposium held in Poznań on December 23–24, 1974. Państwowe Wydawnictwo Rolnicze I Leśne, Warszawa, pp 69–74

Monaghan P, Metcalfe NB (1985) Group foraging in wild brown hares: effects of resource distribution and social status. Anim Behav 33:993–999

Müller-Using D (1974) Diezels Niederjagd [in German]. Paul Parey, Hamburg

Myers K, McInnes CD (eds) (1981) Proceedings of the world lagomorph conference. University of Guelph, Guelph

Mykytowycz R (1966) Observations on odoriferous and other glands in the Australian wild rabbit, *Oryctolagus cuniculus* (L.) and the hare, *Lepus europaeus* P. I. The anal gland. CSIRO Wildl Res 11:11–29

Myrcha A (1968) Studies on the European hare. XXI. Winter food intake in European hare (*Lepus europaeus* Pallas, 1778) in experimental conditions. Acta Theriol 28:453–459

Naldi L, Greco I, Feretti M, Zaccaroni M (2020) Density estimates and habitat preferences of the European hare (*Lepus europaeus*) on mountainous areas in Italy. Mammal Study 45:123–131

Naumova EI, Zharova GK, Chistova TY, Kuznetsova TA (2015) The effect of coprophagy on the size of plant fibers in the digestive tract of hares *Lepus europaeus* and *L. timidus* (Lagomorpha, Leporidae). Biol Bull 42: 426–431

Neumann F, Schai-Braun S, Weber D, Amrhein V (2011) European hares select resting places for providing cover. Hystrix 22:291–299

Nösel H, Ahrens M (1996) Besatzsituation des Feldhasen im Thüringer Becken – Zwischenbilanz des Thüringer Niederwildprojektes [in German with English summary]. Beitr Jagd Wildforsch 21:237–244

Noszczyk-Nowak A, Nicpoń J, Nowak M, Slawuta P (2009) Preliminary reference values for electrocardiography, echocardiography and myocardial morphometry in the European brown hare (*Lepus europaeus*). Acta Vet Scand 51:6

Notini G (1941) Om harens biologi [in Swedish]. Svenska Jägareförbundet Meddelande 4:1–192

Nováková E, Dušek J, Šolc J (1976) Die Wachstumsgeschwindigkeit bei Junghasen in der Kulturlandschaft [in German]. In: Pielowski Z, Pucek Z (eds) Ecology and management of European hare populations. The proceedings of an international symposium held in Poznań on December 23–24, 1974. Państwowe Wydawnictwo Rolnicze I Leśne, Warszawa, pp 85–92

Orthwein L (1984a) Umweltveränderungen in Agrobiotopen Mitteleuropas und ihre Auswirkungen auf die Populationsdichten des Niederwildes. Teil 1 [in German]. Z Angew Zool 71:157–192

Orthwein L (1984b) Umweltveränderungen in Agrobiotopen Mitteleuropas und ihre Auswirkungen auf die Populationsdichten des Niederwildes. Teil 2 [in German with English summary]. Z Angew Zool 71:313–334

Paci G, Bagliacca M, Lavazza A (2006) Stress evaluation in hares (*Lepus europaeus* Pallas) captured for traslocation. Ital J Anim Sci 5:175–181

Paci F, Lavazza A, Ferretti M, Santilli F, Bagliacca M (2011) Relationship between anti-European Brown Hare syndrome serological titers and brown hare (*Lepus europaeus* Pallas) densities. Int J Zool 2011:436193

Packer JP (1989) Annual patterns in reproduction and perirenal fat of hares (*Lepus europaeus*) in sub-alpine Canterbury, New Zealand. J Zool 217:9–21

Pajerský A, Svarc R, Medved'ová M (1992) Loziskové zmeny v pl'úcach zajaca pol'ného vyvolané parazitom *Protostrongylus commutatus* (Diesing, 1851) [in Slovak with English summary]. Veterinárí medicína 37:249–255

Panek M (2013) Long-term changes in the feeding pattern of red foxes *Vulpes vulpes* and their predation on brown hares *Lepus europaeus* in western Poland. Eur J Wildl Res 59:581–586

Panek M (2018) Habitat factors associated with the decline in brown hare abundance in Poland in the beginning of the 21st century. Ecol Indic 85:915–920

Panek M, Kamieniarz R (1999) Studies on the European hare 54. Relationship between density of brown hare *Lepus europaeus* and landscape structure in Poland in the years 1981–1995. Acta Theriol 44:67–75

Pavliska PL, Riegert J, Grill S, Šálek M (2018) The effect of landscape heterogeneity on population density and habitat preferences of the European hare (*Lepus europaeus*) in contrasting farmlands. Mamm Biol 88: 8–15

Pegel M (1986) Der Feldhase (*Lepus europaeus* Pallas) im Beziehungsgefüge seiner Um- und Mitweltfaktoren [in German]. Schriften des Arbeitskreises Wildbiologie und Jagdwissenschaft an der Justus-Liebig-Universität Gießen. F. Enke Verlag, Stuttgart

Pépin D (1977) Phase finale du cycle de reproduction du lièvre, *Lepus europaeus* [in French with English summary]. Mammalia 41:221–230

Pépin D (1986) Spring density and daytime distribution of the European hare in relation to habitat in an open field agrosystem. Z Säugetierkd 51:79–86

Pépin D (1987) Dynamics of a heavily exploited population of European hare in a large-scale farming area. J Appl Ecol 24:725–734

Pépin D (1989) Variation in survival of brown hare (*Lepus europaeus*) leverets from different farmland areas in the Paris basin. J Appl Ecol 23:13–23

Pépin D, Angibault JM (2007) Selection of resting sites by the European hare as related to habitat characteristics during agricultural changes. Eur J Wildl Res 53: 183–189

Pépin D, Cargnelutti B (1994) Studies on the European hare. 51. Individual variations of daily activity patterns in radiotracked European hares during winter. Acta Theriol 39:399–409

Petrak M (1990) Einflüsse landwirtschaftlicher Nutzung auf die Population des Feldhasen (*Lepus europaeus*) [in German with English summary]. Verh Ges Ökol 19: 88–97

Petrov P (1976) Über die Faktoren die den realen Zuwachs des Hasen bestimmen [in German]. In: Pielowski Z, Pucek Z (eds) Ecology and management of European hare populations. The proceedings of an international symposium held in Poznań on December 23–24, 1974. Państwowe Wydawnictwo Rolnicze I Leśne, Warszawa, pp 119–126

Petrovan SO, Ward AI, Wheeler PM (2013) Habitat selection guiding agri-environmental schemes for a farmland specialist, the brown hare. Anim Conserv 16: 344–352

Petrovan SO, Dixie J, Yapp E, Wheeler PM (2017) Bioenergy crops and farmland biodiversity: benefits and limitations are scale-dependent for a declining mammal, the brown hare. Eur J Wildl Res 63:49

Petrović Z, Teodorović V, Djurić S, Milićević D, Vranić D, Lukić M (2014) Cadmium and mercury accumulation in European hare (*Lepus europaeus*): age-dependent relationships in renal and hepatic tissue. Environ Sci Pollut Res 21:14058–14068

Petrusewicz K (1970) Studies on the European hare. XXIV. Dynamics and production of the hare population in Poland. Acta Theriol 15:413–445

Pielowski Z (1966) Forschung über den Feldhasen XII. Die Raumstruktur der Population [in German with English summary]. Acta Theriol 11:449–484

Pielowski Z (1969) Studies on the European hare. XXII. Sex ratio and weight of hares in Poland. Acta Theriol 14:119–131

Pielowski Z (1971a) Studies on the European hare. XXV. The individual growth curve of the hare. Acta Theriol 16:79–88

Pielowski Z (1971b) Studies on the European hare. XXVI. Length of life of the hare. Acta Theriol 16:89–94

Pielowski Z (1972) Studies on the European hare. XXIX. Home range and degree of residence of the European hare. Acta Theriol 9:93–103

Pielowski Z (1976a) Number of young born and dynamics of the European hare population. In: Pielowski Z, Pucek Z (eds) Ecology and management of European hare populations. The proceedings of an international Symposium held in Poznań on December 23–24, 1974. Państwowe Wydawnictwo Rolnicze I Leśne, Warszawa, pp 75–78

Pielowski Z (1976b) The role of foxes in the reduction of the European hare population. In: Pielowski Z, Pucek Z (eds) Ecology and management of European hare populations. The proceedings of an international Symposium held in Poznań on December 23–24, 1974. Państwowe Wydawnictwo Rolnicze I Leśne, Warszawa, pp 135–148

Pielowski Z (1981) Yearly balance of European hare populations. In: Myers K, McInnes CD (eds) Proceedings of the world lagomorph conference. University of Guelph, Guelph, pp 536–540

Pielowski Z (1993) Die Greifvögel. Systematik, Artenbeschreibung, Biologie, Ökologie, Schutz, Falknerei [in German]. Verlag Neumann-Neudamm, Morschen/Heina

Pielowski Z, Pucek Z (eds) (1976) Ecology and management of European hare populations. The proceedings of an international symposium held in Poznań on December 23–24, 1974. Państwowe Wydawnictwo Rolnicze I Leśne, Warszawa

Pielowski Z, Raczyński J (1976) Ecological conditions and rational management of hare populations. In: Pielowski Z, Pucek Z (eds) Ecology and management of European hare populations. The proceedings of an international Symposium held in Poznań on December 23–24, 1974. Państwowe Wydawnictwo Rolnicze I Leśne, Warszawa, pp 269–286

Pierpaoli M, Riga F, Trocchi V, Randi E (1999) Species distinction and evolutionary relationships of the Italian hare (*Lepus corsicanus*) as described by mitochondrial DNA sequencing. Mol Ecol 8:1805–1817

Pilarska J (1969) Individual growth curve and food consumption by European hare (*Lepus europaeus* Pallas 1778) in laboratory conditions. Bull Acad Polonaise Sci 17:299–305

Ponjiger I, Ristić Z, Marković V, Matejević M, Kovačević M (2019) The dynamics of red fox (*Vulpes vulpes*) and brown hare (*Lepus europaeus*) populations in the Vojvodina region (Serbia) in relation to rabies vaccination. Vet Arhiv 896:839–850

Popescu F, Hackländer K, Arnold W, Ruf T (2011) Effects of season and reproductive state on lipid intake and fatty acid composition of gastrointestinal tract contents in the European hare. J Comp Physiol B 181:681–689

Posautz A, Loncaric I, Lundin M, Hoffmann D, Lavazza A, Kelemen Z, Beiglböck C, Walzer C, Kübber-Heiss A (2015) Health screening of free-ranging European brown hares (*Lepus europaeus*) on the German North-Sea island Pellworm. Acta Vet Scand 57:43

Raczyński J (1964) Studies on the European hare. V. Reproduction. Acta Theriol 9:305–352

Rehnus M, Bollmann K (2016) Non-invasive genetic population density estimation of mountain hares (*Lepus timidus*) in the Alps: Systematic or opportunistic sampling? Eur J Wildl Res 62:737–747

Reichholf J (1981) Der Bestandstrend beim Feldhasen (*Lepus europaeus* Pallas 1778) und die jahreszeitliche Verteilung der Verluste im Straßenverkehr [in German with English summary]. Z Jagdwiss 27:240–246

Reichlin T, Klansek E, Hackländer K (2006) Diet selection by hares (*Lepus europaeus*) in arable land and its implications for habitat management. Eur J Wildl Res 52: 109–118

Reitz F, Leonard Y (1993) La survie estivale et automnale du lièvre d'Europe (*Lepus europaeus*) dans une plaine de grande culture du Bassin parisien [in French with English summary]. Gibier Faune Sauvage 10:167–172

Reitz F, Leonard Y (1994) Studies on the European hare. 50. Characteristics of European hare *Lepus europaeus* use of space in a French agricultural region of intensive farming. Acta Theriol 39:143–157

Reynolds JC, Tapper SC (1995) Predation by foxes *Vulpes vulpes* on brown hares *Lepus europaeus* in central southern England, and its potential impact on annual population growth. Wildl Biol 1:145–158

Reynolds JC, Stoate C, Brockless MH, Aebischer NJ, Tapper SC (2010) The consequences of predator control for brown hares (*Lepus europaeus*) on UK farmland. Eur J Wildl Res 56:541–549

Rieck W (1953) Der Feldhase [in German]. Merkblätter des Niederwildausschusses des DJV. Nr. 4. F.C. Mayer, München

Rieck W (1956a) Untersuchungen über die Vermehrung des Feldhasen [in German with English summary]. Z Jagdwiss 2:49–90

Rieck W (1956b) Der Junghasenanteil auf den Strecken 1953–1954 [in German with English summary]. Z Jagdwiss 2:160–164

Rieck W (1966) Zuwachs und Abschuß beim Hasenbesatz des Reviers Biebesheim a. Rh. [in German]. Beitr Jagd Wildforsch 5:95–98

Rieck W (1967) Ergebnis der Untersuchung von Feldhasenstrecken [in German]. In: Isaković I (ed) Les rapports du VII congrès des Biologistes du Gibier, Beograd-Ljubljana, Septembre 1965, pp 383–388

Rijks JM, Kik M, Koene MG, Engelsma MY, van Tulden P, Montizaan MG, Oomen T, Spierenburg MA, Ijzer J, van der Giessen JW, Gröne A, Roest HJ (2013) Tularaemia in a brown hare (*Lepus europaeus*) in 2013: first case in the Netherlands in 60 years. Euro Surveill 18:20655

Ristić Z, Ponjiger I, Matejević M, Kovačević M, Ristić N, Marković V (2021) Effects of factors associated with the decline of brown hare abundance in the Vojvodina region (Serbia). Hystrix 32:67–71

Rödel HG, Dekker JJA (2012) Influence of weather factors on population dynamics of two lagomorph species based on hunting bag records. Eur J Wildl Res 58: 923–932

Rödel HG, Völkl W, Kilias H (2004) Winter browsing of brown hares: evidence for diet breadth expansion. Mamm Biol 69:410–419

Roedenbeck IA, Voser P (2008) Effects of roads on spatial distribution, abundance and mortality of brown hares (*Lepus europaeus*) in Switzerland. Eur J Wildl Res 54: 425–437

Roellig K, Goeritz F, Hildebrandt TB (2010a) Ultrasonographic characterisation of prenatal development in European brown hares (*Lepus europaeus* Pallas, 1778): an evolutionary approach. Reprod Fert Dev 22:448–458

Roellig K, Goeritz F, Fickel J, Hermes R, Hofer H, Hildebrandt TB (2010b) Superconception in mammalian pregnancy can be detected and increases reproductive output per breeding season. Nature Comm 1:78

Roellig K, Menzies BR, Hildebrandt TB, Goeritz F (2011) The concept of superfetation: a critical review on a 'myth' in mammalian reproduction. Biol Rev 86:77–95

Ruf T, Valencak T, Tataruch F, Arnold W (2006) Running speed in mammals increases with muscle n-6 polyunsaturated fatty acid content. PLoS One 1:e65

Rühe F (1999) Effect of stand structures in arable crops on brown hare (*Lepus europaeus*) distribution. Giber Faune Sauvage, Game Wildl 16:317–337

Rühe F, Hohmann U (2004) Seasonal locomotion and home-range characteristics of European hares (*Lepus europaeus*) in an arable region in central Germany. Eur J Wildl Res 50:101–111

Rühe F, Fischbeck I, Rieger A (2000) Zum Einfluss von Habitatmerkmalen auf die Populationsdichte von Feldhasen (*Lepus europaeus* Pallas) in Agrargebieten Norddeutschlands [in German with English summary]. Beitr Jagd Wildforsch 29:333–342

Ruiz-Fons F, Ferroglio E, Gortázar C (2013) *Leishmania infantum* in free-ranging hares, Spain, 2004-2010. Eur Secur 18:20541

Sackmann H-J (1977) Erfahrungen bei der Haltung von Feldhasen (*Lepus europaeus* Pall.) [in German with English summary]. Beitr Jagd Wildforsch 10:267–276
Salvioli M, Pasquali S, Lavazza A, Zanoni M, Guberti V, Chiari M, Gilioli G (2017) EBHS in European brown hares (*Lepus europaeus*): disease dynamics and control. Hystrix 28:202–207
Santilli F, Galardi L (2016) Effect of habitat structure and type of farming on European hare (*Lepus europaeus*) abundance. Hystrix 27:120–122
Santilli F, Paci G, Bagliacca M (2014) Winter habitat selection by the European hare (*Lepus europaeus*) during feeding activity in a farmland area of southern Tuscany (Italy). Hystrix 25:51–53
Schäfers G (1996) Die Jagdstreckenentwicklung des Feldhasen (*Lepus europaeus* Pallas) von 1959 bis 1993 in der alten Bundesrepublik Deutschland in Abhängigkeit von der Landwirtschaftsstruktur [in German with English summary]. Beitr Jagd Wildforsch 21:215–228
Schai-Braun SC, Hackländer K (2014) Home range use by the European hare (*Lepus europaeus*) in a structurally diverse agricultural landscape analysed at a fine temporal scale. Acta Theriol 59:277–287
Schai-Braun SC, Hackländer K (2016) Family Leporidae (Hares and rabbits). In: Wilson DE, Lacher TE Jr, Mittermeier RA (eds) Handbook of the mammals of the world 6. Lynx Edicions, Barcelona, pp 62–149
Schai-Braun SC, Rödel HG, Hackländer K (2012) The influence on daylight regime on diurnal locomotor activity patterns of the European hare (*Lepus europaeus*) during summer. Mamm Biol 77:434–440
Schai-Braun SC, Weber D, Hackländer K (2013) Spring and autumn habitat preferences of active European hares (*Lepus europaeus*) in an agricultural area with low hare density. Eur J Wildl Res 59:387–397
Schai-Braun SC, Peneder S, Frey-Roos F, Hackländer K (2014) The influence of cereal harvest on the home-range use of the European hare (*Lepus europaeus*). Mammalia 78:497–506
Schai-Braun SC, Reichlin TS, Ruf T, Klansek E, Tataruch F, Arnold W, Hackländer K (2015) The European hare (*Lepus europaeus*): a picky herbivore searching for plant parts rich in fat. PLoS One 10: e0134278
Schai-Braun SC, Kowalczyk C, Klansek E, Hackländer K (2019) Estimating sustainable harvest rates for European hare (*Lepus europaeus*) populations. Sustainability 11:2837
Schai-Braun SC, Ruf T, Klansek E, Arnold W, Hackländer K (2020) Positive effects of set-asides on European hare (*Lepus europaeus*) populations: leverets benefit from an enhanced survival rate. Biol Conserv 244:108518
Schai-Braun SC, Jenny H, Ruf T, Hackländer K (2021a) Temperature increase and frost decrease driving upslope elevational range shifts in Alpine grouse and hares. Glob Change Biol 27:6602–6614
Schai-Braun SC, Steiger P, Ruf T, Arnold W, Hackländer K (2021b) Maternal effects on reproduction in the precocial European hare (*Lepus europaeus*). PLoS One 16: e0247174
Schenker L, Bollmann K, Rehnus M, Brodbeck S, Gugerli F (2020) Hare's affairs: Lessons learnt from a noninvasive genetic monitoring for tracking mountain hare individuals. Ecol Evol 10:10150–10166
Schmidt NM, Asferg T, Forchhammer MC (2004) Long-term patterns in European brown hare population dynamics in Denmark: effects of agriculture, predation and climate. BMC Ecol 4:15
Schneider E (1976) Derzeitige Kenntnisse über das Paarungsverhalten des Feldhasen [in German with English summary]. In: Pielowski Z, Pucek Z (eds) Ecology and management of European hare populations. The proceedings of an international Symposium held in Poznań on December 23–24, 1974. Państwowe Wydawnictwo Rolnicze I Leśne, Warszawa, pp 41–57
Schneider E (1977) Verhaltensstudie am Feldhasen (*Lepus europaeus* Pallas) [in German with English summary]. Beitr Jagd Wildforsch 10:277–288
Schneider E (1978) Der Feldhase: Biologie, Verhalten, Hege und Jagd [in German]. BLV Verlagsgesellschaft, München
Schneider E (1981) Studies on the social behaviour of the brown hare. In: Myers K, McInnes CD (eds) Proceedings of the world lagomorph conference. University of Guelph, Guelph, pp 340–348
Schrama M, Kuijper DPJ, Veeneklaas RM, Bakker JP (2015) Long-term decline in a salt marsh hare population largely driven by bottom-up factors. Ecoscience 22:71–82
Schröder J, Soveri T, Suomalainen HA, Lindberg L-A, van der Loo W (1987) Hybrids between *Lepus timidus* and *Lepus europaeus* are rare although fertile. Hereditas 107:185–189
Schröpfer R, Nyenhuis H (1982) Die Bedeutung der Landschaftsstruktur für die Populationsdichte des Feldhasen (*Lepus europaeus* Pallas 1778) [in German with English summary]. Z Jagdwiss 28:213–231
Schwenk S (1985) Österreichische Jagdstatistiken von 1850–1936 [in German]. Homo Venator 13:1–203
Šebová K (1991) Fecundity and prenatal development of hare in the West Slovakian region [in Slovak with English summary]. Folia Venatoria 21:113–122
Sedlák K, Literák I, Faldyna M, Toman M, Benák J (2000) Fatal toxoplasmosis in brown hares (*Lepus europaeus*): possible reasons of their high susceptibility to the infection. Vet Parasitol 93:13–28
Semizorová I (1982) Reproduction indices of hare game in present conditions [in Slovak with English summary]. Folia Venatoria 12:117–125
Severtsov AS, Rosenzweig ML, Shubkina AV (2017) Predators detect the welfare of their potential prey and cull those that are poorly. Evol Ecol Res 18(5):555–569
Siivonen L (1956) The correlation between the fluctuations of partridge and European hare populations and the

climatic conditions of winters in South-West Finland during the last thirty years. Riistatiet Julk – Papers Game Res 17:1–30

Škrivanko M, Hadžiosmanović M, Cvrtila Ž, Zdolec N, Filipović I, Kozačinski L, Florijančić T, Bošković I (2008) The hygiene and quality of hare meat (*Lepus europaeus* Pallas) from Eastern Croatia. Arch Leb 59: 180–184

Slamečka J, Šebová K (1991) Reproductive parameters of brown hare kept in captivity. Transactions of the XXth Congress of the International Union of Game Biologists, Gödöllő. Hungary August 21–26:618–624

Slamečka J, Jurčík R, Hell P, Minárová Z (1997) Weight of inner organs. In: Slamečka J, Hell P, Jurcík R (eds) Brown hare in the Westslovak lowland. Acta Sc Nat Brno 31(3–4):11–12

Slamečka J, Palanská O, Jurčik R, Ondrejička R, Hell P, Mojto J (1998) Meat quality of brown hares (*Lepus europaeus* Pall.) [in German with English summary]. Fleischwirtschaft 78:824–826

Slamečka J, Massányi P, Hell P, Jurčik R, Lukáč N, Sirotkin A (2001) Morphometry and secretory activity of reproductive organs in brown hare *Lepus europaeus*. Folia Zool 50:173–183

Slamečka J, Capcarova M, Jurčik R, Kolesarova A, Ondruska L, Gasparik J, Lukáč N, Mertin D (2012) The occurrence and dynamics of polychlorinated hydrocarbons in brown hare (*Lepus europaeus)* in south-western Slovakia. J Environ Sci Health A 47: 1217–1223

Sliwinski K, Ronnenberg K, Jung K, Strauß E, Siebert U (2019) Habitat requirements of the European brown hare (*Lepus europaeus* Pallas 1778) in an intensively used agriculture region (Lower Saxony, Germany). BMC Ecol 19:31

Smith RK, Jennings NV, Robinson A, Harris S (2004) Conservation of European hares *Lepus europaeus* in Britain: is increasing habitat heterogeneity in farmland the answer? J Appl Ecol 41:1092–1102

Smith RK, Jennings NV, Harris S (2005a) A quantitative analysis of the abundance and demography of European hares *Lepus europaeus* in relation to habitat type, intensity of agriculture and climate. Mamm Rev 35:1–24

Smith RK, Jennings NV, Tataruch F, Hackländer K, Harris S (2005b) Vegetation quality and habitat selection by European hares *Lepus europaeus* in a pastoral landscape. Acta Theriol 50:391–404

Sokolov VE, Ivanitskaya EY, Gruzdev VV, Heptner VG (2009) Lagomorphs. A volume in the series Mammals of Russia and adjacent regions. Smithsonian Institution Libraries, Washington, DC

Soveri T, Valtonern M (1983) Endoparasites of hares (*Lepus timidus* L. and *L. europaeus* Pallas) in Finland. J Wildl Dis 19:337–341

Späth V (1989) Untersuchungen zur Populationsökologie des Feldhasen (*Lepus europaeus* Pallas) in der Oberrheinebene [in German]. Freiburger Waldschutz-Abhandlungen 8. Forstzoologisches Institut der Albert-Ludwigs-Universität, Freiburg i. Br.

Spittler H (1972) Über die Auswirkung der durch die Tollwut hervorgerufenen Reduzierung der Fuchspopulation auf den Niederwildbesatz in Nordrhein-Westfalen [in German with English summary]. Z Jagdwiss 18:76–95

Spyrou V, Stamatis C, Birtsas P, Psychas V, Manolakou K, Billinis C, Mamuris Z (2013) Evidence for European brown hare syndrome virus introduction with translocated brown hares (*Lepus europaeus*): implications for management of restocking operations. Wildl Res 40:545–551

Stamatis C, Suchentrunk F, Sert H, Triantaphyllidis C, Mamuris Z (2007) Genetic evidence for survival of released captive-bred brown hares *Lepus europaeus* during restocking operations in Greece. Oryx 41: 548–551

Stavy M, Terkel J, Kohen F (1978) Plasma progesterone levels during pregnancy and pseudopregnancy in the hare (*Lepus europaeus syriacus*). J Reprod Fert 54: 9–14

Stavy M, Goldblatt A, Terkel J (1985) Home odor preferences in young hares (*Lepus capensis syriacus*): Effects of age and role of maternal presence. Dev Psychobiol 18:125–139

Štěrba O (1981) Prenatal development and growth of *Lepus europaeus*. Folia Zool 30:147–154

Stieve H (1952) Zur Fortpflanzungsbiologie des europäischen Feldhasen (*Lepus europaeus* Pallas) [in German]. Zool Anz 148:101–114

Stoate C, Tapper SC (1993) The impact of three hunting methods on brown hare (*Lepus europaeus*) populations in Britain. Gibier Faune Sauvage 10:229–240

Strauß E, Grauer A, Bartel M, Klein R, Wenzelides L, Greiser G, Muchin A, Nösel H, Winter A (2008) The German wildlife information system: population densities and development of European hare (*Lepus europaeus* PALLAS) during 2002–2005 in Germany. Eur J Wildl Res 54:142–147

Štrbenc M, Fazarinc G, Bavdek SV, Pogačnik A (2003) Apoptosis and proliferation during seasonal testis regression in the brown hare (*Lepus europaeus* L.). Anat Histol Embryol 32:48–53

Stroh G (1931) Zwei sichere Altersmerkmale beim Hasen [in German]. Berlin Tierärztl Wschr 12:180–181

Stubbe I, Stubbe W (1997) Untersuchungen zum Gesundheitsstatus von Hasen (*Lepus europaeus*) in Sachsen-Anhalt [in German with English summary]. Beitr Jagd Wildforsch 22:63–115

Suchentrunk F, Willing R, Hartl GB (1991) On eye lens weight and other age criteria of the brown hare (*Lepus europaeus* Pallas, 1778). Z Säugetierkd 56:365–374

Suchentrunk F, Polster K, Giacometti M, Ratti P, Thulin CG, Ruhlé C, Vasil'ev AG, Slotta-Bachmayr L (1999) Spatial partitioning of allozyme variability in European mountain hares (*Lepus timidus*): gene pool divergence across a disjunct distributional range? Z Säugetierkd 64:308–318

Suchentrunk F, Michailov C, Markov G, Haiden A (2000) Studies on the European hare. 55. Population genetics of Bulgarian brown hares *Lepus europaeus*: allozymic diversity at zoogeographical crossroads. Acta Theriol 45:1–12

Suchentrunk F, Mamuris Z, Sfougaris AI, Stamatis C (2003) Biochemical genetic variability in brown hares (*Lepus europaeus*) from Greece. Biochem Genet 41: 127–140

Suchentrunk F, Ben Slimen H, Stamatis C, Sert H, Scandura M, Apollonio M, Mamuris Z (2006) Molecular approaches revealing prehistoric, historic, or recent translocations and introductions of hares (genus *Lepus*) by humans. Hum Evol 21:151

Suchomel J, Šipoš J, Čepelka L, Heroldová M (2019) Impact of *Microtus arvalis* and *Lepus europaeus* on apple trees by trunk bark gnawing. Plant Protect Sci 55:142–147

Swihart RK (1986) Home range-body mass allometry in rabbits and hares (Leporidae). Acta Theriol 31: 139–148

Szederei A (1959) Beobachtungen über den Feldhasen in Ungarn [in German with English summary]. Z Jagdwiss 5:81–94

Tapper SC, Barnes RFW (1986) Influence of farming practice on the ecology of the brown hare (*Lepus europaeus*). J Appl Ecol 23:39–52

Taylor M (2017) The way of the hare. Bloomsbury, London

Temple HJ, Terry A (2007) The status and distribution of European mammals. Office for Official Publications of the European Communities, Luxembourg

Thulin C-G (2003) The distribution of mountain hares *Lepus timidus* in Europe: a challenge from brown hares *L. europaeus*? Mamm Rev 33:29–42

Thulin C-G, Tegelström H (2002) Biased geographical distribution of mitochondrial DNA that passed the species barrier from mountain hares to brown hares (genus *Lepus*): an effect of genetic incompatibility and mating behaviour? J Zool 258:299–306

Thulin C-G, Fang M, Averianov AO (2006) Introgression from *Lepus europaeus* to *L. timidus* in Russia revealed by mitochondrial single nucleotide polymorphisms and nuclear microsatellites. Hereditas 143:68–76

Toldt K (1935) Aufbau und natürliche Färbung des Haarkleides der Wildsäugetiere [in German]. Deutsche Gesellschaft für Kleintier- und Pelztierzucht, Leipzig

Tomaso H, Otto P, Peters M, Süss J, Karger A, Schamoni H, Zuchantke E, Hotzel H (2018) *Francisella tularensis* and other bacteria in hares and ticks in North Rhine-Westphalia (Germany). Ticks Tick Borne Dis 9:325–329

Topping CJ, Dalby L, Skov F (2016) Landscape structure and management alter the outcome of a pesticide ERA: evaluating impacts of endocrine disruption using the ALMaSS European brown hare model. Sci Total Environ 541:1477–1488

Trouessart È-L (1910) Faune des Mammifères d'Europe [in French]. R. Friedländer & Sohn, Berlin

Tsokana CN, Sokos C, Giannakopoulos A, Mamuris Z, Birtsas P, Papaspyropoulos K, Valiakos G, Spyrou V, Lefkaditis M, Chatzopoulos DC, Kantere M, Manolakou K, Touloudi A, Burriel AR, Ferroglio E, Hadjichristodoulou C, Billinis C (2016) First evidence of *Leishmania* infection in European brown hare (*Lepus europaeus*) in Greece: GIS analysis and phylogenetic position within the *Leishmania* spp. Parasitol Res 115:313–321

Tsokana CN, Sokos C, Giannakopoulos A, Birtsas P, Valiakos G, Spyrou V, Athanasiou LV, Burriel AR, Billinis C (2020) European brown hare (*Lepus europaeus*) as a source of emerging and re-emerging pathogens of public health importance: a review. Vet Med Sci 6:550–564

Ullmann W, Fischer C, Kramer-Schadt S, Pirhofer-Walzl K, Glemnitz M, Blaum N (2020) How do agricultural practices affect the movement behaviour of European brown hares (*Lepus europaeus*)? Agric Ecosyst Environ 292:106819

Valencak TG, Arnold W, Tataruch F, Ruf T (2003) High content of polyunsaturated fatty acids in muscle phospholipids of a fast runner, the European brown hare (*Lepus europaeus*). J Comp Physiol B 173:695–702

Valencak TG, Tataturch F, Ruf T (2009) Peak energy turnover in lactating European hares: the role of fat reserves. J Exp Biol 212:231–237

Valencak TG, Hackländer K, Ruf T (2010) Peak energy turnover in lactating European hares: a test of the heat dissipation limitation hypothesis. J Exp Biol 213: 2832–2839

Valencak TG, Gamsjäger L, Ohrnberger S, Culbert NJ, Ruf T (2015) Healthy n-6/n-3 fatty acid composition from five European game meat species remains after cooking. BMC Res Notes 8:273

Vaughan N, Lucas E-A, Harris S, White PCL (2003) Habitat associations of European hares *Lepus europaeus* in England and Wales: implications for farmland management. J Appl Ecol 40:163–175

Velarde R, Cavadini P, Neimanis A, Cabezón O, Chiari M, Gaffuri A, Lavín S, Grilli G, Gavier-Widén D, Lavazza A, Capucci L (2017) Spillover events of infection of brown hares (*Lepus europaeus*) with Rabbit Haemorrhagic Disease Type 2 Virus (RHDV2) caused sporadic cases of an European Brown Hare syndrome-like disease in Italy and Spain. Transbound Emerg Dis 64:1750–1761

Vidus-Rosin A, Lizier MA, Serrano-Perez S (2012) Habitat selection and segregation by two sympatric lagomorphs: the case of European hares (*Lepus europaeus*) and Eastern cottontails (*Sylvilagus floridanus*) in northern Italy. Acta Theriol 57:295–304

Viviano A, Mori E, Fattorini N, Mazza G, Lazzeri L, Panichi A, Strianese L, Mohamed WF (2021) Spatiotemporal overlap between the European brown hare and its potential predators and competitors. Animals 11:562

Voigt U, Siebert U (2019) Living on the edge – circadian habitat usage in pre-weaning European hares (*Lepus europaeus*) in an intensively used agricultural area. PLoS One 14:e0222205

Voigt U, Siebert U (2020) Survival rates on pre-weaning European hares (*Lepus europaeus*) in an intensively used agricultural area. Eur J Wildl Res 66:67

Wajdzik M, Halecki W, Kalarus K, Gąsiorek M, Pająk M (2017) Relationship between heavy metal accumulation and morphometric parameters in European hare (*Lepus europaeus*) inhabiting various types of landscapes in southern Poland. Ecotoxicol Environ Safety 145:16–23

Watson JS, Taylor RH (1955) Reingestion in the hare *Lepus europaeus* Pal. Science 121:314

Weber D, Roth T, Kohli L (2019) Increasing brown hare (*Lepus europaeus*) densities in farmland without predator culling: results of a field experiment in Switzerland. Eur J Wildl Res 65:75

Weterings MJA, Moonen S, Prins HHT, van Wieren SE, van Langevelde F (2018) Food quality and quantity are more important in explaining foraging of an intermediate-sized mammalian herbivore than predation risk or competition. Ecol Evol 8:8419–8432

Weterings MJA, Ewert SP, Peereboom JN, Kuipers HJ, Kuijper DPJ, Prins HHT, Jansen PA, van Langeveld F, van Wieren SE (2019) Implications of shared predation for space use in two sympatric leporids. Ecol Evol 9:3457–3469

Wibbelt G, Frölich K (2005) Infectious diseases in European brown hare (*Lepus europaeus*). Wildl Biol Pract 1:86–93

Zaccaroni M, Biliotti N, Buccianti A, Calieri S, Ferretti M, Genghini M, Riga F, Trocchi V, Dessì-Fulgheri F (2013) Winter locomotor activity patterns of European hares (*Lepus europaeus*). Mamm Biol 78:482–485

Zachos FE, Ben Slimen H, Hackländer K, Giacometti M, Suchentrunk F (2010) Regional genetic in situ differentiation despite phylogenetic heterogeneity in Alpine mountain hares. J Zool 282:47–53

Zörner H (1977) Erste Ergebnisse dreijähriger Untersuchungen an der Hasenpopulation des Wildforschungsgebietes Hakel [in German with English summary]. Beitr Jagd Wildforsch 8:203–216

Zörner H (1981) Der Feldhase [in German]. Die Neue Brehm Bücherei 169. Ziemsen, Wittenberg, GDR

Iberian Hare *Lepus granatensis* Rosenhauer, 1856

8

Paulo C. Alves, Pelayo Acevedo, and José Melo-Ferreira

Contents

P. C. Alves (✉)
CIBIO, Centro de Investigação em Biodiversidade e Recursos Genéticos, InBIO, Universidade do Porto, Vairão, Portugal

BIOPOLIS Program in Genomics, Biodiversity and Land Planning, CIBIO, Vairão, Portugal

Departamento de Biologia, Faculdade de Ciências, Universidade do Porto, Porto, Portugal

Estação Biológica de Mértola (EBM), Mértola, Portugal
e-mail: pcalves@fc.up.pt

P. Acevedo
Instituto de Investigación en Recursos Cinegéticos IREC (CSIC-UCLM-JCCM), Ciudad Real, Spain

J. Melo-Ferreira
Departamento de Biologia, Faculdade de Ciências da Universidade do Porto, Porto, Portugal

CIBIO/InBIO, Centro de Investigação em Biodiversidade e Recursos Genéticos, Universidade do Porto, Vila do Conde, Portugal

BIOPOLIS Program in Genomics, Biodiversity and Land Planning, CIBIO, Vairão, Portugal

K. Hackländer, P. C. Alves (eds.), *Primates and Lagomorpha*, Handbook of the Mammals of Europe,
https://doi.org/10.1007/978-3-030-34043-8_8

Common Names

English	Iberian hare
German	Iberischer Hase
French	Lièvre Ibérique
Spanish	Liebre Ibérica
Italian	Lepre iberica
Russian	иберийский заяц

Taxonomy, Systematics and Paleontology

The Iberian hare, *Lepus granatensis* Rosenhauer, 1856, was historically classified as a variety of the Cape hare, *Lepus capensis*, under subspecies *L. c. granatensis* (Chapman and Flux 1990), but morphological analyses confirmed the validity of the specific status of the Iberian hare, and the distinction from the Cape hare (Palacios 1983). When compared to the other hare species inhabiting the Iberian Peninsula, the interspecific morphological differences also allow their distinction (Palacios 1983). The Iberian hare shows specific characteristics, indicated by the weight, head, body, hind foot, and ear length ranges, being smallest in the Iberian hare, intermediate in the broom hare *L. castroviejoi*, and largest in the European hare *L. europaeus* (Palacios 1989). Also, *L. granatensis* shows a distinctive white pelage on the inner face of the legs, which extends to the upper side and reaches the tip of the toes. The distinction of this species has also been corroborated by molecular analyses based on an increasingly complex set of genetic markers along the years, such as allozymes (Bonhomme et al. 1986), mitochondrial DNA restriction fragment length polymorphisms (Pérez-Suárez et al. 1994), microsatellites (Estonba et al. 2006; Melo-Ferreira et al. 2014a; Sanz-Martín et al. 2014), single nucleotide polymorphisms (Melo-Ferreira et al. 2009), nuclear DNA sequences (Alves et al. 2003; Melo-Ferreira et al. 2012), whole exomes (Ferreira et al. 2021), and whole genomes (Seixas et al. 2018; Giska et al. 2019). Molecular estimates of species phylogenies based on high throughput DNA sequencing data, using both whole exomes with broad taxonomy coverage (Ferreira et al. 2021) and whole chromosome with more limited taxonomic coverage (Giska et al. 2019), suggested that a clade grouping the mountain hare (*L. timidus*) and the North American arctic hare (*L. arcticus*) and Alaskan hare (*L. othus*), and potentially the Manchurian hare (*L. mandshuricus*), comprise the closest living relatives of the Iberian hare among the analyzed taxa. Divergence time estimation based on full exome and genome sequence data suggested that the Iberian hare diverged from the closest relatives ~2 Mya (Giska et al. 2019; Ferreira et al. 2021).

Three subspecies of the Iberian hare have been proposed, based on morphological characteristics: *L. g. granatensis*, for all Iberian Peninsula, except the northwest; *L. g. gallaecius* Miller 1907, for Galicia and central Asturias, with measurements similar to *L. g. granatensis*, but darker coat coloration; and *L. g. solisi* Palacios and Fernandez 1992, for Mallorca Island on Balearic Archipelago (Spain), somewhat smaller than *L. g. granatensis* and the Anteroflexid of P3 may be absent (Alves and Nithammer 2003; Hoffmann and Smith 2007; Alves and Acevedo 2018). However, the subspecies classification has not been validated by genetic analyses (Seixas et al. 2014; Marques et al. 2017).

Despite difficulties in reconstructing the paleontological records of species of hares, due to overlap of morphological characters and persisting taxonomic uncertainties, the presence of the earliest paleontological records of the Iberian hare date from the Lower Pleistocene in the deposit of Huescar-1, in southern Spain (Mazo et al. 1985; Sesé 1994). An ancient DNA study identified *Lepus* spp. paleontological remains from Les Vautes, southern France (Forest 2003), as belonging to the Iberian hare, showing that the distribution of the species extended beyond the Pyrenees in the Neolithic (Lado et al. 2018). This work suggested that the current range of the species in the Iberian Peninsula, therefore, results from a postglacial retraction of an extended range, possibly due to the recent westwards invasion of *L. europaeus*.

Current Distribution

The Iberian hare is endemic to the Iberian Peninsula (Portugal and Spain). Cabrera (1914) indicated that the species was widely distributed in the Mediterranean areas south of the Ebro River, and also in Galicia, Asturias, and Mallorca Island (Balearic Archipelago, Spain). The following data on species' distribution dates from the 1960s, which maintained the range of the species the same, but some provinces, such as Palencia and Burgos (Northern Plateau) and Cáceres and Sevilla (Southern Plateau), were highlighted for their high population abundance (Ministerio de Agricultura 1968).

The current distribution range can be depicted from efforts to compile biodiversity data, both in Portugal and in Spain. Acevedo et al. (2012a) provided the first detailed map of the species range at 10x10km UTM squares using information from Carro and Soriguer (2007) as well as hunting bags data for Portugal (Almeida et al. 2004). Information for Portugal was recently updated in the Portuguese Mammal Atlas, revealing current absence in several locations in northwestern and central Portugal (Bencatel et al. 2019).

Overall, the Iberian hare naturally occupies most of the Mediterranean region in Iberia, from the Mediterranean coast, south of Ebro River to the Atlantic coast, and from Andalucía coast to the northern mountains of Leon, Palencia, Burgos, and the southern slopes of Navarra and Huesca (Fig. 1), where the broom and the European hare are present (parapatric ranges; see Acevedo et al. 2012b). The natural distribution range of the Iberian hare remain stable in the last decades; nevertheless, in some locations, it has been extirpated by humans, as mentioned for northwestern and central Portugal, namely near the Atlantic coast.

The Iberian hare was introduced in Mallorca Island (Balearic Archipelago, Spain) more than ~4000 years ago (Seixas et al. 2014), and traces of DNA from this species has been found in Corsica, predominantly as hybrids with the Apennine hare, thus most likely there are no pure individuals of Iberian hare in Corsica (Pietri et al. 2011). A more recent introduction (1980s) has been done in the French Pyrenees (Perpignan, southwestern France), for hunting purposes (Bordes et al. 2007). In the French Pyrenees, the Iberian hare is in sympatry with the European hare in some localities, and has been reaching high-density populations in some areas, according to the French Office for Biodiversity (unpub. data).

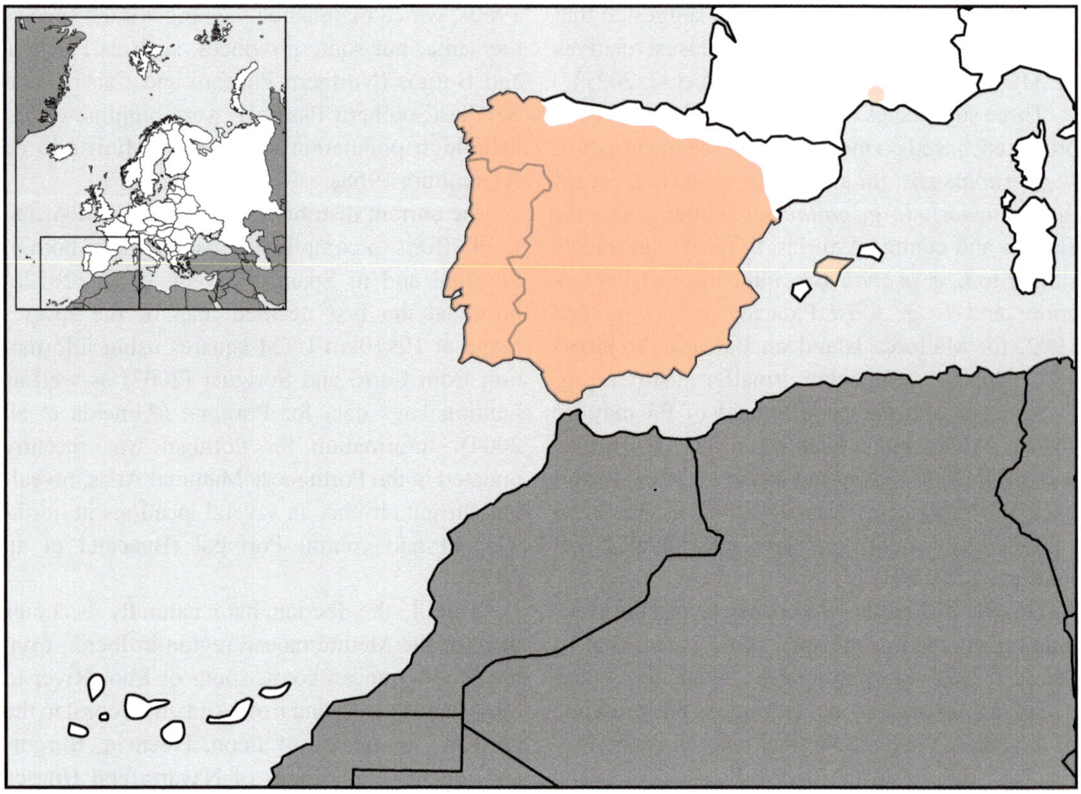

Map template: © Getty Images/iStockphoto

Fig. 1 Distribution of the Iberian hare (*L. granatensis*) showing endemic area (apricot) and where it was introduced (light apricot). Raw data were obtained from Carro and Soriguer (2007) and Bencatel et al. (2019) for Spain and Portugal, respectively. (Map template: © Getty Images/iStockphoto)

Description

In contrast to the general uniformity of the pelage pattern of the other hares in Europe, in the Iberian hare, the white ventral area is extensive, and there is a clear contrast between the ochraceous brown (gray-brown) color of the back and the white belly pelage that extends in a white strip to the forefeet and hindfeet (Fig. 2). It is the only species with white spots on the upper fore and hind feet. Pelage has an ochraceous brown tone with basically white long guard hairs as opposed to the gray hairs in the other two species occurring in the Iberian Peninsula. The external faces of the hips are bright reddish in contrast with the gray brown of the dorsal part of the back. The white coloration of the inner face of the legs is continued with the upper side of the legs and reaches to the tip of the toes. This pelage pattern is quite typical of *L. granatensis*, not occurring in the other hare species (Fig. 2). According to Palacios (1989), in spite of the color pelage uniformity, lightness differences are seen from one population to another, being darker in the northern range. The black area on the upper tail surface is never rounded posteriorly like in *L. europaeus*, but ends on the tip. The long guard hairs are very conspicuous, being white at the top and black at the base. The length of these guard hares range between 40 and 57 mm, and the longest whiskers range between 67 and 91 mm. No white stripes on the face, as in the *L. castroviejoi*.

The Iberian hare is smaller than the other two hare species that occur in Iberia, mean body weight ranging from 2.0 kg to 2.5 kg, and hind feet smaller than 130 mm (Table 1). However, independently of the body size, the Iberian hare

Fig. 2 Iberian hare (*Lepus granatensis*). © J Jambas with kind permission

has the longest tail length of all hare species in Europe.

From the cranial characters (Fig. 3), *L. granatensis* is more similar to *L. europaeus* than to *L. castroviejoi*. It has a very developed posterior arm of the supraorbital process, which reaches the temporal tubercle. However, craniometrical measurements reflect the same size relationship as the body measurements, with *L. granatensis* being the smallest (Table 2). According to Palacios (1989), the aboral zygomatic width (AZW) was the only discriminating character found (of 46 cranial measurements) being significantly smaller than that observed in the other two hare species in Iberia (AZW: *L. granatensis* 38–43 mm, n = 155; *L. castroviejoi* 43.7–46.2 mm, n = 14; *L. europaeus* 43.5–48.4 mm, n = 23). The shape is similar to *L. europaeus,* but the posterior arm of the processus supraorbitalis does not diverge and the sutura coronalis is straight (Palacios 1977). Some skull indices described by Palacios (1983) allow the differentiation of the three species. Cranial dimensions of *L. granatensis* across the range show no important differences (Palacios 1989). The teeth are similar to *L. europaeus*, but the anterior fold of the first lower premolar (P3) is comparatively larger; see Palacios and Lopez-Martinez (1980) for detailed information on dental morphology. Regarding the postcranial skeletal, the relationship of the humerus length by femur length and of tibia to femur length is almost identical in the three species occurring in Iberia, but *L. granatensis* has the longer radius as related to the humerus.

There is no significant variation of measurements across diverse Spanish and Portuguese regions. Nevertheless, the ear length is somewhat greater in the south and hindfoot length is greater in the north. Iberian hares from Andalusia and the Guadalquivir valley are particularly lighter in color, whereas those from the Northwest, Coruña, are particularly darker. Thus, there is a geographical variation of coat color, with Iberian hares with the darkest coat found in areas with particularly high levels of annual precipitation, c. 2000 mm (Palacios 1989). Although there is no sexual dimorphism in the coloration pattern, females are in general heavier than males (average around 2.3 kg in males and 2.5 kg in females, Alves and Niethammer 2003, Table 1).

Physiology

There are almost no studies focusing on aspects of the physiology of the Iberian hare, namely on metabolism, body condition, hormones, etc. However, a recent study has described the reference intervals for 22 hematological and serum biochemistry parameters for this species (Alves and Santos 2021). In general, the values are in accordance with those reported for European hares, except for higher number of white blood cells and urea concentration (Table 3). There are no differences when considering the scale mass index, despite differences among some parameters were observed between sex and age classes, as well between seasons.

Table 1 Body measurementsof adult *Lepus granatensis* in the Iberian Peninsula (Adapted from Alves and Niethammer 2003). The numbers of specimens (within brackets), mean ± Standard deviation (SD) of each parameter are indicated

	Spain[a,b]							Portugal[c]				
	Leon	Burgos	Valladolid	Madrid	Toledo	Albacete	Granada	Pancas	Santarém	Idanha	Males[d]	Females[d]
Weight (g)	2335 ± 238 (11)	2541 ± 238 (11)	2365 ± 315 (17)	2189 ± 206 (9)	2339 ± 317 (9)	2060 ± 145 (8)	2315 ± 277 (6)	2417 ± 247 (157)	2321 ± 226 (45)	2269 ± 194 (9)	2329 ± 200 (154)	2554 ± 278 (57)
Head and body length (mm)	462 ± 16 (9)	465 ± 18 (9)	459 ± 14 (16)	456 ± 16 (7)	461 ± 15 (9)	446 ± 18 (6)	470 ± 14 (4)					
Tail length (mm)	102 ± 6.9 (10)	102 ± 13 (7)	100 ± 11 (15)	105 ± 7.1 (8)	103 ± 9.1 (10)	104 ± 8.9 (7)	109 ± 11 (4)					
Hind feet length (mm)	116 ± 3.2 (16)	117.9 ± 3.6 (12)	117 ± 4.9 (24)	117 ± 4.5 (14)	114 ± 3.9 (13)	114 ± 4.3 (11)	114 ± 4.0 (8)	121 ± 5.6 (215)	124 ± 4.0 (57)	120 ± 4.7 (13)	122 ± 5.4 (160)	122 ± 5.4 (125)
Ear length (mm)	96 ± 3.6 (14)	96.8 ± 4.9 (17)	96 ± 5.0 (28)	101 ± 3.2 (17)	101 ± 4.6 (24)	100 ± 5.2 (11)	100 ± 4.4 (8)	97.1 ± 4.9 (215)	99 ± 3.5 (58)	102.3 ± 4.7 (13)	97.6 ± 5.1 (160)	97.8 ± 4.5 (126)
Total length[e](mm)	564 (9)	567 (7)	559 (15)	561 (7)	564 (9)	550 (6)	550 (4)	551 ± 21.4 (130)	556 ± 17.8 (58)	547 ± 19.7 (13)	545 ± 19.4 (110)	560 ± 18.35 (91)

[a]Palacios (1989)
[b]Names of the Spanish locations are the approximate cities of the region described by Palacios (1989)
[c]Alves and Niethammer (2003)
[d]Values for females and males were calculated with the data from the three Portuguese populations
[e]For the Spanish locals was calculated summing the head and body length with tail length

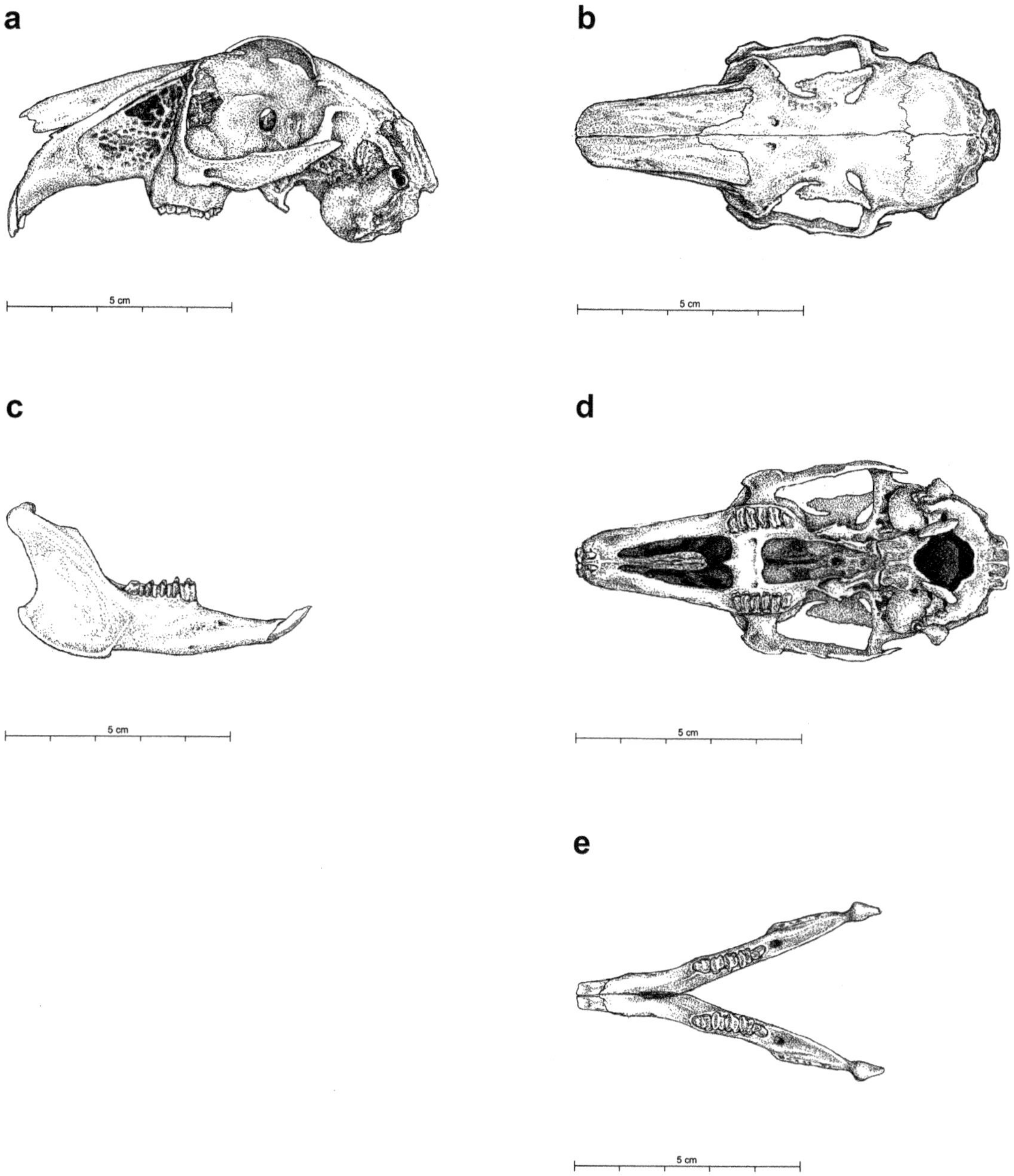

Fig. 3 Skull of *Lepus granatensis*: (**a, b, d**) lateral, dorsal, and ventral view of skull; (**c**) lateral view of left mandible (view from outside), and (**e**) dorsal view of mandible. (© Franz Müller with kind permission)

Table 2 Skull measurements (mm) of adult *Lepus granatensis, L. castroviejoi, and L. europaeus* in the Iberian Peninsula (Palacios 1983). The number of specimens (n), mean ± standard deviation (SD) are indicated

	L. granatensis		*L. castroviejoi*		*L. europaeus*	
	n	*Mean ± SD*	*n*	*Mean ± SD*	*n*	*Mean ± SD*
Occipital length	161	87.6 ± 2.7	13	95.0 ± 1.9	29	99.3 ± 3.1
Total length	158	88.7 ± 2.5	13	95.4 ± 2.1	30	96.6 ± 2.5
Condylobasal length	155	78.1 ± 2.3	13	84.6 ± 2.0	30	89.2 ± 2.2
Palatal length	161	35.5 ± 1.3	14	37.6 ± 0.9	30	48.8 ± 1.4
Internal nasal length	164	31.2 ± 1.9	14	32.0 ± 1.1	30	37.2 ± 2.3
External nasal length	168	39.2 ± 1.4	14	41.9 ± 0.9	30	44.7 ± 2.2
Zygomatic width	156	40.9 ± 1.1	14	45.0 ± 0.7	29	45.9 ± 1.0
Upper dental series	171	15.8 ± 0.6	14	17.4 ± 0.3	131	18.3 ± 0.6
Alveolus height	168	15.4 ± 0.7	14	15.3 ± 0.5	31	15.9 ± 0.8
Width between supraorbital processes	114	30.3 ± 1.5	12	31.7 ± 0.6	23	37.4 ± 1.7
Postpalatal width	158	9.7 ± 0.5	14	11.4 ± 0.5	31	11.2 ± 0.7
Tympanic bulla length	162	11.7 ± 0.5	13	11.3 ± 0.6	30	11.5 ± 0.6
Vertical diameter of the foramen magnum	148	11.9 ± 0.5	13	11.8 ± 0.5	27	12.7 ± 0.6
Radius of incisors curvature	50	10.2 ± 0.5	14	10.9 ± 0.3	27	10.1 ± 0.6

Table 3 Hematology and serum biochemistry values for the Iberian hare, obtained in live animals (Adapted from Alves and Santos 2021). The number of specimens (n), mean ± standard deviation (SD), reference interval (using iterative processes for identifying the location of the median and values distribution), and range are indicated

Parameter	Units	*n*	Median	SD	Reference interval	Minimum–maximum
WBC	$10^3/\mu L$	100	6.73	2.66	2.76–13.29	2.28–16.90
RBC	$10^6/\mu L$	86	8.72	1.06	6.30–10.63	5.78–11.10
HG	g/dL	99	16.80	1.63	13.12–19.81	11.50–20.40
HTC	%	102	48.25	5.22	36.21–57.11	27.2–58.5
MCV	fL	100	55.25	3.85	49.49–63.98	48.80–65.70
MCH	pg	102	19.35	1.29	17.19–22.42	16.9–22.9
MCHC	g/dL	101	34.70	2.00	30.09–38.03	28.6–38.3
PLAT	$10^3/\mu L$	95	386.00	136.81	131.10–673.28	118.00–783.00
TP	g/dL	101	5.08	0.90	3.72–7.37	3.54–7.52
ALB	g/dL	99	2.85	0.30	2.30–3.47	2.17–3.64
CREA	mg/dL	103	1.17	0.80	0.83–2.19	0.80–2.95
Urea	mg/dL	100	45.00	11.18	25.69–69.89	21.00–81.00
URIC	mg/dL	98	1.20	1.46	0.22–5.59	0.20–7.90
GLU	mg/dL	101	223.00	76.47	99.39–493.00	77.00–459.00
TRIG	mg/dL	101	110.00	70.31	49.20–360.29	39.00–386.00
CHOL	mg/dL	101	20.00	5.79	12.32–37.63	10.00–43.00
Ca	mmol/L	96	2.97	0.22	2.55–3.46	2.33–3.40
Na	mmol/L	97	138.00	2.68	132.98–143.82	133.00–146.00
K	mmol/L	101	4.12	1.00	2.70–6.63	2.76–7.13
Cl	mmol/L	100	105.00	4.12	96.87–113.33	97.00–121.6
Mg	mmol/L	100	1.66	0.40	1.19–2.98	1.10–3.47
P	mmol/L	102	1.78	0.87	0.77–8.90	0.68–8.90

WBC = white blood cell count; RBC = red blood cell count; HG = hemoglobin concentration; HTC = hematocrit; MCV = mean corpuscular volume; MCH = mean corpuscular hemoglobin; MCHC = mean corpuscular hemoglobin concentration; PLAT = platelet count; TP = total protein; ALB = albumin; CREA = creatinine; URIC = uric acid; GLU = glucose; TRIG = triglycerides; CHOL = cholesterol; Ca = calcium; Na = sodium; K = potassium; Cl = chloride; Mg = magnesium; P = phosphorus.

Genetics

Karyotype: 2n = 48 (assuming conservation across hares; Palacios 1979; Robinson et al. 1983).

Genetic and genomic analyses and ecological niche modelling suggest that the Iberian hare has colonized northern Iberia after the last glacial maximum, possibly from the southwest (Acevedo et al. 2012a, 2015; Marques et al. 2017; Seixas et al. 2018). Analyses of genetic variation showed that the species has relatively shallow population structure, and that the distribution of genetic variation is mostly determined by the demographic dynamics of the northwards range expansion (Melo-Ferreira et al. 2011; Marques et al. 2017; Seixas et al. 2018). The exception is the Northwest Iberia, where a genetic cluster has been found with single nucleotide polymorphism (SNP) data, suggesting that population might have escaped the expansion dynamics (Marques et al. 2017), and the Central and Northwest Iberia where an Iberian hare mitochondrial DNA (mtDNA) sublineage was found (Melo-Ferreira et al. 2011).

During the northwards postglacial expansion, the Iberian hare encountered and hybridized with Iberian populations of the mountain hare *L. timidus*, a species that was then present in the region, as attested by paleontological records (Altuna 1970). The Iberian hare replaced the mountain hare, and the latter species disappeared from the Iberian Peninsula. From these hybridization events, mtDNA haplotypes of *L. timidus* origin are now predominant in the northern range of the Iberian hare (Fig. 4), but are absent from the South (Melo-Ferreira et al. 2005; Alves et al. 2008; Acevedo et al. 2015). These ancient introgression events from *L. timidus* also affected *L. castroviejoi* and the Iberian populations of *L. europaeus*, where the introgressed mtDNA lineage is also found (Alves et al. 2003; Melo-Ferreira et al. 2005), explaining the high mtDNA diversity detected in *L. europaeus* from the Iberian Peninsula (Pérez-Suárez et al. 1994). Among the introgressed mtDNA variants of *L. timidus* origin, three sublineages were found in the Iberian hare, structured from Northeast to Northwest Iberia, which may represent a remnant of preexistent structure in the replaced *L. timidus* populations, and/or expansion of distinct haplotypes in different regions, "surfing" on the northwards expansion of the species (Melo-Ferreira et al. 2011). A study focused on patterns of evolution along the complete mtDNA sequence of several hare species showed that the variant of *L. timidus* may have evolved under positive selection, affecting several codons of mtDNA-encoded genes (Melo-Ferreira et al. 2014b). Whether the high frequencies of mtDNA introgression result from a selective advantage, from the demographic dynamics during and after the hybridization events, or both, remains being debated (Melo-Ferreira et al. 2011, 2014a, b; Seixas et al. 2018; Carpio et al. 2021). An ancient DNA study showed that the northwards expansion of the Iberian hare reached southern France in the Neolithic (Lado et al. 2018). These analyzed individuals carried the mitochondrial DNA haplotype of *L. timidus*, in keeping with the high proportion of introgression found in the northern Iberian populations.

The ancient hybridization between the Iberian and mountain hares also affected the nuclear genome, with inferences based on genome-wide data of 10 individuals suggesting that at least 1.3–2.4% of the Iberian hare genome is of mountain hare origin (Seixas et al. 2018). These analyses also proposed that the introgressive hybridization events occurred between 24 thousand years ago, based on identity by state tracts, and 7000 years ago, based on the introgressed tract length distribution. Nuclear introgression was found to be higher in the north but, contrary to mtDNA, also affects southern populations (Seixas et al. 2018). These patterns are compatible with a northwards range expansion of the species with hybridization and flow of introgressed variants towards the south, possibly driven by male-mediated dispersal. Introgressed fragments of *L. timidus* origin are less prevalent close to the center of the chromosomes and in the X-chromosome, which is consistent with an increased effect of low recombining regions and the X-chromosome in genomic incompatibilities between species. The inspection of the genome of

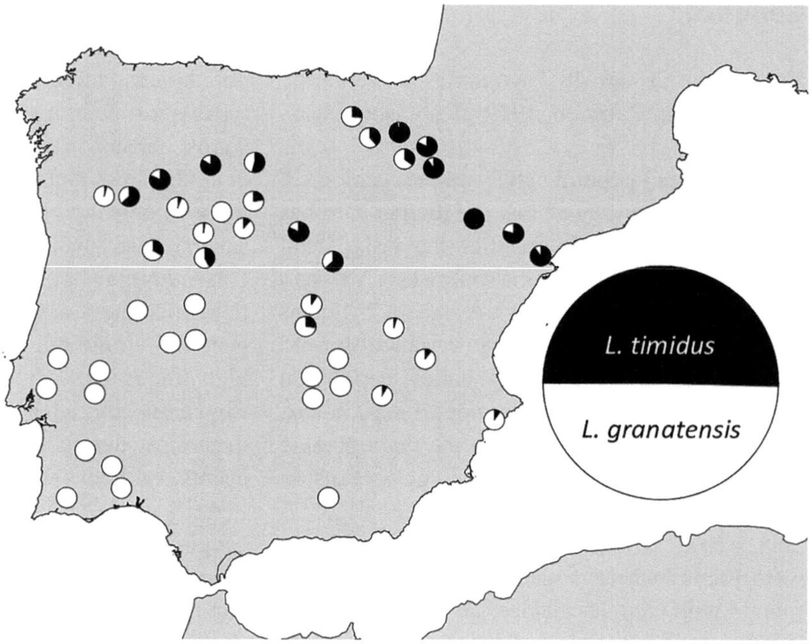

Fig. 4 Distribution of mountain hare (*Lepus timidus*) mitochondrial DNA (mtDNA) introgression Into the Iberian hare (*L. granatensis*) populations. Pie charts show the proportion of analyzed Iberian hares with mountain hare mtDNA (in black) (Adapted from Acevedo et al. 2015)

the Iberian hare identified genomic regions with high frequencies of introgression that could not be predicted by a neutral model of demographic replacement with hybridization and are compatible with selection-driven introgression from *L. timidus* into *L. granatensis*. These genomic fragments include genes with immune functions and related with male fertility, suggesting that introgression in these cases was driven by adaptation to the environment or genomic conflicts (Seixas et al. 2018). Also, patterns of nuclear DNA introgression similar to the geographic distribution of mtDNA introgression affecting nuclear DNA-encoded genes with mtDNA functions suggest co-introgression and coevolution during and after the ancient hybridization events.

Given the presence of *L. granatensis* in southern France in the Neolithic (Lado et al. 2018), the natural current range of the species confined to the Iberian Peninsula may reflect a more recent retraction, possibly caused by the westward invasion of *L. europaeus* and interspecific competition (Acevedo et al. 2012a, b). SNP (Melo-Ferreira et al. 2009) and microsatellite data (Melo-Ferreira et al. 2014a) showed that the Iberian hare and the European hare hybridize in the area of contact in northern Iberia, but no traces of introgressed variants were found away from the contact zone, suggesting that genetic admixture between the species is rather limited (Melo-Ferreira et al. 2014a). Whole genome data estimates found that one Iberian hare from Navarra has 0.39% of the genome introgressed from *L. europaeus*, while for nine other individuals from the rest of the range the estimates varied between 0.01% and 0.03% (Seixas 2017). Analyses of genomic variation inferring the genomic tracts of introgression of different origins also suggested that the Iberian hare first contacted and hybridized with the mountain hare, and then more recently with the European hare (Seixas 2017). Minor traces of genetic admixture with the broom hare have been suggested from analyses of molecular variation (Melo-Ferreira et al. 2009, 2012). Also, a genetic variant originating from ancient hybridization with the Iberian hare has been found to cause a winter gray pelage in the mountain hare from Faroe Islands (Giska et al. 2019).

Studies comparing genetic diversity estimates in hare species from Europe show that levels of diversity are lower in the Iberian hare when compared to *L. timidus* and *L. europaeus* and larger than in *L. castroviejoi* (Estonba et al. 2006; Melo-Ferreira et al. 2009, 2014a; Seixas et al. 2018), in keeping with expected differences in effective population sizes across species. Estimates of long-term effective population size of *L. granatensis* vary between ~100,000 (Melo-Ferreira et al. 2012; Seixas et al. 2018) and 180,000 individuals (Giska et al. 2019), based on inferences using genetic variation data.

A study analyzing patterns of synonymous and non-synonymous substitutions in protein-coding genes based on transcriptome sequencing data, including the Iberian hare, other vertebrates and invertebrate species, suggested that the non-synonymous to synonymous ratio (π_N/π_S) did not differ significantly between vertebrates and invertebrates, contrary to expectations of higher adaptive rate in species with larger effective population sizes (Gayral et al. 2013). This study showed that even though neutral genetic diversity (π_S) is lower in the Iberian hare than in the European wild rabbit *Oryctolagus cuniculus*, π_N/π_S is much higher in the Iberian hare, which is compatible with the hypothesis that smaller effective population sizes leads to a reduction in the efficiency of purifying selection.

The extensive studies using genetic information from the Iberian hare produced important genetic resources. These include partial mtDNA sequences (Alves et al. 2003; Melo-Ferreira et al. 2007, 2011), complete mtDNA sequences (Melo-Ferreira et al. 2014b), microsatellites (Estonba et al. 2006; Sanz-Martín et al. 2014; Melo-Ferreira et al. 2014a, b), SNPs (Marques et al. 2017), transcriptomes (Cahais et al. 2012; Gayral et al. 2013; Marques et al. 2017), exomes (Ferreira et al. 2021), and whole-genomes (Seixas et al. 2018). However, an Iberian hare reference genome assembly is currently lacking.

Life History

The reproductive characteristics of the Iberian hare do not vary significantly among seasons, despite a lower activity observed in autumn, especially in males. This continuous pattern of breeding activity, without summer/autumn interruption, is different from the seasonal pattern reported in most populations of other hare species occurring in Europe (Palacios 1981; Lopez et al. 1996; Alves et al., 2002; Alves and Rocha 2003; Fernández et al. 2008, 2010), despite being in line with the observed in some hare species dwelling in the Mediterranean region, namely *L. europaeus* (Antoniou et al. 2008). Nevertheless, the peak of breeding occurs in spring. Mean litter size (based on embryo counts) observed in different populations from Portugal was 1.56 ± 0.10 (range between 1 and 4; Alves et al. 2002). However, higher average litter size (2.1) has been reported for Navarra and Granada (Farfán et al. 2004; Fernández et al. 2008). A maximum litter size of 7 was observed in Spain (Duarte 2000). Ovary mean weight 1.39gr ±0.08 (range between 0.64 and 2.20gr), mean testis weight 7.93gr ± 0.21 (range between 2.15 and 12.81gr), and epididymis weight 1.54gr ± 0.04 (range between 0.43 and 2.28gr) (Alves et al. 2002). The average daily sperm production and extra-gonadal sperm reserves was estimated as 353 x 10^6 and 1762 x 10^6 sperm cells, respectively, and the mean basal testosterone 7.28 ng/ml ± 1.08 (Alves et al. 2002). Evidence for precocious sexual maturity, pregnant females with 2307 g and males with descended testis with only 1811 g were detected in Portugal in March and June, respectively (Alves and Niethammer 2003). In Spain, males with 1.5 kg and females with 1.75 kg have been observed to be reproductively active (Fernández et al. 2010), suggesting reproduction in the year of birth. Prenatal mortality (due to loss of ova and/or resorption of embryos) in pregnant females was estimated as 21%, and the proportion of pregnant females with higher number of corpora lutea than visible embryos, as 26% (Alves et al. 2002). The gestation period is around 42 days. The annual potential production of young per adult female of 9.8 is in the range of values observed for other hare species (Alves et al. 2002). In Granada (southern Spain), the productivity of female per year has been estimated at 7.21 leverets, with an estimated average number of 3.48 litters per adult female and per year (Farfán et al. 2004). In Navarra (Northern Spain), a productivity of 6.2 litters per year was reported, indicating an average potential of 16.2

leverets per adult female and year (Fernández et al. 2008, 2010).

Mean weight of newborn leverets in captivity is 128.6gr ± 6.2 (n = 6, range between 123 and 140, litter size 1 leveret/female) (Alves and Niethammer 2003). However, smaller sizes of newborns (62 and 82gr) have been reported in Spain (Rodriguez et al. 1997). There are indications for superfetation as some observations of small embryos and fetuses in the final stage of the development in the uteri have been done (PC Alves unpublished data). It is also common to observe lactating pregnant females around all the year (Farfán et al. 2004). In general, females both being pregnant and lactating are heavier and bigger (Fernandez et al. 2010). The life span of the Iberian hare in captivity can reach at least 10 years (PC Alves unpublished data).

A study conducted in southern Spain (Doñana) estimated that the survival of the Iberian hare ranged from 38% (marshland) for a period of 141 days to 22% (ecotone) for a period of 220 days (Carro et al. 2002). It has been estimated in Granada a 27.91% minimum annual survival for leverets (Farfán et al. 2004). The age structure is not well known, but the existent data from Navarra suggest that the majority of the population is around 1 year of life and few specimens exceed that age (Fernández et al. 2010).

Habitat and Diet

Habitat

The Iberian hare occupies heterogeneous habitats in agroecosystems (Acevedo et al. 2012a). In general, the species prefers open landscapes and areas with sparse scrubland, avoiding close woodlands. However, in northern Iberia, it appears in humid mountains areas with oceanic climate (annual precipitation 1500–2000 mm), meadows, pastures, agricultural fields (wheat, etc.), and deciduous forests (Palacios and Meijide 1979; Duarte 2000; Tapia et al. 2010). In central Iberia (annual precipitation 300–800 mm), the species occupies mainly open areas of agricultural landscapes such as wheat, barley, oats, vineyards, and alfalfa. Calzada and Martínez (1994) described seasonal differences in the habitat preferences for this region, selecting agricultural fields in winter and fallows and alfalfa in summer. In southern Iberia, hares inhabit dry area (annual precipitation 300–600 mm, with some places lower than 200 mm) with olive groves, vineyards, wheat, barley, and natural scrubland vegetation (Duarte and Vargas 1998). Olive groves should be highlighted within the preferences of the species in southern range since the highest population densities are achieved in this woody crop (Duarte 2000). Farfán et al. (2012) described changes in the habitat preferences of the species between 1960s and 1990s in this region, selecting natural vegetation in the earlier period and agricultural lands in the latter one. Overall, this species is favored by patchy agrosystems within mosaics with well-developed natural vegetation giving a high structural diversity of the landscapes (Acevedo et al. 2012a; Farfán et al. 2012). Regarding resting site selection, hares mainly select uncultivated lands and pastures that are areas with a low rate of urban disturbance (Bartolomé et al. 2004; Sánchez-García et al. 2012). Hares are looking for sites with vegetation providing coverage (except in the front) that is characteristic of species in which hideout and scape is their antipredator mechanism. Other patterns of resting site selection were described in southern Iberia where the olive groves and neighboring agricultural fields are preferred (Duarte 2000).

In southern Iberia, marginally significant differences were reported for the home range size (90% minimum convex polygon) between VHF (very high frequency) radio-collared males (28 ha, n = 5) and females (24 ha, n = 9; Carro et al. 2011). Data from central Spain provided similar values to those reported in southern populations (35 and 19 ha, for 3 males and 3 females, respectively) using GPS (Global Positioning System) collars and 95% kernels for home range estimation (M. Martinez-Haro, unpublished data). In northern Iberian, the average home range size (95% minimum convex polygon) was 24 ha (Jaramillo-Fayad et al. 2015) and 39.6 ha (Sánchez-García et al. 2012), in Galicia (n = 14) and Valladolid (n = 23), respectively, from VHF radio-collared animals. Finally, Rodríguez et al. (1997) reported higher values (100–300 ha; 95% minimum convex polygon) in translocated VHF radio-collared individuals that can be not directly generalized to natural populations. In general, home range values for natural populations are equivalent to those reported for *L. europaeus*.

Diet

Hares are herbivorous, mainly eating on grasses with spatiotemporal variations in diet depending on resources availability. Studies on Iberian hare feeding ecology are limited to Pauperio and Alves (2008) in central Iberia and to Carro and Soriguer (2010) in southern Iberia. These authors found a wide range of plant species in Iberian hare's diet (higher than reported for other hare species; see, e.g., Dingerkus and Montgomery 2001), being grasses the most represented ones. The species with higher rates of intake were *Anthoxanthum odoratum*, *Secale cereale*, and *Agrostis* spp. in central Iberia (Pauperio and Alves 2008), and *Cynodon dactylon*, *Hordeum* sp., *Aeluropus littoralis* and *Paspalum* sp. in southern Iberia (Carro and Soriguer 2010). In summer, when grass availability is lower, the ingestion of other groups increases, like shrubs, inflorescences (Pauperio and Alves 2008), and even cereal crops (Carro and Soriguer 2002), being generally associated with the higher nutritive value of alternative groups and reproductive plant parts. Overall, the information available for Iberian hare suggests that its diet did not differ substantially of that described for other hare species (Peroux 1995; Ballesteros 2003). According to Alves and Rocha (2003), the vegetation available for the Iberian hare in southern Portugal acquires a maximum protein content between January and April, where fiber reach the minimum value; on the contrary, from July to December, the trend is reversed and the protein content acquires its minimum (less than 10%) and that of fiber its maximum. The biomass of the vegetation acquires high values between January and April and its minimum between July and October.

Behavior

In contrary to wild rabbits, hares do not use burrows for breeding neither to hide from predators. Usually they lay down in vegetation, making shelters and breeding sites (dens). Leverets born with open eyes and cover by hair, typical from nidifugous species. Parental care is limited (usually one or two visits a day when lactating), and leverets with 2 weeks are feeding grass and with 3 weeks do no longer depend of the mother's milk (Duarte 2000).

Social Behavior

Iberian hares are generally solitary animals appearing sometimes in small groups to feed during the night, presumably as antipredator strategy. Social hierarchy is established by means of intimidating poses and pursuits. The hierarchy level is used for selecting feeding sites; dominant animals displace to subordinary ones to suboptimal areas (Rodríguez et al. 1997). Iberian hares are not territorial (Ballesteros 1998) but fights are frequent among males during the mating period. Fights are frequent when animals are maintained in captivity that limits the success of the intensive breeding programs. In particular, female seems to be more aggressive than males in captivity conditions.

Activity

The activity patterns of Iberian hare were studied in Doñana National Park, southern Iberia (Carro and Soriguer 2010) and in Zamora, northern Iberia (Rodríguez et al. 1997). In general, these authors, both for southern and northern populations, reported that the Iberian hare is mainly nocturnal, spending most of the daytime in resting places and there is no substantial differences in the circadian rhythms between males and females. Males and females are mostly active slightly before dusk until a bit after dawn (sun rising), with slight interindividual variability. During its activity period (approx. 12 h), the hares search food resources and establish social relationships with conspecifics. Resting period starts earlier in the morning by entering forms. This period is only interrupted by short grooming periods. There are no robust data for assessing intra-annual variation in the activity pattern, but the activity in spring peaked in the sunset and is more variable during the night (Rodríguez et al. 1997).

Parasites and Diseases

Health status is one the subject more broadly assessed in the Iberian hare, mainly in the last years and likely due to their potential relevance in the species' mortality, and as reservoir for zoonosis (Alzaga et al. 2008; Alzaga 2010; Carro and Soriguer 2010; Sánchez-García et al. 2012). For example, Alzaga (2010) studied the causes of mortality in 75 Iberian hares and found that bacterial diseases caused 45% of the total deaths (similar rate to traumatism), and parasites caused 8% of the deaths; no viral diseases were reported in this study.

Among bacterial diseases, tularemia (*Francisella tularensis*) is one of the most important, but other diseases such as pseudotuberculosis (*Yersinia pseudotuberculosis*), pasteurellosis (*Pasteurella multocida* and *P. haemolytica*), salmonellosis (*Salmonella typhimurium*), tuberculosis (*Mycobacterium tuberculosis* complex) were described (Rodríguez et al. 1997; García Pena et al. 1998; Aranaz et al. 2004; Carro and Soriguer 2010). But other bacterial diseases, namely caused by *Pseudomonas aeruginosa*, *Fusobacterium necrophorum*, *Streptococcus* sp., *Arcanobacterium pyogenes*, *Pseudomonas aeruginosa*, *Haemophilus* sp., *Clostridium perfringens*, *Staphylococcus aureus*, have been reported by Alzaga (2010).

Hares are considered to be very susceptible to tularemia, which may cause important epizootics, but they do not seem to be the usual reservoir. In Iberia Peninsula, tularemia has been related to common voles (*Microtus arvalis*; e.g., Rodríguez-Pastor et al. 2017), but it was suggested that Iberian hares can be associated in the transmission to humans (Luque-Larena et al. 2017).

Alzaga (2010) examined 170 Iberian hare sera widely distributed across the species range in Spain from 2003 to 2006, and no specific antibodies against *F. tularensis*, *Brucella* sp., and *Leptospira interrogans* were found. However, she found a high seroprevalence of *Toxoplasma gondii* (47%), a parasitic zoonosis, and concluded that Iberian hares could represent a relevant intermediate host for *T. gondii* in south-central Iberia. A moderate seroprevalence was also obtained in a posterior study in northern Iberia (Fernández-Aguilar et al. 2013). These authors studied the epidemiology of *T. gondii* and described evidence supporting that most juvenile seropositive hares did not survive as adults and/or there is a short-term humoral immune response against *T. gondii* in Iberian hares.

Iberian hare was described as an important reservoir for *Leishmania infantum*. In central Iberia, hares achieved high prevalence (74%) during an epidemic outbreak between 2009 and 2012 (Molina et al. 2012). In a retrospective study between 2004 and 2010, Ruiz-Fons et al. (2013) described a widespread presence of the parasite in Iberia and suggested that hares may have and unexpected role in the epidemiology of this zoonosis.

The parasite communities of the Iberian hare were studied by Alzaga (2010). Using more than 300 hunted animals widely distributed across the species range in Spain, this author described the parasite communities comparatively among the three hare species inhabiting the Iberian Peninsula (*L. castroviejoi* and *L. europaeus*, in addition to Iberian hare). In Table 4, a list of reported parasites is described for the Iberian hare (Sánchez-Covisa et al. 1999; García-Romero et al. 2001; Alzaga 2010; Segovia et al. 2014; Carvalho-Varela 1971; Cordero-del-Campillo 1994; Martínez et al. 2011; Rodríguez et al. 1973; Romero-Rodriguez 1976; Romero-Rodríguez et al. 1973; Vicente 1969).

The most well-known viral disease occurring in hares is provoked by the European brown hare syndrome virus (EBHSV), which causes the European brown hare syndrome (EBHS) in

Table 4 Parasites reported in the Iberian hare (*Lepus granatensis*)

Trematoda	*Dicrocoelium dendriticum*
	Fasciola hepatica
Nematoda	*Dermatoxys hispaniensis*
	Graphidium strigosum
	Micipsella numidica
	Micipsella indica
	Nematodiroides zembrae
	Nematodirus sp.
	Passalurus ambiguus
	Passalurus nonanulatus
	Protostrongylus commutatus
	Protostrongylus terminalis
	Trichuris sp.
	Trichuris leporis
	Trichostrongylus retortaeformis
	Trichostrongylus colubriformis
Cestoda	Multiceps *serialis*
	Cittotaenia denticulata
	Leporidotaenia pseudowimerosa
	Taenia pisiformis
	Anoplocephalidae
	Mosgovoyia viscaciae
	Mosgovoyia pectinata
	Taenia pisiformis
Protozoa	*Eimeria* sp.
	Eimeria leporis
	Eimeria europaea
	Eimeria magna
	Eimeria media
	Eimeria septentrionalis
	Sarcocystis sp.
	Toxoplasma gondii
	Cryptosporidium parvum
Ticks	*Rhipicephalus* sp.
	Ixodes sp.
	Haemaphysalis sp.
	Dermacentor sp.
	Hyalomma sp.

brown and mountain hares. Despite causing severe deaths in those two species, interestingly it was never reported in the Iberian hare. Nevertheless, a virus from the same genus *Lagovirus* (family Caliciviridae), the rabbit hemorrhagic virus (RHDV), which is causing high mortalities in the European wild rabbit *Oryctolagus cuniculus*, have been detected in from dead Iberian hares collected in the 1990s in Portugal, where clinical signs were compatible with a *Lagovirus* infection (Lopes et al. 2014).

More recently, a new viral outbreak was detected in the Iberian hare. During 2018, widespread mortalities in the Iberian hare with myxomatosis-like clinical signs were reported in Portugal and Spain, which supported the potential species jump of the myxoma virus (MYXV) from the European rabbit to the Iberian hare (Águeda-Pinto et al. 2019; Dalton et al. 2019; García-Bocanegra et al. 2019; Carvalho et al. 2020). Molecular studies revealed that an insertion or recombination event with respect to the MYXV Lausanne reference strain may have been involved in the cross-species jump and increased virulence in its new host, the Iberian hare (Águeda-Pinto et al. 2019; Dalton et al. 2019). A comprehensive study between 2018 and 2020 in Spain demonstrates that spatial distribution of this viral disease is not homogeneous, with most outbreaks concentrated in the southern and central parts of this country. Nevertheless, consecutive outbreaks reported after 2018 suggest endemic circulation of this new emerging virus (García-Bocanegra et al. 2021). Moreover, a retrospective serological study shows that the Iberian hare may have been in contact with myxoma like virus between 1994 and 1999, since antibodies were detected in animals from that period (Abade dos Santos et al. 2022). The past contact of the Iberian hare with MYXV may have occurred with strains that circulated in wild rabbit, or unnoticed strains circulating in Iberian hare populations, suggesting the potential circulation of the MYXV (or similar virus) at least 20 years before the severe virus outbreaks observed in 2018. It should be noticed that hare myxomatosis outbreaks are causing severe mortalities in the Iberian hare, causing strong populations declines in Portugal and Spain.

Recently, a new leporid herpesvirus (LeHV-5) was detected in Iberian hares. Herpesvirus DNA was detected in hares with myxomatosis, where, in most cases, herpetic-like skin vesicles were present in the nostrils and lips along with necrosis of the genitalia, most evident in males affecting the penile glans but also observed in females (Abade dos Santos et al. 2020).

Population Ecology

Population Dynamics

The population abundance of the Iberian hare is widely variable in response to factors such as habitat quality, hunting pressure, natural predation, and diseases (Rodríguez et al. 1997; Gortázar et al. 2007; Lázaro et al. 2019). Available data on population density are scarce, old and mostly local, and results in diverse density estimates across regions (Table 5).

Studies on population age structure are very scarce in the Iberian hare, even considering them pivotal to understand population trends. In this species, the population trends are unclear. In the 1990s, numbers declined in northern Iberia (Duarte 2000) but increased in southern Iberia (Duarte and Vargas 1998). On the one hand, Gortázar et al. (2007) described significant interannual differences in abundance (period 1992–2002), both at local and regional scales, that produced a general positive trend suggesting that the species slightly increased in numbers during the study period in Aragón (northern Iberia). Carro and Soriguer (2017) showed that the species in Doñana National Park (southern Iberia) moderately declined between 1995 and 2012. Results from game bags in Spain (Garrido et al. 2019) suggest a 50% reduction between 2000 (1400.000) and 2017 (700.000). Moreover, from September 2018 populations are being affected by myxomatosis (Garcia-Bocanegra et al. 2019; see also section "Parasites and Diseases"). The MYXV is causing a high mortality and therefore is expected to contribute to hare population dynamics (but see Lázaro et al. 2019). More studies are needed to understand the current population trend of Iberian hare at both local and global scales, as well as to evaluate the impact of recent hare myxomatosis outbreaks on population dynamics.

Competition with other Leporids

Parapatry is a biogeographical term used to refer to species whose distribution ranges do not overlap but are immediately adjacent to each other. No environmental barriers usually exist in the contact zones between the Iberian hare and the other hare species occurring in Iberia, hence competitive exclusion is advocated as a main factor that modulates species ranges (Bull and Possingham 1995). Thus, the Iberian hare distribution range (see section "Current Distribution") might be explained by occurrence of the *Lepus* species inhabiting the Iberia (Acevedo et al. 2012a, b), representing, therefore, a clear example of parapatry. The Iberian hare seems to have advantage over *L. europaeus*, as their contact zone is more favorable for the former (Acevedo et al. 2012b). This result is consistent with positive population trends described for *L. granatensis* and negative trends for *L. europaeus* in a region where the species coexist in northern Spain (Gortázar et al. 2007). Similarly, a macroecological study showed the existence of some areas with a high environmental potential for *L. castroviejoi,* but unoccupied by this species, that are currently within the area currently occupied by Iberian hare (Acevedo et al. 2014). This result

Table 5 Summary of the values of population density reported for Iberian hare (*Lepus granatensis*)

Region	Density (ind/km^2)	References
Navarra	5.8	Carro and Soriguer (2010)
León	13	Calzada and Martínez (1994)
León	22.1	López et al. (1996)
Zamora	2–10	Rodríguez et al. (1997)
Valladolid	11–15.3	Sánchez-García et al. (2012)
Castilla-La Mancha	1.1–2.7	Herranz (2000)
Castilla-La Mancha	7.6	Lázaro et al. (2019)
Évora	13.7–37.4	Batista and Cruz de Carvalho (1996)
Granada	32	Palomares and Ruiz-Martinez (1994)
Huelva	1–31	Carro and Soriguer (2010)

suggests that competition with the Iberian hare could be an important factor constraining the expansion process of *L. castroviejoi* to these favorable territories (Ballesteros et al. 1986). Summarizing, the Iberian hare seems to have adaptive advantages for the possible interspecific relationships that could be established with the other hare species inhabiting Iberia.

Effects of Global Change

The effects of land use changes on the Iberian hare were studied by Farfán et al. (2012). These authors showed that the requirement of the species changed substantially from a highly association with natural vegetation in the 1960s to one linked to cultivated lands in the 1990s. In this case, changes in land use did not suppose relevant shifts in the species distribution range but in the habitat availability for the species. The Iberian hare has benefited from new habitat since the cultivated areas provide year-round cover and food. Regarding climate change, Acevedo et al. (2012a) forecasted an increase in the species distribution range in a close future. According to these authors, it is predicted a higher probability of occurrence of the species in its current range, and a northward expansion (see also Araujo et al. 2011). Moreover, results from the previous study suggested favorable conditions for the Iberian hare presence in southern France, a region where there are paleontological evidence suggesting its presence since 5.5–7.5 thousand years before present (Lado et al. 2018), and where it has been recently distributed (Bordes et al. 2007). Overall, and despite the current negative population trend (namely due to diseases and change in habitat use), given the ecological requirements of the species, climate change is not expected to be a relevant threat for the species.

Conservation Status

It is considered *Least Concern* in the IUCN Red List of Threatened Species (Soriguer and Carro 2019). It is also considered *Least Concern* in the regional IUCN red list assessments of Portugal (Cabral et al. 2005) and Spain (Palomo et al. 2007). Nevertheless, it is possible that the new information on population trends might change this status, due to the population decline observed in the last years, which has been dramatically accelerated by hare myxomatosis outbreaks since 2018.

Management

The Iberian hare is highly associated with open agricultural landscapes. Nevertheless, it is fundamental to have shelters for refugee, namely for avian predators, namely Spanish imperial eagle (*Aquila adalberti*), golden eagle (*Aquila chrysaetos*), Bonelli's eagle (*Aquila fasciata*), booted eagle (*Hieraaetus pennatus*), red kite (*Milvus milvus*), black kite (*Milvus migrans*), common buzzard (*Buteo buteo*), goshawk (*Accipiter gentilis*), and the Eurasian eagle-owl (*Bubo bubo*). The red fox (*Vulpes vulpes*) is the main mammal predator, namely for juveniles. The Iberian hare is an important small game species in Portugal and Spain. It represents an important social, cultural, and economic value for rural areas. The main hunting activities are done traditional by shooting, but dog hare coursing is also very popular in some regions of Portugal and Spain. Hunting is only allowed in hunting estates, and the open season is between September and December (in some regions may go to end of February). A limited number of animals are allowed per hunter. Translocations are common, namely for hare dog coursing, although they are not recommended by the national entities, and in several regions are not allowed. In some regions, the Iberian hare can cause agricultural damages (crops, vineyards, etc.) and need to be controlled when densities are high, or the fields be protected using fences. Nevertheless, hare control may be also done when cases of zoonosis outbreaks are detected, as it happened last decade in Madrid region, Spain. Leishmaniasis was detected in 2009 and, by December 2012, human infections had increased significantly in several cities of this region (Carrillo et al. 2013). While dogs are the main reservoir host, the Iberian hare were suspected as a potential reservoir for *Leishmania* infections (Molina et al. 2012), and thus Iberian hare populations were controlled by the authorities and/or hunted.

The main factors affecting the Iberian hares are: loss of habitat due to the disappearance of traditional crops and intensification of farming (Palacios and Ramos 1979; Carro and Soriguer 2007); transformation of Mediterranean landscapes, with the abandonment of traditional agricultural practices, associated with the scrub and forest regeneration, which has increased the abundance of the wild boar (*Sus scrofa*), and thus reducing the Iberian hare namely in mountain areas, as well as the implementation of intensive crops (Delibes-Mateos et al. 2009); unsustainable hunting management (Palacios and Ramos 1979; Carro and Soriguer 2007); abusive and indiscriminate use of phytosanitary products (Rodríguez et al. 1997; Carro and Soriguer 2007) and rodenticides in agricultural areas (Olea et al. 2009); poaching (Palacios and Ramos 1979; Carro and Soriguer 2007); death by run over on roads; increase in opportunistic predators (Herranz 2000; Carro and Soriguer 2007; Ballesteros 2003); and, last but not least, emerging diseases (Molina et al. 2012; García-Bocanegra et al. 2021).

Future Challenges for Research and Management

The Iberian hare is endemic in the Iberian Peninsula, and its populations have the ancestral footprints of past hybridization with the mountain hare, already extinct in the region: a ghost of a hybrid past. The several genetic and genomic studies conducted in the Iberian hare make the reticulate evolutionary history of this species a textbook example in evolutionary biology (Arnold 2015). The patterns of ancient introgression from the mountain hare and more recent introgression from the brown hare show that the range shifts since the last glacial maximum promoted contacts between species and hybridization. Thus, the presence of portions of genetic variation originating from other species, in particular the mountain hare and affecting massively mtDNA and some parts of the nuclear genome, make the Iberian hare a particularly suited model to study the impact of animal hybridization in the evolutionary trajectories of species. Research has shown that possibly a combination of demographic circumstances and natural selection contributed to the current admixed gene pool of the species. Further research is needed to clarify the physiological and ecological impacts of introgression in the Iberian hare populations.

Moreover, despite the increasing knowledge in the evolutionary history of the Iberian hare, basic ecological and physiological studies are still very scarce. The population dynamics of the Iberian hare is poorly known, and there is no information on the body metabolism and hormone function of this species. Finally, this species has recently been affected by a virus jump from the wild rabbit, and so far the consequences of the recent outbreaks caused by the myxoma virus are still not well known, thought the observed population decline and extirpation of populations in some regions is a main concern for conservation of this endemic species. The absence of management and conservation programs in Spain and Portugal for this species, associated with the fact that is an important game and simultaneously prey species, elevated the concern regarding its threatened status. Definitely, more attention should be given for preserving the natural populations of the Iberian hare.

Acknowledgments P. Acevedo is partially supported by MICINN-FEDER (grant: PID2019-111699RB-I00). JMF is supported by Fundação para a Ciência e a Tecnologia (2021.00150.CEECIND).

References

Abade dos Santos FA, Monteiro M, Pinto A, Carvalho CL, Peleteiro MC, Carvalho P, Peleteiro P, Mendonça P, Carvalho T, Duarte MD (2020) First description of a herpesvirus infection in genus *Lepus*. PLoS One 15: e0231795

Abade dos Santos FA, Santos N, Carvalho CL, M. H., M., Gortázar C, García-Bocanegra I, Capucci L, Duarte M, Alves PC (2022) Retrospective serological and molecular survey of myxoma or antigenically-related virus in the Iberian hare, *Lepus granatensis*. Transboundary and Emerging Diseases. https://doi.org/10.1111/tbed.14734

Acevedo P, Melo-Ferreira J, Real R, Alves PC (2012a) Past, present and future distributions of an Iberian endemic, *Lepus granatensis*: ecological and evolutionary clues from species distribution models. PLoS One 7(12):e51529

Acevedo P, Jiménez-Valverde A, Melo-Ferreira J, Real R, Alves PC (2012b) Parapatric species and the implications for climate change studies: a case study on hares in Europe. Glob Change Biol 18:1509–1519

Acevedo P, Melo-Ferreira J, Real R, Alves PC (2014) Evidence for niche similarities in the allopatric sister species *Lepus castroviejoi* and *Lepus corsicanus*. J Biogeogr 41:977–986

Acevedo P, Melo-Ferreira J, Farelo L, Beltrán-Beck B, Real R, Campos R, Alves PC (2015) Range dynamics driven by quaternary climate oscillations explain the distribution of introgressed mtDNA of *Lepus timidus* origin in hares from the Iberian Peninsula. J Biogeogr 4:1727–1735

Águeda-Pinto A, Lemos de Matos A, Abrantes M, Kraberger S, Risalde MA, Gortázar C, McFadden G, Varsani A, Esteves PJ (2019) Genetic characterization of a recombinant myxoma virus in the Iberian hare (*Lepus granatensis*). Viruses 11:530

Almeida J, Santos E, Bio A (2004) Characterization of population and recovery of Iberian hare in Portugal through direct sequential co-simulation. Quantitative Geol Geostat 13:127–138

Altuna J (1970) Hallazgo de una liebre ártica (*Lepus timidus*) en el yacimiento prehistórico de Urtiga (Guipúzcoa). Munibe (San Sebastián) 22:165–168

Alves PC, Acevedo P (2018) *Lepus granatensis* Rosenhauer, 1856 Iberian Hare. In: Smith AT, Johnston CH, Alves PC, Hackländer K (eds) Lagomorphs: Pikas, rabbits, and hares of the world. Johns Hopkins University Press, Baltimore, pp 193–195

Alves PC, Niethammer J (2003) Iberische Hasen *Lepus granatensis* und *Lepus castroviejoi*. In: Niethammer J, Krapp F (eds) Handbuch der Säugetiere Europas. Band 3/II: Hasentiere Lagomorpha. Aula Verlag, Wiebelsheim, pp 137–154

Alves PC, Rocha A (2003) Environmental factors have little influence on the reproductive activity of the Iberian hare (*Lepus granatensis*). Wild Res 30:639–647

Alves PC, Santos N (2021) Sex and Age-Specific Hematology and Biochemistry Reference Intervals of Live Iberian Hares (*Lepus granatensis*) and Comparison with Postmortem Sampling. J Wildl Dis 57:178–183

Alves PC, Goncalves H, Santos M, Rocha A (2002) Reproductive biology of the Iberian hare, *Lepus granatensis*, in Portugal. Mamm Biol 67:358–371

Alves PC, Ferrand N, Suchentrunk F, Harris DJ (2003) Ancient introgression of *L. timidus* mtDNA into *L. granatensis* and *L. europaeus* in the Iberian Peninsula. Mol Phyl Evol 27:70–80

Alves PC, Melo-Ferreira J, Freitas H, Boursot P (2008) The ubiquitous mountain hare mitochondria: multiple introgressive hybridization in hares, genus *Lepus*. Philos Trans R Soc Lond, B, Bio Sci 363:2831–2839

Alzaga V (2010) Ecología de la parasitación de las liebres (*Lepus* spp.) de la Península Ibérica. Tesis doctoral. Universidad de Castilla-La Mancha, España

Alzaga V, Vicente J, Villanua D, Acevedo P, Casas F, Gortázar C (2008) Body condition and parasite intensity correlates with escape capacity in Iberian hares (*Lepus granatensis*). Behav Ecol Sociobiol 62: 769–775

Antoniou A, Kotoulas G, Magoulas A, Alves PC (2008) Evidence of autumn reproduction in female European hares (*Lepus europaeus*) from southern Europe. Eur J Wildl Res 54:581–587

Aranaz A, de Juan L, Montero N, Sánchez C, Galka M, Delso C, Alvarez J, Romero B, Bezos J, Vela AI, Briones V, Mateos A, Domínguez L (2004) Bovine tuberculosis (*Mycobacterium bovis*) in wildlife in Spain. J Clin Microbiol 42:2602–2608

Araújo MB, Guilhaumon F, Rodrigues Neto D, Pozo Ortego I, Gómez Calmaestra R. (2011) Impactos, vulnerabilidad y adaptación de la biodiversidad española frente al cambio climático. 2. Fauna de vertebrados. Dirección general de medio Natural y Política Forestal. Ministerio de Medio Ambiente, y Medio Rural y Marino, Madrid

Arnold ML (2015) Divergence with genetic exchange. Oxford University Press, Oxford, UK

Ballesteros F (1998) Las Especies de caza en España. Biología, ecología y conservación. Estudio y gestión del medio, Oviedo

Ballesteros F (2003) Liebre de piornal, *Lepus castroviejoi* (Palacios, 1976). Galemys 15:3–13

Ballesteros F, Benito JL, González-Quirós P (1996) Situación de las poblaciones de liebres en el norte de la península Ibérica. Quercus 128:12–17

Bartolomé DJ, Alonso ME, Pérez JA, Díez C, Olmedo JA (2004) Habitat selection of the Iberian hare (*Lepus granatensis*) in a plain agro environment depending on the time of the year. In: Alves PC, Ferrand N, Suchentrunk F, Hackländer K (eds) 2nd World lagomorph conference, Vairão, Portugal, 26–31 July 2004. Abstract Book. CIBIO, Research Center in Biodiversity and Genetic Resources, University of Porto, Portugal, p 156

Batista T, Cruz de Carvalho E (1996) Dinâmica de uma população de lebres (estudo de caso). Rev Flor 9: 175–188

Bencatel J, Sabino-Marques H, Álvares F, Moura AE, Barbosa AM (2019) Atlas de Mamíferos de Portugal, 2ª edição. Universidade de Évora, Évora

Bonhomme F, Fernandez J, Palacios F, Catalan L, Machordon A (1986) Caractérisation biochimique du comples d'espéces du genre Lepus en Espagne. Mammalia 50:495–506

Bordes F, Langand J, Feliu C, Morand S (2007) Helminths communities of an introduced hare (*Lepus granatensis*) and a native hare (*Lepus europaeus*) in southern France. J Wild Dis 43:747–751

Bull CM, Possingham H (1995) A model to explain ecological parapatry. Am Natur 145:935–947

Cabral MJ, Almeida J, Almeida PR, Delliger T, Ferrand de Almeida N, Oliveira ME, Palmeirim JM, Queirós AI, Rogado L, Santos-Reis M (eds) (2005) Livro Vermelho dos Vertebrados de Portugal. Instituto da Conservação da Natureza, Lisboa, p 659p

Cabrera A (1914) Fauna Ibérica. Mamíferos. Museo Nacional de Ciencias Naturales, Madrid

Cahais V, Gayral P, Tsagkogeorga G, Melo-Ferreira J, Ballenghien M, Weinert L, Chiari Y, Belkhir K, Ranwez V, Galtier N (2012) Reference-free transcriptome assembly in non-model organisms from next generation sequencing data. Mol Ecol Res 12:834–845

Calzada E, Martínez FJ (1994) Requerimientos y selección de hábitat de la liebre mediterránea (*Lepus granatensis*, Rosenhauer, 1856) en un paisaje agrícola mesetario. Ecología 8:381–394

Carpio AJ, Melo-Ferreira J, Martinez-Haro M, Farelo L, Campos R, Alves PC, Acevedo P (2021) Iberian hares with anciently introgressed mitochondrial DNA express a marginal environmental niche. J Biogeogr. https://doi.org/10.1111/jbi.14155

Carrillo E, Moreno J, Cruz I (2013) What is responsible for a large and unusual outbreak of leishmaniasis in Madrid? T Trends Parasitol 29:579–580

Carro F, Soriguer C (2002) *Lepus granatensis* Rosenhauer, 1856. In: Palomo JL, Gisbert J (eds) Atlas de los mamíferos terrestres de España. Dirección General de Conservación de la Naturaleza- SECEM-SECEMU, Madrid, pp. 452–455

Carro F, Soriguer RC (2007) *Lepus granatensis* Rosenhauer, 1856. In: Palomo J L, Gisbert J, Blanco JC (Eds.). Atlas y libro rojo de los mamíferos terrestres de España. Dirección General para la Biodiversidad-SECEM-SECEMU, Madrid, pp. 476–478

Carro F, Soriguer RC (2010) La liebre ibérica. Serie técnica Naturaleza y Parques Nacionales. Ministerio de Medio Ambiente y Medio Rural y Marino, Madrid, Spain

Carro F, Soriguer RC (2017) Long-term patterns in Iberian hare population dynamics in a protected area (Doñana National Park) in the southwestern Iberian Peninsula: effects of weather conditions and plant cover. Integr Zoo 12:49–60

Carro F, Beltrán JF, Márquez FJ, Pérez JM, Soriguer RC (2002) Supervivencia de la Liebre Ibérica (*Lepus granatensis* Rosenhauer, 1856) en el Parque Nacional Doñana durante una época de inundaciones. Galemys 14:31–38

Carro F, Soriguer RC, Beltrán JF, Andreu AC (2011) Heavy flooding effects on home range and habitat selection of free-ranging Iberian hares (*Lepus granatensis*) in Doñana National Park (SW Spain). Acta Theriol 54:375–382

Carvalho Varela M (1971) Nouvelle contribution a l'etude des helminthocenoses des lievres au Portugal. Anais da Escola Superior de Medicina Veterinaria 13:165–171

Carvalho CL, dos Santos FA, Monteiro M, Carvalho P, Mendonça P, Duarte MD (2020) First cases of myxomatosis in Iberian hares (*Lepus granatensis*) in Portugal. Vet Rec Case Rep 8:e001044

Chapman J, Flux J (1990) Introduction and overview of the lagomorphs. In Rabbits, Hares and Pikas, Status survey and conservation action plan. (Eds) J. Chapman & J. Flux. UICN/SSC Lagomorph Specialist Group, Switzerland. Gland, pp 1–6

Cordero del Campillo M, Castañón Ordóñez L, Reguera Feo A (1994) Índice – catálogo de zooparásitos ibéricos. Segunda edición. Secretariado de publicaciones, Universidad de León

Dalton KP, Martín JM, Nicieza I, Podadera A, de Llano D, Casais R, Gimenez S, Badiola I, Aguero M, Duran M, Romero LJ, Garcia E, Parra F (2019) Myxoma virus jumps species to the Iberian hare. Transbound Emerg Dis 66:2218–2226

De la Calzada E, Martínez FJ (1994) Requerimientos y selección de hábitat de la liebre mediterránea (*Lepus granatensis* Rosenhauer, 1856) en un paisaje agrícola mesetario. Ecología 8:381–394

Delibes-Mateos M, Farfán MA, Olivero J, Márquez AL, Vargas JM (2009) Long-term changes in game species over a long period of transformation in the Iberian Mediterranean landscape. Environ Manag 43: 1256–1268

Dingerkus SK, Montgomery WI (2001) The diet and landclass affinities of the Irish hare *Lepus timidus hibernicus*. J Zool 253:233–240

Duarte J (2000) Liebre ibérica *Lepus granatensis* (Rosenhauer 1856). Galemys 12:3–14

Duarte J, Vargas JM (1998) La perdrix rouge et le lièvre ibérique dans les oliveraies du sud de l'Espagne. Perspectives de gestion de ce type d'habitat. Bulletin Mensuel de l'Office National de la Chasse 236:14–23

Estonba A, Solís A, Iriondo M, Sanz-Martín J, Pérez-Suárez G, Markov G, Palacios F (2006) The genetic distinctiveness of the three Iberian hare species: *Lepus europaeus*, *L. granatensis* and *L. castroviejoi*. Mamm Biol 71:52–59

Farfán MA, Vargas JM, Real R, Palomo LJ, Duarte J (2004) Population parameters and reproductive biology of the Iberian hare *Lepus granatensis* in southern Iberia. Acta Theriol 49:319–335

Farfán MA, Duarte J, Vargas JM (2012) Effects of human induced landuse changes on the distribution of the Iberian hare. J Zool 286:258–265

Fernández A, Soriguer R, Castién E, Carro F (2008) Reproduction parameters of the Iberian hare *Lepus granatensis* at the edge of its range. Wild Biol 14: 434–443

Fernández A, Soriguer R, Carro F, Castién E (2010) Relation between physical condition and reproductive activity in a population of Iberian hares, *Lepus granatensis* in northern Iberian Peninsula. Fol Zool 59:1–8

Fernández-Aguilar X, Alzaga V, Villanúa D, Cabezón O, García-Bocanegra I, Dubey JP, Almería S (2013) Epidemiology and prevalence of *Toxoplasma gondii* infection in the Iberian hare (*Lepus granatensis*). Vet Parasit 196:194–198

Ferreira MS, Jones MR, Callahan CM, Farelo L, Tolesa Z, Suchentrunk F, Boursot P, Mills LS, Alves PC, Good JM, Melo-Ferreira J (2021) The legacy of recurrent introgression during the radiation of hares. Syst Biol 70:593–607

Forest V (2003) Les données archéozoologiques. In: Guilaine J, Escallon G (eds) Les Vautes (Saint-Gély-du-Fesc, Hérault) et la fin du Néolithique final en Languedoc oriental),. Archives d'Ecologie Préhistorique, Toulouse, pp. 147–169

García Pena FJ, Suárez Mayoral P, Cogolludo Cogolludo C, Arriola Garrote C, Anadón Navarro E (1998) Brote de tularemia en la comunidad autónoma de Castilla-León. Primer aislamiento en España de Francisella tularensis. Med Vet 15:418–423

García-Bocanegra I, Camacho-Sillero L, Risalde MA, Dalton KP, Caballero-Gómez J, Agüero M, Gómez-Guillamón F (2019) First outbreak of myxomatosis in Iberian hares (*Lepus granatensis*). Transbound Emerg Dis 66:2204–2208

García-Bocanegra I, Camacho-Sillero L, Caballero-Gómez J, Agüero M, Gómez-Guillamón F, Manuel Ruiz-Casas J, Díaz-Cao JM, Garcia E, Ruano MJ, de la Haza R (2021) Monitoring of emerging myxoma virus epidemics in Iberian hares (*Lepus granatensis*) in Spain, 2018–2020. Transbound Emerg Dis 68:1275–1282

García-Romero C, Corchero J, Valcárcel F (2001) Prevalencia de las helmintosis digestivas y cisticercosis de la liebre ibérica *Lepus granatensis* en la zona central de España. Ecología 15:225–232

Garrido JL, Gortázar C, Ferreres J (2019) Las especies cinegéticas españolas en el siglo XXI. Independently published, Spain

Gayral P, Melo-Ferreira J, Glémin S, Bierne N, Carneiro M, Nabholz B, Lourenco JM, Alves PC, Ballenghien M, Faivre N, Belkhir K, Cahais V, Loire E, Bernard E, Galtier N (2013) Reference-free population genomics from next-generation transcriptome data and the vertebrate-invertebrate gap. PLoS Genet 9:e1003457

Giska I, Farelo L, Pimenta J, Seixas FA, Ferreira MS, Marques JP, Miranda I, Letty J, Jenny H, Hackländer K, Magnussen E, Melo-Ferreira J (2019) Introgression drives repeated evolution of winter coat color polymorphism in hares. PNAS 116:24150–24156

Gortázar C, Millán J, Acevedo P, Escudero MA, Marco J, Fernández de Luco D (2007) A large-scale survey of brown hare *Lepus europaeus* and Iberian hare *L. granatensis* populations at the limit of their ranges. Wild Biol 13:244–250

Herranz J (2000) Efectos de la depredación y del control de depredadores sobre la caza menor en Castilla – La Mancha. PhD Thesis, Universidad Autónoma de Madrid, 338 pp

Hoffmann RS, Smith AT (2007) Lagomorphs. In: Wilson DE, Reeder DM (eds) Mammal speciesof the world, 3rd edn. Smithsonian Institution Press, Washington DC

Jaramillo-Fayad JC, Pérez-Torres J, Pardavila Rodríguez X, Lamosa Torres A (2015) Dinámica espacial de la liebre ibérica (*Lepus granatensis*) en una zona sometida a restauración vegetal (escombrera de lignitos Meirama, A Coruña-España). Acta Zool Mex 31:380–390

Lado S, Farelo L, Forest V, Acevedo P, Dalén L, Melo-Ferreira J (2018) Post-glacial range revolutions in South European hares (*Lepus* spp.): insights from ancient DNA and ecological niche modelling. J Biogeogr 45:2609–2618

Lázaro C, Guzmán JL, Sánchez-García C (2019) Situación de las poblaciones de liebre ibérica (*Lepus granatensis*) tras el brote de mixomatosis en Castilla-La Mancha. I Congreso Ibérico de Ciencia Aplicada a los Recursos Cinegéticos

Lopes AM, Marques S, Silva E, Magalhaes MJ, Pinheiro A, Alves PC, Le Pendu J, Esteves PJ, Thompson G, Abrantes J (2014) Detection of RHDV strains in the Iberian hare (*Lepus granatensis*): earliest evidence of rabbit lagovirus cross-species infection. Vet Res 45:94

López J, Hernandez A, Purroy F, Robles J (1996) Datos sobre la biologia de la reproduction de la liebre iberica (*Lepus granatensis* Rosenhauers, 1856) en agrosistemas cerealistas de la provincia de Leon (NW de España). Revista Florestal 9(1):49–60

Luque-Larena JJ, Mougeot F, Arroyo B, Vidal MD, Rodríguez-Pastor R, Escudero R, Anda P, Lambin X (2017) Irruptive mammal host populations shape tularemia epidemiology. PLoS Pathog 13:e1006622

Marques JP, Farelo L, Vilela J, Vanderpool D, Alves PC, Good JM, Boursot P, Melo-Ferreira J (2017) Range expansion underlies historical introgressive hybridization in the Iberian hare. Sci Rep 7:40788

Martínez R, García A, Blanco JE, Blanco J, Rey J, Alonso JM, Gómez L, Sánchez S (2011) Occurrence of verocytotoxin-producing *Escherichia coli* in the faeces of free-ranging wild lagomorphs in Southwest Spain. Eur J Wildl Res 57:187–189

Mazo AV, Sesé C, Ruiz-Bustos A, Peña JA (1985) Geologia y paleontologia de los yacimientos Plio-Pleistocenos de Huescar (Depresion de Guadix-Baza, Granada). Estud Geol 41:467–493

Melo-Ferreira J, Boursot P, Suchentrunk F, Ferrand N, Alves PC (2005) Invasion from the cold past: extensive introgression of mountain hare (*Lepus timidus*) mitochondrial DNA into three other hare species in northern Iberia. Mol Ecol 14:2459–2464

Melo-Ferreira J, Boursot P, Randi E, Kryukov A, Suchentrunk F, Ferrand N, Alves PC (2007) The rise and fall of the mountain hare (*Lepus timidus*) during Pleistocene glaciations: expansion and retreat with hybridization in the Iberian Peninsula. Mol Ecol 16: 605–618

Melo-Ferreira J, Alves PC, Freitas H, Ferrand N, Boursot P (2009) The genomic legacy from the extinct *Lepus timidus* to the three hare species of Iberia: contrast between mtDNA, sex chromosomes and autosomes. Mol Ecol 18:2643–2658

Melo-Ferreira J, Alves PC, Rocha J, Ferrand N, Boursot P (2011) Interspecific x chromosome and mitochondrial dna introgression in the Iberian hare: selection or allele surfing? Evolution 65:1956–1968

Melo-Ferreira J, Boursot P, Carneiro M, Esteves PJ, Farelo L, Alves PC (2012) Recurrent introgression of mitochondrial DNA among Hares (*Lepus* spp.) revealed by species-tree inference and coalescent simulation. Syst Biol 61:367–381

Melo-Ferreira J, Farelo L, Freitas H, Suchentrunk F, Boursot P, Alves PC (2014a) Home-loving boreal hare mitochondria survived several invasions in Iberia: the relative roles of recurrent hybridisation and allele surfing. Heredity 112:265–273

Melo-Ferreira J, Vilela J, Fonseca MM, da Fonseca RR, Boursot P, Alves PC (2014b) The elusive nature of adaptive mitochondrial DNA evolution of an arctic lineage prone to frequent introgression. Genome Biol Evol 6:886–896

Ministerio de Agricultura (1968) Mapa Cinegético Nacional. Ministerio de Agricultura, Madrid

Molina R, Jiménez MI, Cruz I, Iriso A, Martín-Martín I, Sevillano O, Melero S, Bernal J (2012) The hare (*Lepus granatensis*) as potential sylvatic reservoir of *Leishmania infantum* in Spain. Vet Parasit 190:268–271

Moreno I, Alvarez J, García N, de la Fuente S, Martínez I, Marino E, Torano A, Goyache J, Vilas F, Domínguez L, Domínguez M (2014) Detection of anti-*Leishmania infantum* antibodies in sylvatic lagomorphs from an epidemic area of Madrid using the indirect immunofluorescence antibody test. Vet Parasit 199:264–267

Moreno-Montañez T, Becerra Martell C, Navarrete López-Cozar I (1979) Contribución al conocimiento de los parásitos de la liebre *Lepus capensis*. Rev Iber Parasitol 39:383–393

Olea PP, Sánchez-Barbudo IS, Viñuela J, Barja I, Mateo-Tomás P, Piñeiro A, Mateo R, Purroy FJ (2009) Lack of scientific evidence and precautionary principle in masive release of rodenticides threatens biodiversity: old lessons need new reflections. Environ Conserv 36:1–4

Palacios F (1977) Description de una nueva especie de liebre (*Lepus castroviejoi*), endemica de la Cordillera Cantábrica. Doñana Acta Vertebrata 3(2):205–223

Palacios F (1979) Análisis cromosómico, carga de DNA y electroforesis de las liebres españolas. Doñana, Acta Vertebrata 6(2):203–215

Palacios F (1980) Notas sobre la reproduction en libertad de las especies del género *Lepus* Linne, 1758, en España. In Actas II Reunion Iberoamericana de Zoologia e la Conservacion de la Naturaleza. Caceres, Spain, pp 350–358

Palacios F (1981) Sobre los periodos de reproducción de las liebres ibéricas. Proceedings of Reunión Iberoamericana de Conservación de Zoología de Vertebrados, p 94

Palacios F (1983) On the taxonomic status of the genus *Lepus* in Spain. Acta Zool Fennica 174:27–30

Palacios F (1989) Biometric and morphological features of the species of the genus *Lepus* in Spain. Mammalia 53: 227–264

Palacios F, Fernández J (1992) A new subspecies of hare from Majorca (Balearic Islands). Mammalia 56:71–85

Palacios F, Lopez N (1980) Morfologia dentaria de las liebres europeas (Lagomorpha, Leporidae). Doñana Acta Vertebrata 7:61–91

Palacios F, Lopez-Martinez N (1980) Morfologia dentária de las liebres europeas (Lagomorpha, Leporidae). Doñana Acta Vertebrata 7:61–81

Palacios F, Meijide M (1979) Distribución geográfica y hábitat de las liebres en la península ibérica. Naturalia Hispanica 19:1–40

Palacios F, Ramos B (1979) Situación actual de las liebres en España y medidas para su conservación. Boletín de la Estación Central de Ecología 8:69–75

Palomares F, Ruiz-Martínez I (1994) Die Dichte des Rotfuchses und die Beute an Niederwild wahrend der Periode der Jungenaufzucht im Sudosten Spaniens. Zeitschrift fur Jagdwissenschaft 40:145–155

Palomo LJ, Gisbert J, Blanco JC (2007) Atlas y Libro Rojo de los mamíferos terrestres de España. Dirección General para la Biodiversidad- SECEM-SECEMU, Madrid

Pauperio J, Alves PC (2008) Diet of the Iberian hare (*Lepus granatensis*) in a mountain ecosystem. Eur J Wild Res 54:571–579

Pérez-Suárez G, Palacios F, Boursot P (1994) Speciation and paraphyly in western Mediterranean hares (*Lepus castroviejoi*, *L. europaeus*, *L. granatensis*, and *L. capensis*) revealed by mitochondrial DNA phylogeny. Biochem Genet 32:423–436

Peroux R (1995) Le lievre d'Europe. Número spéecial du Bull. Mens Office National de la Chasse 204:1–96

Pietri C, Alves PC, Melo-Ferreira J (2011) Hares in Corsica: high prevalence of *Lepus corsicanus* and hybridization with introduced *L. europaeus* and *L. granatensis*. Eur J Wildl Res 57:313–321

Robinson TJ, Elder FB, Chapman JA (1983) Karyotypic conservatism in the genus *Lepus* (order Lagomorpha). Can J Genet Cytol 25:540–544

Rodríguez JA, Zapatero-Ramos LM, Sánchez-Covisa A (1981) Variación estacional de Ixodidae y Siphonaptera de lepóridos en la provincia de Ciudad Real. Rev Iber Parasitol 41:527–538

Rodríguez M, Palacios J, Martín JA, Yanes T, Martín P, Sánchez C, Navesco M, Muñoz R (1997) La liebre. Mundiprensa, Barcelona

Rodríguez-Pastor R, Escudero R, Vidal D, Mougeot F, Arroyo B, Lambin Z, Vila-Coro AM, Rodriguez-Moreno I, Anda P, Luque-Larena J (2017) Density-dependent prevalence of *Francisella tularensis* in fluctuating vole populations, northwestern Spain. Emerg Infect Dis 23:1377–1379

Romero-Rodríguez J (1976) Contribución al conocimiento de las coccidiopatías de los Lagomorpha: estudio de los Protozoa - Eimeridae, *E. leporis* y *E. europaea*, parásitas de *Lepus granatensis* (= *L. capensis*). Rev Iber Parasitol 36(1–2):131–134

Romero-Rodríguez J, Guevara Pozo D, Lizcano-Herrera J (1973) Estudios sobre el género Passalurus, Dujardin 1845, parasitando al *Oryctolagus cuniculus domestica* (L.) y *Lepus granatensis*. Rev Iber Parasitol 33: 315–329

Ruiz-Fons F, Ferroglio E, Gortázar C (2013) *Leishmania infantum* in free-ranging hares, Spain, 2004-2010. Eur Secur 18:5–39

Sánchez-Covisa A, Manzanaro M, Vitutia MM, Briones I, Rodríguez-Rodríguez JA (1999) Garrapatas (Acarina: Metastigmata) parásitas de conejos y liebres silvestres en Castilla La Mancha: variación estacional. Med Vet 16:420–423

Sánchez-García C, Alonso MA, Bartolomé DJ, Pérez JA, Larsen RT, Gaudioso VR (2012) Survival, home range patterns, probable causes of mortality, and den-site selection of the Iberian hare (*Lepus*, Leporidae, Mammalia) on arable farmland in north-west Spain. Ital J Zool 79:590–597

Sanz-Martín MJ, Estonba A, Manzano C, Iriondo M, Pérez-Suárez G, Fernández A, Palacios F (2014) Genetic structure of brown and Iberian hare populations in northern Iberia: implications for conservation of genetic diversity. J Wildl Manag 78:632–644

Segovia JM, Vila T, Vargas JM, Fuentes MV, Feliu C (2014) The helminth community of the Iberian hare, *Lepus granatensis* (Lagomorpha: Leporidae), in the province of Granada, Spain. Helminthologia 51: 281–287

Seixas FA (2017) Genome admixture with massive mitochondrial DNA introgression in hares (*Lepus* spp.): the relative roles of demography and natural selection. PhD thesis. University of Porto and University of Montpellier

Seixas FA, Juste J, Campos PF, Carneiro M, Ferrand N, Alves PC, Melo-Ferreira J (2014) Colonization history of Mallorca Island by the European rabbit, *Oryctolagus cuniculus*, and the Iberian hare, *Lepus granatensis* (Lagomorpha: Leporidae). Biol J Linnean Soc 111: 748–760

Seixas FA, Boursot P, Melo-Ferreira J (2018) The genomic impact of historical hybridization with massive mitochondrial DNA introgression. Genome Biol 19:91

Sesé C (1994) Paleoclimatical interpretation of the quaternary small mammals of Spain. Geobios 27(6):753–767

Soriguer R, Carro F (2019) *Lepus granatensis*. The IUCN Red List of Threatened Species 2019: e. T41306A2953195

Tapia L, Domínguez J, Rodríguez L (2010) Modelling habitat use by Iberian hare *Lepus granatensis* and European wild rabbit *Oryctolagus cuniculus* in a mountainous area in northwestern Spain. Acta Theriol 55: 73–79

Vicente FS (1969) On *Dermatoxys hispaniensis* n.sp. (Nematoda: Oxyuridea), from *Oryctolagus cuniculus* and *Lepus timidus* of Spain. J Helminthol 43:417–426

Mountain Hare *Lepus timidus* Linnaeus, 1758

9

Anders Angerbjörn and Stéphanie Claire Schai-Braun

Contents

A. Angerbjörn (✉)
Department of Zoology, Stockholm University, Stockholm, Sweden
e-mail: Anders.angerbjorn@zoologi.su.se

S. C. Schai-Braun
Department of Integrative Biology and Biodiversity Research, Institute of Wildlife Biology and Game Management, University of Natural Resources and Life Sciences, Vienna, Vienna, Austria
e-mail: stephanie.schai-braun@boku.ac.at

K. Hackländer, P. C. Alves (eds.), *Primates and Lagomorpha*, Handbook of the Mammals of Europe,
https://doi.org/10.1007/978-3-030-34043-8_12

Common Names

English: Mountain hare, varying hare, arctic hare, snow hare, Irish hare (*L. t. hibernicus*), blue hare, tundra hare, Alpine mountain hare (*L. t. varronis*), white hare. Since the name "arctic hare" also refers to *L. arcticus*, it should not be used for *L. timidus*. "Tundra hare" is mostly used for *L. othus* and should also be avoided.

English	Mountain hare
German	Schneehase
French	Lièvre variable
Spanish	Liebre variable
Italian	Lepre bianca
Russian	За́яц-беля́к, беляк

Taxonomy and Systematics

Lepus timidus Linnaeus, 1758. Linnaeus did not specify a type locality but it has later been decided to be "Uppsala" Sweden. Other synonyms are *Lepus algidus* Pallas, 1778; renaming of *alpinus* (=*timidus*); *Lepus borealis* Pallas, 1778; renaming of *alpinus* (=*timidus*); *Lepus variabilis* Pallas, 1778; renaming of *timidus*; *Lepus septentrionalis* Link, 1795; renaming of *variabilis* (=*timidus*); *Lepus hibernicus* Bell, 1837, type locality "Ireland." *Lepus sclavonius* Blyth, 1842, based on Polish or Russian trade skins; *Lepus canescens* Nilsson, 1844; renaming of *borealis sylvaticus* Nilsson, 1831 (but *sylvaticus* predates *canescens* and is therefore preferred, see further below). *Lepus lugubris* Kastschenko, 1899, type locality "Onguday in Altai," Russia. *Lepus varronis* Miller, 1901, type locality "Grisons, Heinzenberg, Canton of Graubünden, Switzerland." *Lepus gichiganus* Allen, 1903, type locality "Gichiga, west coast of Okhotsk Sea."

Corbet and Hill (1987) suggested a circumpolar distribution of *L. timidus* with *L. arcticus* and *L. othus* as conspecific with *L. timidus* based on morphological characteristics. This is still (partially) supported by genetic studies based on mitochondrial DNA (mtDNA) even though evidences based only on mtDNA should be treated carefully (Waltari et al. 2004; Wu et al. 2005). A phylogenetic study of the genus *Lepus* based on mtDNA and nuclear DNA (nDNA) suggests that *L. timidus*, *L. arcticus*, and *L. othus* may have started diverging at about 270,000 years ago in a presumably strict allopatric speciation process (Melo-Ferreira et al. 2012). Another phylogenetic study of *Lepus* analyzing whole exome data revealed a potential paraphyly of *L. timidus* with *L. othus*. As sister taxa they likely shared a very recent common ancestor and, hence, the classification as separate species has been debated (Alves et al. 2008; Melo-Ferreira et al. 2012). Although closely related (Ferreira et al. 2021; Melo-Ferreira et al. 2012; Waltari and Cook 2005), the three species are treated nowadays as different species. However, the border between the mountain hare and the Alaskan hare are not clear and might be either in Bering Strait or in Kolyma.

Historically, there have been discussions whether or not the mountain hare includes different subspecies. Some authors defined different subspecies based on morphological characteristics. Ognev (1940) recognized 16 subspecies in Asia based mainly on pelage characteristics including colors, and Hoffmann and Smith (2005) listed 15 subspecies. Corbet (1980) included most of the subspecies in *L. t. timidus*. The European populations of *L. timidus* in Ireland (*L. t. hibernicus*), Scotland (*L. t. scoticus*), and the Alps (*L. t. varronis*), are geographically separated, have some morphological distinct characteristics, and can be treated as subspecies. The other proposed subspecies should be treated as preliminary hypotheses. *L. t. hibernicus* Bell, 1837 (*lutescens* Barrett-Hamilton is a synonym) was considered a species by Barrett-Hamilton (1912) because it remained distinct in the presence of *L. t. scoticus* on the island of Mull since 1860 and no interbreeding between the two subspecies had been known. *L. t. kozhevnikovi* Ognev, 1929, type locality "near the village Tschernaja, district Bogorodski, Mosk. government" *L. t. scoticus* Hilzheimer, 1906, type locality "northern Scotland." *L. t. sylvaticus* Nilsson, 1831 (*canescens* Nilsson is a synonym), type locality "Mörkö, Södermanland," Sweden. *L. t. timidus* Linnaeus, 1758 (*abei* Kuroda, 1938, *algidus* Pallas, *alpinus* Erxleben, *borealis* Pallas, *collinus* Nilsson, *septentrionalis* Link, *sclavonicus* Blyth, *typicus* Barrett-Hamilton, *variabilis* Pallas, are all synonyms). *L. t. varronis* Miller, 1901 (*breviauritus* Hilzheimer is a synonym).

Taxonomy and Geographic Variation

There are preliminary 15 recognized subspecies: *Lepus timidus ainu*, *L. t. begitschevi*, *L. t. gichiganus*, *L. t. hibernicus*, *L. t. kolymensis*, *L. t. kozhevnikovi*, *L. t. lugubris*, *L. t. mordeni*, *L. t. orii*, *L. t. scoticus*, *L. t. sibiricorum*, *L. t. sylvaticus*, *L. t. timidus*, *L. t. transbaicalicus*, and *L. t. varronis*. However, this is mostly based on fur characteristics and no detailed revision has been made with modern methods, namely using genetic markers. Genetic analyses have confirmed support for the European subspecies (Hamill et al. 2006; Thulin et al. 1997) *L. t. timidus, L. t. hibernicus*, *L. t. scoticus,* and *L. t. varronis*. The two subspecies on the islands of Sakhalin and Hokkaido, *L. t. ainu* and *L. t. orii*, are probably also valid due to the geographical isolation. However, since there are no further genetic analyses for the other suggested subspecies, they remain. The main subspecies *L. t. timidus* has a distribution through all Norway, Sweden south to 59 N, all Finland, Russia south to 57 N and east of the Ural mountains. For the hare of southern Sweden, there is no genetic support for a subspecies (*L. t. sylvaticus*); however, there are suggestions that a difference to *L. t. timidus* could be connected to different immigration history with one route from the north and another route from the south (Thulin 2003). In the Scottish Highlands, the distribution of *L. t. scoticus* has been extended by introduction to southwestern Scotland in 1834–62, but also to the English Peak District, north Wales, and to many islands. In Ireland, *L. t. hibernicus* occurs over the whole of island on moorland and pasture down to sea level. *L. t. hibernicus* was introduced to southwestern Scotland, Mull, and the English Lake District about 1890. *L. t. varronis* consists of relict populations in the Alps above ca. 1,300 m in Germany, France, Switzerland, Italy, Austria, and Slovenia. *L. t. kozhevnikovi* is found in Central Russia, from 57 to 58 N, where it apparently intergrades with *L. t. timidus*, and south to 53 N. The western boundary is unclear; most authors refer to eastern Baltic hares as *L. t. timidus*.

Paleontology

The hares probably arose in the Pliocene, and *Lepus* in the early Middle Pleistocene (Kurtén 2007). The first appearance of *L. timidus* was in the Eemian (e.g., Lambrecht Cave, Hungary) and is common in Würm, with numerous Magdalenian sites from the Pyrenees (Altuna 1971) to southeastern Europe (Malez 1972). *L. timidus* was present in southern Poland (the Krakow-Wielun Uplands) in beds from the last glaciation. Hares from the Sudetic Caves in Pleistocene might also be *L. timidus* (Kowalski 1959). Mountain hares were in general common in the Pleistocene (Koby 1960) with remains from Bohemia (Mostecky 1964, 1969) and Germany (Heller 1960; Luttschwager 1956). In central Europe, *L. timidus* appears at the end of the Riss glaciation in the Bocksteinhöhle in Württemberg, Germany of Acheulian age. *L. t. varronis* can be recognized from the Mousterian and became abundant in the upper Palaeolithic, especially the Magdalenian (Couturier 1981). The general pattern of distribution suggests an origin in Asia or even North America, spreading eastward to Greenland, and westward to Britain. There are many remains in southern England, Italy, Spain, and in glacial deposits in Crimea (Dahl 1928). The genetic differentiation in Europe suggests an origin from a panmictic population during Weichsel (Suchentrunk et al. 1999). *L. timidus* is clearly cold adapted and the distribution over large parts of Siberia and Europe, including Iberia, during cold periods was probably related to an expansion at about 164,000 years BP, that is, before the last glaciation (Melo-Ferreira et al. 2007). After the last glacial maximum, *L. timidus* expanded north and to alpine regions to previously glaciated areas tracking its habitat (Smith et al. 2017), for example, Scandinavia (Thulin 2003) and the Alps, but was replaced in southern areas by *L. europaeus* and *L. granatensis* (Melo-Ferreira et al. 2007). Contrary to the once continuous distribution, at present, Central European mountain hares are restricted to the Alps and secluded forests in Poland. Furthermore, mountain hares occur in Ireland and in the Scottish Highlands.

L. timidus was eaten by humans 28,000 years BP in Belgium (Gautier 1973) and remains are common in Pleistocene middens (Adam 1982). During the Mesolithic period, hares on Gotland in the Baltic Sea are genetically related to modern hares from Russia, whereas hares from the Neolithic and onwards (2848 BCE–1641 CE) cluster with Neolithic hares from the Swedish mainland (Ahlgren et al. 2016). This suggests that humans reintroduced hares to Gotland already this early.

Current Distribution

The geographic distribution of the mountain hare in tundra and taiga habitats spans from northern Europe through Russia and Siberia to Japan in the east (Fig. 1). It occurs all over Norway, Finland, and Sweden, including the Baltic islands Öland, Gotland, and Åland. It has also been introduced to many small islands. It is distributed in Estonia, including Hiiumaa and Saaremaa, and Latvia and Lithuania. In Poland, it is found only in the isolated Augustow and Rominty forests (54 N). Isolated alpine populations occur above 1,300 m in the mountains of southern Germany, with natural postglacial occurence also in Austria, France, Italy, Slovenian Alps, and all Switzerland except Jura. It has never been present on Iceland. *L. arcticus groenlandicus* and *L. t. timidus* were introduced into Spitzbergen in 1930 and 1931 (Myrberget 1972), but are gone now. A hybrid of *L. t. timidus* and *L. t. sylvaticus* was introduced into the Faroe Islands in 1854. Mountain hares have a natural origin in Scotland, Ireland, and the Isle of Man. There have been many introductions (mainly in the 19th century) to for example the Shetland Islands (Mainland 1907; Vaila 1900), Orkney, Outer Hebrides, Skye, Raasay, Scalpay, Mull, Arran, Bute, Jura, Eigg, and Islay (now extinct). However, the distribution has decreased in several areas, especially at the southern range all through Europe and Russia (Thulin 2003). In Russia, it is found all through the taiga and tundra from Kola Peninsula to Kamchatka southward to 53°N with isolated pockets to 49°N.

The range of the mountain hare in Scandinavia, the subspecies *L. t. timidus*, includes all Norway, Sweden south to 59°N, all Finland, Russia south to 57 to 58°N, and east of Ural mountains. Polish hares may belong to this subspecies but exact determination is not possible (Cabon-Raczynska 1963). *L. t. sylvaticus* is the hare of southern Sweden, with hybrid zones with *L. t. timidus* to west Latvia, south coast of Norway, and Faroes (introduced). *L. t. kozhevnikovi*, the Central Russian hare, ranges from 57 to 58°N, where it apparently intergrades with *L. t. timidus*, south to 53°N. The western boundary is unclear and most authors refer to eastern Baltic hares as *L. t. timidus*, although Ognev (1940) suggests *L. t. kozhevnikovi* reaches St. Petersburg (formerly Leningrad).

The original distribution of *L. t. scoticus* in the Scottish Highlands has been extended by introduction to southwestern Scotland in 1834–62 (Ritchie 1929), the English Peak District, north Wales, and many islands (Yalden 1984). *L. t. hibernicus* occurs over the whole of Ireland on moorland and pasture down to sea level. *L. t. hibernicus* was introduced to southwestern Scotland, Mull, and the English Lake District about 1890 (Corbet and Southern 1977). *L. t. varronis* consists of relict populations in the Alps above ca. 1,300 m in Germany, France, Switzerland, Italy, Austria, Liechtenstein, and Slovenia.

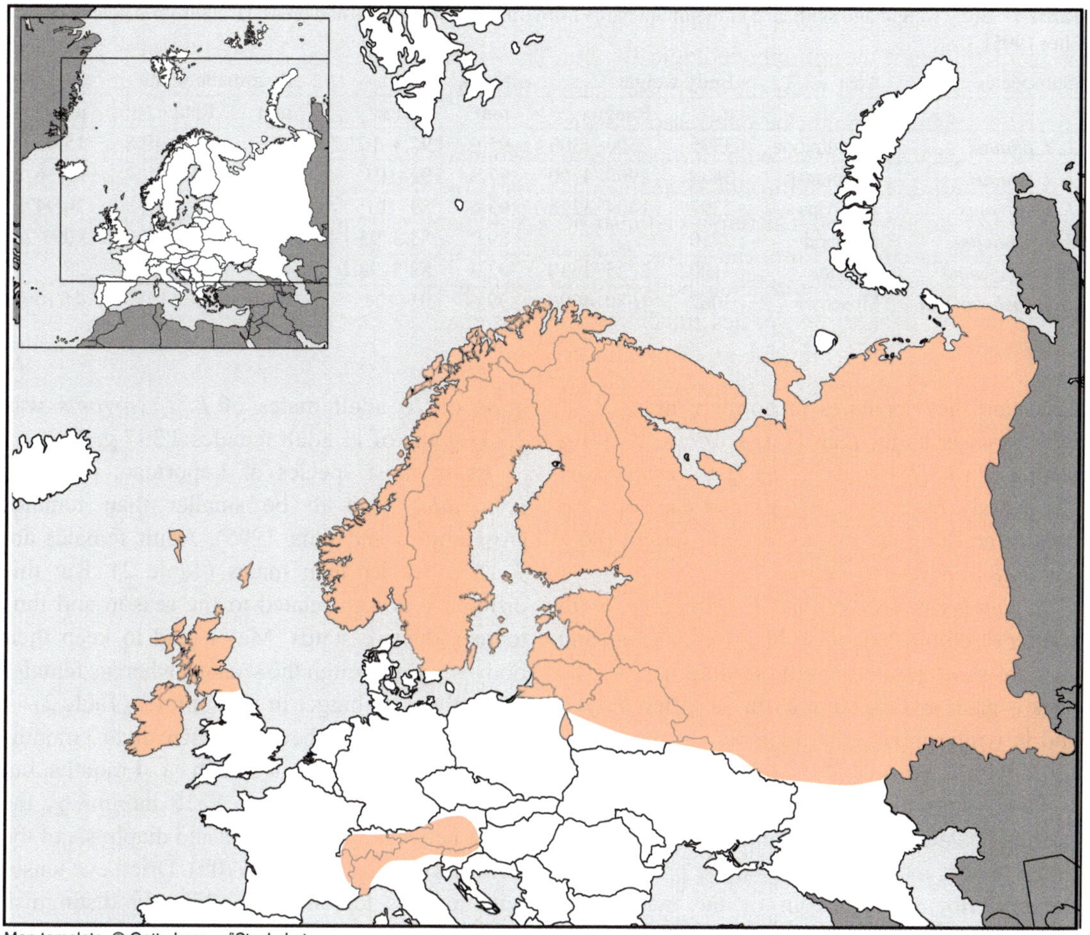

Map template: © Getty Images/iStockphoto

Fig. 1 Distribution map *Lepus timidus* – for source of data please see text. (Map template: © Copyright Getty Images/iStockphoto)

Description

General Characters

The mountain hare is large, but not as large as *L. europaeus* especially where they are sympatric. The relatively large head has prominent eyes with a yellowish iris in the adult but dark brown in the young. They are larger in northern populations and subspecies, supporting Bergmann's Rule. The basal length of the skull is greater in northern populations, as indicated by the following measurements (in mm): Ireland, 73; Scotland, 70; Scandinavia, 73.2; northern Scandinavia and Russia, 77.8; northern Siberia, 87.5 (Barrett-Hamilton 1900). The European subspecies show a similar trend where the smallest subspecies *L. t. varronis* has a mass of 2,397 g (1,404–3,228); length of head and body, 509 mm (460–560); tail length, 53–65 mm; hind foot length, 159 mm (145–172); ear length, 98–106 mm, n = 94. This can be compared to the largest subspecies *L. t. timidus* (Angerbjörn and Flux 1995), which at Murmansk had a mass of 3,438 g (2,900–4,700); length of head and body, 546 mm (430–610); length of tail, 63 mm (40–70); hind foot length, 164 mm (150–180); ear length, 88 mm (76–100, n = 29). There are several studies reporting on body weight, skull length, and zygomatic width (Table 1). This demonstrates the basic pattern with

Table 1 Body weight and skull size in mountain hares from different subspecies and areas. (Data from Angerbjörn and Flux 1995)

Subspecies	Area	Body weight		Skull length		Zygomatic width		
		g	Range	mm	Range	mm	Range	n
L. t. timidus	Scandinavia	3395	2500–4300	95.6	92.3–103.2	47.5	45.3–49.8	15/68
L. t. timidus	Murmansk	3438	2900–4700	97.8	92–103	48.9	46–52	29/36
L. t. varronis	The Alps	2397	1404–3228	93.6	89–101	46.1	43–49	20/94
L. t. scoticus	Scotland	2750		89.2	83.3–95.1	44.2	42.4–45.8	1097/26
L. t. hibernicus	Ireland	3190	2725–3630	93.3	88.5–98.1	45.2	43.1–47.7	28
L. t. kozhevnikovi	Moscow	3062	1680–4590	99.6	95–104	48.7	45–51	41/100

larger and heavier hares in northern areas. *L. t. hibernicus* is larger than *L. t. scoticus*, 3,190 g compared to 2,750 g (Table 1). *L. t. kozhevnikovi* has a body mass of 3,062 g. The ears of *L. t. timidus* are long relative to its skull length (86.7 and 94.8 mm, respectively).

L. timidus is smaller than *L. europaeus*, with shorter forelimbs but longer hind feet. The tail and ears are shorter, and the tail is white below. The back is paler and the summer fur is softer, grayer, and in winter most individuals turn white. Compared with *L. europaeus*, *L. timidus* has a shorter facial cranium, more concave frontal region, longer bony palate; a stronger mandible, deeper in the molar region, more vertical condylar process, and the posterior inner margin of the symphysial region is not as sharp; anterior branch of supraorbital process is less distinct; foramen mesopterygium often is broader than incisive foramen; zygomatic arch is not less deep toward the front, and anterior process is less developed (Fig. 2). The upper incisors are squarer in cross section and less curved, thus the roots rise between the premaxilla and maxilla, farther posterior than in *L. europaeus*. The third lower premolar is erect in *L. timidus*, but curved backward in *L. europaeus*. The tibia and phalanges of the hind foot are longer in *L. timidus* than in *L. europaeus*, and the radius of the forearm is shorter (Angermann 1967). The scapula is broader (especially so in *L. t. hibernicus*) than in *L. europaeus* (Barrett-Hamilton 1912).

L. t. sylvaticus has shorter ears (86.2 mm) relative to skull length (97.7 mm). The adult mass of *L. t. timidus* in Norway was 3,395 g. In Russia, the mass of *L. t. timidus* was 3,438 g. The average mass of 20 adult males of *L. t. varronis* was 2,337 g and of 11 adult females 2,397 g.

As in most species of Leporidae, males of *L. timidus* tend to be smaller than females (Angerbjörn and Flux 1995). Adult females are 7–15% heavier than males (Table 2). But this difference is also related to the season and thus to reproductive status. Males tend to keep their body weight through the seasons whereas females naturally grow bigger in the summer (Table 2).

Young hares can be aged by size up to 3 months of age. They reach adult size in ca. 4 months, but can be distinguished up to 8–10 months by the notch between the apophysis and diaphysis of the tibia and humerus (Flux 1970b). Dried eye lenses with a mass less than ca. 250 mg distinguish young up to 1 year old from older individuals (Walhovd 1965). However, this depends on subspecies. For instance, in the smallest subspecies *L. t. varronis* the distinction between subadult and adults is made at 210 mg (Schai-Braun et al. 2017). Growth of the supraorbital process with age also has been used (Walhovd 1965), but annual increments in the mandible provide the most accurate technique (Ohtaishi et al. 1976).

Fat deposits in both sexes are very low from June to October, increasing to a peak in January–March, and declining when breeding starts. Fat and body mass are positively correlated for all age and sex groups (van der Merve and Racey 1991). Artificial winter feeding in Finland allows hares in poor habitats to put on more fat in winter (Soveri and Aarnio 1983). In Sweden, body condition affects the number of young weaned (Angerbjörn 1986b).

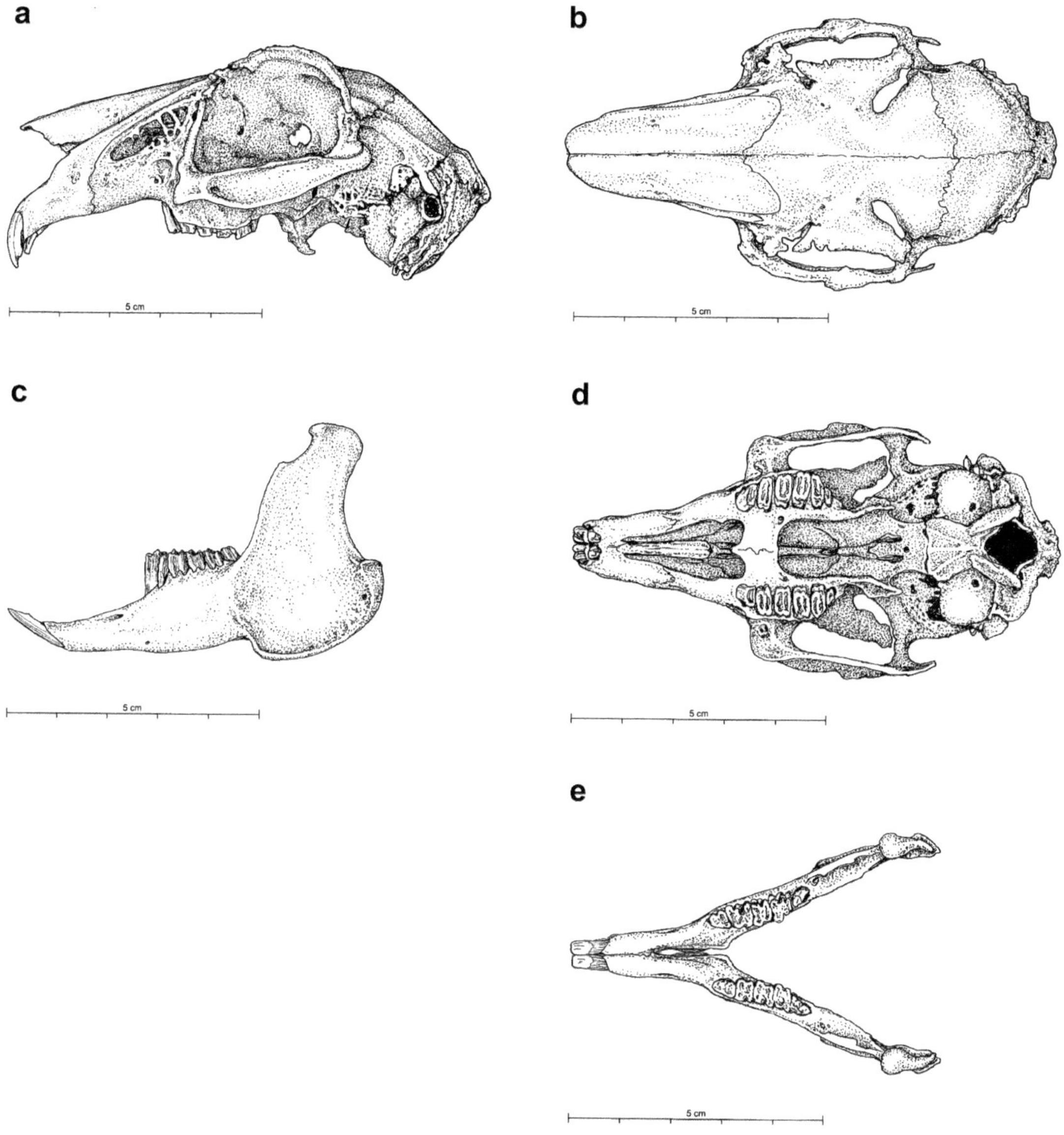

Fig. 2 Skull of *Lepus timidus*. (**a**, **b**, **d**) lateral, dorsal, and ventral view of skull, (**c**) lateral view of left mandible (view from outside), (**e**) dorsal view of mandible. (© Franz Müller with kind permission)

The dental formula is: I^2, C^0, P^3, M^3 and I_1, C_0, P_2, M_3, total 28. Mammary glands are 2 pectoral +6 abdominal, total 8.

The ears are short and narrow with a black tip 30 mm long on the exterior surface and ca. 5 mm long on the interior surface. The hind feet are long, with hairy soles and widely spreading toes. The head and body are dusky brown in summer, and often richer brown on head, with gray-blue underfur showing through especially on the flanks. The limbs, lower throat, and upper breast are lighter than the back. The chin, upper throat, and insides of limbs are dirty white; the abdomen and tail are white. The line of demarcation is not abrupt and varies with season (Barrett-Hamilton 1912). The coat tends toward reddish brown at low latitudes (*L. t. hibernicus*, and *L. t. kozhevnikovi* in the region of Moscow) but is

Table 2 Body size (g) in female and male mountain hares and sexual dimorphism from different subspecies

Subspecies	Area, sex	Body weight (g)	n	Sex dimorphism	Ref
L. t. scoticus	Scotland female	2890	590	15%	Flux (1970b), Hewson (1968)
L. t. scoticus	Scotland male	2510	507		Flux (1970b), Hewson (1968)
L. t. scoticus	Scotland female winter	2850	218	8%	Flux (1970b), Hewson (1970)
L. t. scoticus	Scotland female summer	3000	372	15%	Flux (1970b), Hewson (1968)
L. t. scoticus	Scotland male winter	2650	206		Flux (1970b), Hewson (1968)
L. t. scoticus	Scotland male summer	2610	301		Flux (1970b), Hewson (1968)
L. t. timidus	Norway female	3504	37	8%	Walhovd (1965)
L. t. timidus	Norway male	3257	30		Walhovd (1965)
L. t. timidus	Norway northern	3345	47		Walhovd (1965)
L. t. timidus	Norway southern	3063	37		Walhovd (1965)
L. t. hibernicus	Ireland female	3289	13	7%	Barrett-Hamilton (1912)
L. t. hibernicus	Ireland male	3070	14		Fairley (1974)
L. t. kozhevnikovi	Moscow female	3758	158	13%	Ognev (1940)
L. t. kozhevnikovi	Moscow male	3336	217		Ognev (1940)

grayish brown at high altitudes and toward the arctic, probably matching the color of the habitat (Ognev 1940).

Molting processes in hares and other mammals have been reviewed by Zimova et al. (2018). The white winter fur is related both to thermoregulation, with better insulation, and to camouflage. Most scientists have reported two molts per year, but a detailed study showed three molts in *L. t. scoticus*: spring, white to brown, February–May; autumn, brown to brown, June–September; winter, brown to white, October–February (Hewson 1958). There are two morphs of winter coat, white = W (dominant) and blue = w (Bergengren 1969). In white morphs, *L. t. timidus* and *L. t. varronis*, the coat is finer, woollier, and the external appearance is almost pure white except for a small brown spot on the nose and above the eye, and black ear tips. In blue forms, *L. t. scoticus* and *L. t. sylvaticus*, the fur is 15 mm long, supporting fibers of winter underfur 25 mm at mid-back, black underfur hair 25 mm long with white or pale buff subterminal bands, white or white-tipped guard hairs, and white or black and white vibrissae. The front of the ears is gray-brown, with a whitish back, the nose and forehead are rust brown with a brown spot over each eye, and little rust-brown below it. The outside of the forelimbs and the top of the paws are rust brown; the hind feet are white, but rust brown between the toes; back, sides, and flanks are pale blue-gray with few black hairs, and the base of the fur is ash-blue with a light brown layer above (Bergengren 1969). In European Russia, the underfur of *L. t. timidus* is snow white, and in *L. t. kozhevnikovi* it is straw white with a blue form occasionally present in the southwest (Ognev 1940); only 10% are pure white in Belarus (Gaiduk 1975). The winter fur of *L. t. hibernicus* is typically brown, but with white rump, flanks, and legs. A sprinkling of reddish hair is always present.

The summer coat in *L. t. scoticus* consists of gray hairs that are brown toward the tips, ca. 15 mm long. The underfur hair is ca. 25 mm long, dark gray with 2–4 mm brown subterminal band, and a black tip. The guard hairs are ca. 40 mm long and black, although some white ones may persist from the winter coat (Hewson 1958). The head is rich brown, and the nape is gray-brown and woolly. There are variations in color of the upper parts from warm brown overlying dark gray-black to pale brown, due to difference in color and length of the subterminal band of underfur hair. In *L. t. sylvaticus*, the summer coat is yellowish gray-brown, blending evenly with an off-white abdomen. The dark upper parts of *L. t. timidus* contrast sharply with the white underside (Bergengren 1969). The head of *L. t. kozhevnikovi* is rich rust-brown, and the body is the same with grayish-black mottling; the underfur has gray roots, and dull rust tips (Ognev 1940). The summer coat of *L. t. hibernicus* is russet, not smoky brown or "blue," and the coat of *L. t. varronis* is dull smoky gray mixed with brown, faint red and pale yellow, with the extremities lighter and more reddish; the chest is yellowish gray (Couturier 1981). The coat of juveniles is softer and duller than that of adults, and has fewer guard hairs. Growth of late-season young may be stunted, and they may not acquire a full winter coat. Albino, melanic, and buff variations in coat color have been reported from several areas. Black melanistic hares appear regularly in both southern and northern Sweden (Angerbjörn unpubl.). All Faroese hares are winter-gray, a morph that occurs in the source population from Fennoscandia at low frequency and prevailed by positive selection due to local adaptation within ~65 y on the Faroe Islands (Giska et al. 2019). The variant causing this winter-gray phenotype introgressed from the Iberian hare (*L. granatensis*) (Giska et al. 2019).

The molt to the white winter coat (Fig. 3) begins on the dorsum and is relatively rapid; the spring molt, in contrast, begins ventrally. The color change is slower in cold springs. Females turn brown ca. 2 weeks before males (Flux 1970a). Males also start the molt earlier than females during the autumn. Young hares delay their molt and those born later in the season molt even later (Bergengren 1969). Hares at higher altitudes and latitudes remain white longer, in correlation with the average ambient temperature (Bergengren 1969), and hence with the duration of snow cover. The timing of molt, however, is strongly influenced by photoperiod (Zimova et al. 2018) with an endocrine regulation. Dominant males may molt earlier than less aggressive (Verkaik and Hewson 1986) but sick animals show a delayed molt during the autumn. The spring color change is initiated by daylength and its rate is influenced by temperature (Flux 1970a). Molt for hares within a population tend to have an identical onset each year but in cold temperature, and there will be an accelerated growth of white coat (Zimova et al. 2018). But typically there will be hares of all molting stages on the same day (Bergengren 1969). A similar process occurs in autumn, but the extent of late winter whitening is more closely correlated with snow cover (Jackes and Watson 1975).

Physiology

Mountain hares may eat snow in the winter but are rarely seen to drink. The energy requirement for maintenance in winter is about 105 kcal/kg/day, with a maximum intake of 150 kcal (Pehrson 1984). At least 500 g/day of *Betula* and *Salix* food is eaten (Sperber 1974). The feces, pellets, are small round "pearls" with mostly fibers about 1 cm round deposited while hares are feeding and therefore often spread out. They have many nicknames, for example, turds. The hares produce two different sources of feces where the first sort is black and mucous. These soft pellets are consumed directly during daytime while the hare is resting in the lair. The hard pellets are from the second sort. The reingestion is important for the nitrogen uptake (Pehrson 1984). Reingestion increases digestibility up to 25%, the soft pellets containing three times as much protein (Pehrson 1983a), and takes place from 0900 to 1600 h; 208 hare pellets (33–450, Flux 1970b), or 200–450 (Lampio 1952), are produced daily. Alpine mountain hares did not have higher fecal glucocorticoid metabolites (GCM) values in winter than in the other seasons. This suggests that *L. timidus varronis* copes with harsh

Fig. 3 Mountain hares from different areas. (**a**) From northern Sweden. (© Per Klaesson). (**b**) From southern Sweden, in winter pelage. Hållö, Bohuslän, Sweden, 1977. (© A. Angerbjörn). (**c**) From the Alps. (© Rolf Giger). (**d**) From Ireland. (© Mike Brown). (**e**) From Scotland. (© Mills lab research photo)

environmental conditions by reducing metabolic rate instead of depleting fat reserves as elevated glucocorticoid secretion facilitates catabolism (Rehnus et al. 2010). GCM concentrations were influenced by season and sex. Within the mating period, females had higher GCM levels than males, but individual differences were more expressed in males. In the post-reproductive period, GCM concentrations were neither influenced by sex nor individual (Rehnus and Palme 2017).

Genetics

The karyotype is 2n = 48. *Lepus* is highly conservative karyotypically, with G-banding patterns identical in most species (Robinson et al. 1983; Schröder et al. 1987). Marques et al. (2017a) characterized and annotated the first hare transcriptome, belonging to *L. timidus*. Polymorphism was detected in two analyzed populations (*L. t. varronis* and *L. t. hibernicus*) and 4672 putative diagnostic sites were identified for the two hare species *L. timidus* and *L. europaeus*. A subset of 85 random independent Single Nucleotide Polymorphisms (SNPs) was successfully validated. These genomic resources can be used to design tools to assess population status and monitor hybridization between species. Marques et al. (2020) reported the first de novo draft reference whole genome for a hare species, from an Irish *L. timidus*. A total of 24,578 protein coding genes were annotated, of which 683 were solely derived from hare-specific transcriptome data. The hare reference genome is a resource to discover and investigate hare-specific variation. The results suggest that the chromosome-scale hare draft genome should enable chromosome-wide analyses and genome scans on hares. *L. timidus* has a high contemporary genetic diversity with no clear differentiation between geographically isolated populations. Neither habitat shifts nor demographic fluctuations at the last glacial/interglacial transition have had any substantial impact on the genetic diversity of mountain hares inhabiting Europe (Smith et al. 2017). This remarkable resilience, which is in contrast to a majority of other cold-adapted species, is likely due to its generalist nature that may make it less vulnerable to environmental changes (Smith et al. 2017).

An investigation of the molecular regulation of the color molts in *L. timidus* found 632 differentially expressed genes, with a major pulse of expression early in the molt, followed by a milder one in late molt (Ferreira et al. 2020). The functional makeup of the differentially expressed genes anchored the sampled molt stages to the developmental timeline of the hair growth cycle. The progression of color change was characterized by differential expression of genes involved in pigmentation, circadian, and behavioral regulation. Significant overlap existed between differentially expressed genes across the seasonal molts of *L. timidus* and the snowshoe hare (*L. americanus*), particularly at molt onset, suggesting a conservatism of gene regulation across species and seasons (Ferreira et al. 2020). Additionally, the transcriptomic work showed that genes with functions in pigmentation and circadian rhythm are involved in the regulation of the color molt in *L. t. varronis*.

Hybrids with *L. europaeus* occur in Sweden where predominately mountain hare females mate with European hare males (Thulin 2003). The reciprocal hybridization is also possible, but is very rare due to the typical mating behavior of both hare species in natural surroundings (Flux 1970b; Hewson 1990; Holley 1986; Wiley and Poston 1996). This results in unidirectional introgression where DNA from mountain hares will be present in the DNA from European hares. The hybrids are morphologically intermediate between the species and are fertile (Lönnberg 1905). Analysis of single nucleotide polymorphic markers supports that the introgression is highly asymmetrical from mountain hare to European hare, but the levels of nuclear gene introgression are independent of mtDNA introgression (Levänen et al. 2018b). Hybrids' distribution in Scandinavia has been studied based on mtDNA (Levänen et al. 2018a) and SNPs (Levänen et al. 2018b). About 15% of the genetically analyzed European hares in areas of current sympatry in Sweden were hybrids (Thulin and Tegelström 2002), whereas about 4% of the analyzed Alpine mountain hares in Grisons were hybrids (Zachos et al. 2010). There is genetic evidence of similar hybridization in the Iberian Peninsula where ancient introgression of mountain hare DNA into *L. granatensis, L. europaeus* and the broom hare, *L. castroviejoi* has been described (Melo-Ferreira et al. 2005). Analyses of both nuclear and mtDNA show that *L. granatensis* has extensive mitochondrial DNA introgression from *L. timidus* (Alves et al. 2003; Melo-Ferreira et al. 2005). *L. granatensis* replaced *L. timidus* after the last glacial maximum demonstrating that the cold adapted *L. timidus* was replaced by the more warm adapted *L. granatensis* (Marques et al. 2017b; Seixas et al. 2018). An introgression of *L. timidus* DNA into *L. castroviejoi* and *L. europaeus* has also been detected, but its pattern is more unclear. Outliers of elevated introgression include several genes related to immunity, spermatogenesis, and mitochondrial metabolism. Introgression is reduced on the X chromosome and in low recombining regions (Seixas et al. 2018). Some slight introgression of individual haplotypes from *L. timidus* to

L. europaeus in Central and Eastern European countries (Hungary, Poland, Serbia, Lithuania, Romania, Georgia, and Italy) are also documented (Ashrafzadeh et al. 2018).

Life History

The gestation period of *L. t. timidus* in captivity is 50.3 days (47–55, n = 192) with no difference between first (50.8) and second (50.0) litters (Höglund 1957a). There is no difference in gestation period between northern and southern Sweden (50.0, 47–54, n = 50 compared to 50.1, 47–53, n = 35) (Angerbjörn 1986b; Bergengren 1969), and it does not vary with nutritional levels (Pehrson and Lindlöf 1984). Postpartum estrus, with copulation a few hours after parturition, is the rule as in many hares (Höglund 1957a). Superfetation where a female carries embryos of two different development stages is not as common as in *L. europaeus* (Borg et al. 1952), but does occur (Collett 1912). Differences in breeding exist between populations and/or subspecies. As an example, *L. t. timidus* has normally one or two litters per year in northern Norway and Sweden, but can have three in years with an early spring (Angerbjörn 1986b; Myrberget 1983). *L. t. timidus* in Finland has most frequently two litters (Kauhala et al. 2005b). *L. t. sylvaticus* of southern Sweden regularly has three litters, and starts breeding in the first half of March. Another difference is the length of the reproductive period with *L. t. hibernicus* having a longer one than *L. t. scoticus* (Barrett-Hamilton 1912; Fairley 1974). The photoperiodic control of reproduction acts through the pineal organ and its secretion of mela-tonin (Iason 1987).

In mountain hares, the testes are abdominal and in a regressed state during the non-reproductive season, from July to December, but are scrotal and well developed (7–9 g) from February to May. Ovaries size of 34 pregnant females in Scotland were 20.8 by 9.8 mm. Graafian follicles reach 3–5 mm in diameter before rupturing at estrus, and corpora lutea grow from 7.4 mm (5–10, n = 23) to 11.0 mm (8–14, n = 27) during gestation. Start of breeding varies with temperature and gestation has been recorded from January/February until August/September. Pregnant females were found during February–August and all adults were pregnant March–June. However, juvenile females did not breed in the year of their birth. The average number of corpora lutea, 2.4 (1–5), increased with age, body weight, and seasonally from 1.8 (February) to 3.7 (June), followed by a decline to 2.7 (August). The mean monthly number of embryos per female in Scotland was 2.1, and the average number of litters was 2.6, giving an annual production of 5.4–6.3 young per female despite 47% prenatal mortality (Flux 1970b). In Finland the average number of young per female was 6.1 (Kauhala et al. 2005b). The weather and climate have an important influence on the number of litters produced within one reproductive season. On less productive habitats, litters were smaller and breeding began later. Among 1,331 females, pregnancies in Scotland occurred January–September, with over 50% pregnant March–July (Hewson 1970). But only 64% of females were fertile for the first litter, but fertility rose to 95.8 and 87% for second and third litters, respectively. Hares with a large first litter had small third litters and vice versa (Iason 1990).

The second litter is on average larger than the first one. In Sweden, the first litters averaged 2.15 young (1–5, n = 342) and second litters 3.24 (1–8, n = 311). The mean number of litters, and of young per litter, was highest for hares 3–6 years old and declined at 8–10 years (Höglund 1957a). In northern Russia (Yakutia), there is only one litter per year (from 6.4 to 6.9 young) (Naumov and Shatalova 1974), but in the southwest (Belarus) three or four litters are usual. First litters averaged 2.0 young, second, 3.1; third, 3.0; and fourth, 2.7 (Gaiduk 1973). Near Moscow, first litters averaged 5.0 corpora lutea and 3.5 embryos, and second litters 4.5 and 3.8, respectively (Naumov and Shatalova 1974). For *L. t. varronis,* there are usually two or three litters of 2–5 and the young are born April to July (Baumann 1949). *L. t. varronis* has on average 8.6 young per female per reproductive season (Schai-Braun et al. 2017). Elevation affected litter size in *L. t. varronis*, whereas the elevation did not influence the number of litters. Hence, some reproductive parameters seem to indicate that the Alpine mountain hare changes the reproductive strategy in relation to elevation similar to the mountain hares living further north do in relation to latitude (Schai-Braun et al. 2017).

At birth, the young are fully furred with their eyes open, and they start suckling at once (Höglund 1957a). In *L. t. scoticus*, the birth-mass of nine full-term embryos averaged 87.3 g (71–106, Flux 1970b); *L. t. hibernicus*, 167 g and 182 g (n = 2, Fairley 1974); *L. t. timidus*, 61–83 g (n = 5, Walhovd 1965), 109–150 g, depending on quality of food fed to the mother, and 80–130 g depending on litter size (Pehrson and Lindlöf 1984), and on mother age and mass (Angerbjörn and Pehrson 1986); *L. t. kozhevnikovi* 84–140 g (Ognev 1940). The growth rate is 14 g per day (7–21, n = 9) in *L. t. scoticus* (Flux 1970b; Hewson 1968), 23 g (14–30, n = 4) in *L. t. timidus* in Norway (Walhovd 1965), and 30 g per day (n = 11) in captivity (Pehrson and Lindlöf 1984). In captivity, the young hares depend on milk for 10–20 days (Borg et al. 1952) but those of the final litter may continue to suckle for 6 weeks (Flux 1970b). Young born early in the breeding season grow for a longer period, and reach a larger size (hind foot length and body mass) compared to young that are born later. These show instead a more rapid growth rate but for a shorter period (Iason 1989a, b). In captivity, growth of *L. t. timidus* during lactation was related to birth mass, and juvenile females grew faster than males (Angerbjörn and Pehrson 1986). Over a longer period of growth (up to 320 days), juvenile males had a faster growth rate as indicated by hind foot length, than females; females continued to grow for a longer period, however, and ended up averaging larger (Iason 1987).

Sex can be determined at birth by the circular genital aperture in males (Höglund and Viklund 1953). The sex ratio is equal, with 49.3% male (n = 140) in captive Swedish hares (Bergengren 1969). Among adults, females often predominate slightly but this varies with local conditions (Angerbjörn 1986a).

A sample from hunting statistics in *L. t. scoticus* from August to January showed that the percentage of young was 43%, increasing seasonally from 22% young in August to 55% young in December. Based on lens weights, 48% were young, 32% 1 year old, and the remaining 29% 2 years or older indicating ca. 50% juvenile mortality per year (Flux 1970b). In Norway, 63% were young (Walhovd 1965), and on a Swedish island the ratio of young to adults in autumn reached 84% and the ratio was negatively correlated with spring population density (Angerbjörn 1986b). A few *L. t. scoticus* reach an age of 9 years in the wild (Hewson 1976), with a maximum of 18 years (Angerbjörn and Flux 1995). The recruitment of juvenile hares can be related to population density but fecundity seems to be independent of density (Knipe et al. 2013).

Both adult and juvenile survival varies considerably between populations from 6 to 88% annually (Dahl 2005; Flux 1970b; Hewson 1976). Among captive hares, there was an average mortality of 30% for hares 1–8 years old (Höglund 1957a). In Sweden, the average winter survival for adult hares was 0.42 at low predation pressure, and 0.19 under high predation. Juvenile survivorship was 0.75 from weaning to autumn, 0.36 during a winter with low predation pressure, and 0.16 during a winter with high predation. Overall juvenile survival from birth to the following spring averaged 0.20 (Angerbjörn 1992). However, on islands with no or only few red foxes (*Vulpes vulpes*) survival rates were between 14 and 86% (Angerbjörn 1989; Marcström et al. 1989).

Habitat and Diet

The mountain hare is predominantly a species of mixed and coniferous forest in Europe (Naumov and Shatalova 1974) with the highest densities in transition zones with open clearings, in swamps, river valleys, and patches of regrowth among fallen trees. In Finland, bogs with *Salix* and *Vaccinium* are favored, followed by stands of spruce (*Picea*) and birch (*Betula*). Fewer hares live in less favorable habitats such as in pine forests or birch forests on windy slopes. Such habitats may be occupied when populations are high (Pulliainen 1983). Hares selected thickets with dense understories as habitat, probably because they offer both food and shelter (Hiltunen et al. 2004). In winter, *L. timidus* preferred mixed sapling stands over coniferous sapling stands (Hiltunen and Kauhala 2006). A second major habitat is tundra, where hares can survive to the limits of vegetation if there is cover such as cliffs or rocks. In southern Russia, forests with aspen (*Populus*) and shrub thickets are the normal habitat (Ognev 1940). Female hares in Britain selected grassland-mire habitat in summer and showed no strong selection for any habitat in winter, whereas males selected heather moorland in both summer and winter (Rao et al. 2003). Hares in Scotland reach high densities (50/km^2) on well-managed heather moorland in the north-east at 300–900 m a.s.l., and on arctic-alpine ground over rocks rich in nutrients (Hewson 1984). Hares in Ireland occupy moorland and agricultural land at all altitudes even down to sea level (Dingerkus and Montgomery 2001; Wolfe et al. 1996). This is exceptional as mountain hares in temperate climate zones are usually restricted to high elevation areas. Rushes and hedgerows are important diurnal resting areas for the Irish hare (Dingerkus and Montgomery 2002). Groups of 100 may be seen on airfields. *L. t. varronis* lives on alpine pastures up to 4,000 m elevation in summer, and at 3,200 m in January, but most move to shelter in woodland escarpments (Bille 1974). This subspecies prefers forests rich in structure and thickets at the forest boundary as habitat (Rehnus et al. 2013). The hares strongly use heterogeneous habitats with a high diversity of vegetation layers, abundance of saplings and storeyed vegetation structures (Rehnus et al. 2016). In spring, the importance of meadows augments (Gamboni 1997) as herbaceous food is abundant. When the habitat offers both food and shelter against predators, *L. t. varronis* uses the same habitat for the resting and the active phase (Bisi et al. 2013). On agricultural land, *L. timidus* tends to be replaced by *L. europaeus* in Europe and *L. capensis* further east.

The literature shows that mountain hares are grazers in summer and browsers in winter (Hulbert et al. 1996). The diet of hares varies with the habitat. The diet of hares in forest habitats includes leaves and twigs of *Salix*, *Sorbus*, *Betula*, *Juniperus*, and *Populus*. In tundra habitats, alpine plants are preferred, especially dwarf *Salix*, which is rich in nutrients. On Scottish moorlands and Swedish islands, *Calluna* is the most important food (Angerbjörn and Pehrson 1987). Palatable grasses and clovers are eaten when available, and hares prefer plants growing on fertilized soil (Miller 1968), or having a high nutrient content (Angerbjörn and Pehrson 1987). Snow cover restricts the diet to twigs, bark, moss, and lichens. The diets of both sexes are similar (September–November), although reproductively active females eat more grasses and less heather than males (Iason and Waterman 1988), and young eat more grass (*Agrostis* and less *Nardus*) than adults (Flux 1970b). The plant groups used by the Alpine mountain hare are 47.0% gymnosperms, 32.5% graminoids, dwarf shrubs (8.5%), forbs (2.8%), and bryophytes (0.6%) (Rehnus et al. 2013). *L. t. varronis* preferred ligneous plants, whereas grasses were neutrally or negatively selected (Schai-Braun et al. 2020). In low-diversity habitats and at higher elevations, hares exhibited greater diet breadth compared to high-diversity habitats or at lower elevations, even if they had access to the same plant types (Schai-Braun et al. 2020). This is likely due to the hares' need to diversify their diet when plant quantity is low. Studies on captive hares (*L. t. timidus*) demonstrated that dry matter digestibility is the basic factor in meeting energy requirements. Secondary plant compounds in the twigs of *Calluna* or *Betula* may reduce digestibility by causing hares to excrete sodium, but they maintained a sodium balance on a mixed diet (Pehrson 1983b). Plant phenolic compounds have been suggested to be involved in this reduced digestibility.

The winter food of mountain hares in Scotland and Ireland is dominated by heather, 30 to 90% frequency from stomach analyses, whereas grasses (especially *Agrostis* and *Deschampsia*), *Eriophorum*, and dicotyledons like *Galium*, are more frequent in the summer diet (Flux 1970b; Iason 1987). Mountain hares in Ireland consume more grass and herbs (Dingerkus and Montgomery 2001; Wolfe et al. 1996) than in areas with sympatric European hares (Hewson 1962; Hulbert et al. 1996). In Fennoscandia, the winter diet (from browsed plants) is dominated by *Betula* (15–62%), *Sorbus* (5–72%), and *Salix* (10–34%, Lindlöf et al. 1974), whereas in Russia *Salix* (20–36%), *Populus* (6–20%), and *Vaccinium* (5–25%) are more important (Naumov and Shatalova 1974). Little *Pinus* and *Picea* are eaten (0–3% in all areas) but more *Juniperus* (3–10% in all areas). In summer, captive hares in Finland avoided many of the winter foods, instead preferring *Corylus avellana*, *Carex echinata*, *Elymus repens*, *Lathyrus pratensis*, *Vicia cracca*, *Trifolium pratense*, and *Equisetum* spp. (Helminen et al. 1966); winter trials indicate the preference order of *Salix*, *Betula*, *Populus*, and *Sorbus* (Aarnio 1983; Karlsen 1983) but this is mediated through food availability. Other species eaten in low frequencies include *Molina*, *Nardus*, *Scirpus*, *Carex*, *Juncus*, *Alnus*, *Rubus*, *Ribes*, *Prunus*, *Rosa*, *Hippophae*, *Erica*, *Arctostaphylos*, *Empetrum*, *Myrica*, *Ledum*, *Potentilla*, *Filipendula*, *Chamenerium*, *Saxifraga*, *Gentiana*, *Ranunculus*, *Achillea*, *Rumex*, *Fragaria*, *Taraxacum*, *Quercus*, *Acer*, *Fagus*, *Larix*, and *Rhododendron* (Angerbjörn and Flux 1995). Hares feed with their back to the wind, and can clear snow from vegetation with forepaws, but when the snow is hard, they can move 5–10 km to lower ground to feed (Hewson 1962). Long migrations have been recorded from Russia (Ognev 1940). In open country, hundreds of hares may gather on good feeding areas.

Behavior

The mountain hare is primarily nocturnal, but shows increased daylight activity in summer when nights are short and in winter when food is scarce. Activity patterns change seasonally in parallel with circadian rhythms (Bisi et al. 2013; Gamboni 1997). The hare rests by day in a "form" with ears back and eyes half closed, but sleeps for only a few minutes. They groom themselves carefully, especially in early morning and evening, and can be seen rolling in dust in dry weather. When disturbed, *L. t. scoticus* rises from its form at ca. 20 m but this is variable and affected by weather. *L. t. timidus* in Finland rises at longer distances at cold weather (43 m at –20 °C). From April to June more hares are active by day than in other months. The main feeding period begins ca. 2 h before sunset in June, 1 h before sunset in March, and around sunset in September–February, and ends 2–3 h after sunrise in June, but about sunrise in winter (Hewson 1962). They confuse tracks and jump to the side before resting for the day. Mountain hares move, rest, and feed often together in parties of 2–5 individuals (Hewson and Hinge 1990). Groups of 20–100 individuals have been recorded to feed in the same place (Flux 1970b).

From January to July, male hares often follow 2–20 m behind a female for hours. Males obviously use scent to track females; the activity of inguinal glands increases during reproduction (Sokolov et al. 1987). When courtshipping, several hare males chase the female, and male mountain hares can guard females and fight each other (Angerbjörn 1986b and unpubl., but see Flux 1970b). Females dominate males, and may strike with their forepaws with ears laid backward if the male approaches too closely. Copulation can involve five or more males and a female (Flux 1970b), and males can be seen in long fights (Angerbjörn 1986b, and unpublished) as in *L. europaeus* (Hackländer, this volume).

Mountain hares move on average 1400 m per night, but may travel up to 2400 m in one night (Flux 1970b). Mountain hares are not territorial and have overlapping home ranges. Home range sizes vary due to different climate. In the moorland and pastures of Scotland, the home range size was for males 64 ha (n = 16) and for females 62 (90% minimum convex polygons, MCP; Hewson and Hinge 1990), or 21 and 17 ha for males and females, respectively (95% MCP; Hulbert et al. 1996). On heather moorland and grassland-mire in Britain, mean home ranges were 10 ha (multinuclear probability polygon technique, CLUST; Rao et al. 2003). On the coastal grasslands of Ireland, home ranges were bigger; 100% MCP was 46 and 21 ha for males and females, respectively (Wolfe and Hayden 1996). In the boreal coniferous forests of

Fennoscandia, the home ranges were even larger; 206 ha in Finland (Kauhala et al. 2005a) and in northern Sweden 280 and 116 ha for males and females, respectively (Dahl and Willebrand 2005). But home range sizes were larger in winter and smaller in summers according to local food availability. In the Alps, mean home ranges vary from 38 ha (95% kernel density estimator, KDE; Genini Gamboni et al. 2008), 54 ha (95% KDE; Bisi et al. 2011), 99 ha (100% MCP; Gamboni 1997), 136 ha (females) and 195 ha (males, 100% MCP; Nodari 2006), to 551 ha (95% KDE; Slotta-Bachmayr 1997). Usually, the home ranges do not differ between sexes and are smaller in winter than in the other seasons (Gamboni 1997; Genini Gamboni et al. 2008). Night home ranges are larger than day home ranges. Night and day home ranges overlap around 35% (Hewson and Hinge 1990). The daytime range for adult males was 16 ha (n = 18), and 10.1 (n = 17) for adult females but only 3.7 for 14 breeding females (February–July); and 1.6 ha for small young (n = 14). In Finland, home range sizes varied with food supply, 72 and 305 ha for two hares tracked January–March (Seiskari 1957), but only 6.6 ha (n = 9) on a 92-ha island (Häkkinen and Jokinen 1974). Mountain hares showed strong site fidelity with little adult dispersal but juvenile hares showed a relative short dispersal during their first winter (Dahl and Willebrand 2005). But distances moved by 99 hares recovered after game-stocking in Finland, Norway, and Sweden were: 54 hares 0–4 km; 20, 5–9 km; 15, 10–29 km; 9, 30–100 km; and 1 over 200 km (Höglund 1957a, b; Myrberget 1965; Vuolanto 1972). Individual home ranges and core areas of *L. timidus* and *L. europaeus* living in sympatry assessed by GPS telemetry in eastern Finland overlapped. Interspecific interaction seems to be frequent during the breeding season, which also enables cross-breeding (Levänen et al. 2019).

Hares make several forms to sit in by day, trimming protruding vegetation but adding no lining. Forms can reduce wind speeds as much as 90% (Thirgood and Hewson 1987). If undisturbed, the same form may be used for weeks, or may be changed depending on the weather. On Scottish moors, forms may last 25 years. Hares use deep vegetation for concealment in summer, and exposed positions or rocks in winter (Flux 1970b; Häkkinen and Jokinen 1974). In Finland, of 287 forms in forest, 189 were in spruce, 51 below fallen trees, 46 in juniper, and 1 in willow; of 72 in the open, 27 were among ice, 14 in rocks, 12 were unsheltered, 11 were beside fences, and 8 were in reeds (Nyholm 1968). In snow, hares burrow to reach or make forms and enter rock crevices or marmot burrows for shelter or escape (Thirgood and Hewson 1987). They sometimes dig permanent burrows 1–2 m long in the ground, often used by young but seldom by adults. Hares in Russia can build complex burrows up to 7 m long (Ognev 1940). Hares in Ireland and the Alps do not burrow.

In a study on captive Alpine mountain hares, the animals (n = 2) had a low vocalization rate of one vocalization every 33 night hours. All 20 recorded vocal events corresponded with intraspecific communication events such as warning others about danger and defending resources (Rehnus et al. 2019).

Parasites and Diseases

Mountain hares carry both several species of fleas, ticks and lice, but also intestinal and lung parasites. Parasites include the flea, *Spilopsyllus cuniculi*; lice, *Haemodipsus lyriocephalus*, *H. leporis*, and *Listrophorus gibbus* (Sosnina and Novozhilova 1985); ticks, *Ixodes ricinus*, *I. hexagonus*, *I. persulcatus*, *Rhipicephalus sanguineus*, and *Haemaphysalis punctata* (Galimov and Galimova 1988; Muroma 1951; Raitis and Ermala 1972); trematode *Fasciola hepatica* (Ognev 1940); the pentastomid, *Linguatula taenioides* (Ognev 1940); cestodes, *Dicrocoelium lanceolatum*, *D. dendriticum*, *Cittotaenia pectinata*, *Taenia serrata*, *T. pisiformis*, *Multiceps serialis*, *Paranoplocephala wimerosa*, *Andrya cuniculi*, and *A. rhopalocephala* (Burgaz 1970; Mead-Briggs and Page 1967); nematodes, *Protostrongylus pulmonalis*, *P. commutatus*, *Trichostrongylus retortaeformis*, *T. instabilis*, *Graphidium strigosum*, *Passalurus ambiguus*, *Trichuris leporis*, and *Microfilaria* spp.; coccidia, *Eimeria leporis*, *E. robertsoni*, *E. townsendi*, *E. semisculpta*, and *E. hungarica* (Burgaz 1973); and microorganisms and bacteria such as *Toxoplasma gondii*, *Pasteurella pseudotuberculosis*, *P. lepiseptica*, *P. multocida*, *Francisella tularensis*, *Listeria monocytogenes*, *Bordetella bronchiseptica*, and *Fusobacterium necrophorus* (Borg et al. 1952; Couturier 1981). Inkoo arbovirus has been recorded (Brummer-Korvenkontio 1973), *Campylobacter jejuni* (Rosef et al. 1983), *Cynomyia mortuorum* (Itamies and Koskela 1980), and isolated cases of myxomatosis (Thompson 1957). Heavily parasitized hares are more frequent in high-density populations, and have enlarged spleens, less coronary fat, and less body mass (Berg 1981); these populations might be limited by parasitism (Angerbjörn 1983; Newey and Thirgood 2004), with a relationship between prevalence of *Trichostrongylus retortaeformis* and body condition, and fecundity of hares in Scotland (Newey and Thirgood 2004). Tularemia has been reported in mountain hares central Sweden (Mörner 1999). In the French and Italian Alps, the most frequent parasite of *L. t. varronis* was the tapeworm *Paranoplocephala wimerosa* (Bouche 1989: 78%, n = 43; Meneguz and Rossi 1990: 57%, n = 60). In the Austrian Alps, most frequent was the infection by Coccidia (46.2%), whereas stomach intestine strongylids, *Trichuris* spp., and Cestoda were only found in 9.6% of all feces (n = 52; Schai-Braun et al. 2019). Parasitic incidence in fecal pellets of British *L. timidus* was high (≥ 93%, n = 14, Irvin 1970), whereas in *L. t. varronis* half of the fecal pellets were free of parasites (46.2%, n = 52, Schai-Braun et al. 2019). The Calicivirus of the European Brown Hare Syndrome (EHBS) have been identified to affect also mountain hares (Mörner 1999). There is also a strain in the related RHD2 (Rabbit Haemorrhagic Disease) that might affect mountain hares. However, the consequences of these Calicivirus for mountain hares are unknown, but the spread and risk are alarming. Through their ticks, mountain hares can also affect the spread of tick-borne encephalitis (Jaenson et al. 2018).

Population Ecology

The mountain hare is an important prey to several predators (namely mammalian carnivores and birds of prey). Of 84 *L. t. scoticus* killed by predators, red fox took 63, wild cat (*Felis silvestris*) 14, dog 2, stoat (*Mustela erminea*) 2, golden eagle (*Aquila chrysaetos*) 1, buzzard (*Buteo buteo*) 1, hen harrier (*Circus cyaneus*) 1 (Flux 1970b; Hewson 1976). In Finland, of 18 *L. t. timidus* killed, pine marten (*Martes martes*) took 6, wolf (*Canis lupus*) 4, golden eagle 4, goshawk (*Accipiter gentilis*) 2, eagle owl (*Bubo bubo*) 1, stoat 1 (Nyholm 1968). Mountain hares represented 4–10% by volume of the diet of red foxes in Sweden in winter, as estimated from stomach analyses, and during summer from 1 to 14% by volume depending on vole abundance (Englund 1965). Mountain hares formed 38 to 94% by volume of the winter diet for red foxes on islands (Angerbjörn 1989). The diet for wolf (Rukovsky and Kupriyanov 1972) and lynx (Pulliainen 1981) can be up to 70% of mountain hares, and from 1 to 33% for pine marten (Danilov and Ivanter 1967; Lockie 1961; Parovschikov 1961). In northern Fennoscandia, mountain hare remains were found in 5% of Arctic fox (*Vulpes lagopus*) fecal droppings during the winter diet. Some mountain hares were found in the stomachs or fecal droppings of wolverine (*Gulo gulo*) and mink (*Neovison vison*). There are recent concerns that the raccoon dog (*Nyctereutes procyonoides*) could be a serious predator on juvenile mountain hares in Belarus (Sidorovich et al. 2008).

For golden eagle, mountain hares have been found in 21–26% of pellets in Finland; in Norway, 28–34%; in Scotland, 50%; in north Russia, 35%; in the Alps, 10%; and in Estonia, 19% (Sulkava et al. 1984). The proportion of mountain hare in the diet of eagle owls can also be substantial (for Norway, 2–7%; Finland, 4%; Sweden, 4%; Germany, 7%; Sulkava 1966). Fewer hares were taken by goshawks and hen harriers in Norway, but in Scotland hares made up 11% of the diet of harriers (Picozzi 1978). Mountain hares are taken in small amounts (less than 1%) by white-tailed eagle (*Haliaeetus albicilla*), peregrine falcon (*Falco peregrinus*), gyr falcon (*F. rusticolus*), buzzard, snowy owl (*Bubo scandiacus*), ural owl (*Strix uralensis*), and long-eared owl (*Asio otus*).

Mountain hare densities are variable: 1 hare/km^2 over large parts of Sweden (Bergengren 1969); 1–2/km^2 in European Russia (Novikov et al. 1970); 1–15/km^2 in Japan (Abe and Ota 1987); 2–6/km^2 in central Finland (Lind 1961); higher on Baltic islands, 25/km^2 on Krunnian (Nyholm 1968), and ca. 60/km^2 on Signilskär (Häkkinen and Jokinen 1974); up to 245/km^2 on the best areas in northeastern Scotland, but only 0.14/km^2 in western Scotland (Watson and Hewson 1973); and 400/km^2 on Vedholmen, Sweden (Angerbjörn 1986b). Evidence suggests that *L. t. hibernicus* abundance in Northern Ireland has declined substantially, with present total population estimates for Northern Ireland ranging from 8,250 to 21,000 individuals (0.65 hares/km^2; Dingerkus and Montgomery 2002). Long-term field counts (1954–2008) of *L. t. scoticus* in the UK indicated that intensification of game bird management (red grouse, *Lagopus l. scotica*) has resulted in recent severe declines of the mountain hare, exacerbating longer term declines associated with land-use change. Management practices founded on misinterpretation of earlier research (i.e., hare culls affect grouse surpluses) are the probable cause (Watson and Wilson 2018). Population density estimates for *L. t. varronis* based on a spatially explicit capture-recapture approach using nDNA of fecal pellets ranged from 3.2 to 3.6 mountain hares/km^2 in the Swiss Alps (Rehnus and Bollmann 2016). Models predicting the distribution of *L. t. timidus* anticipated an average habitat loss of 35% by the year 2100 in the Swiss Alps, mainly due to an increase in temperature during the reproductive season (Rehnus et al. 2018). Trends in *L. timidus* abundance indices in Scottish moorland varied with region and grouse management intensity. Hare indices were higher and relatively stable on moors where driven grouse shooting was practiced relative to lower indices and declines on moors where grouse were either walked-up or not shot (Hesford et al. 2019). Despite increases in numbers of *L. timdus* killed in Scotland over the last 20 years, it appears that range contraction may be attributed to factors other than culling, such as changes in habitat and management (Hesford et al. 2020). The population fluctuations vary through the distribution range where about 50% of the populations show some type of cycles and the remaining are noncyclic (Newey et al. 2007). In taiga and tundra areas in Russia and Siberia, there are peaks every 8–12 years apart and in Scotland time series analyses showed similar patterns with quasi-cycles of 9 years (4 to 15 years; Newey et al. 2007). But in Fennoscandia, populations in northern areas typically peak 3–4 years apart (Lindén 1991). This is connected to the dominating vole cycle and predators would shift their diet from voles to hares and thereby forcing the hares into the vole (*Microtus*) cycle with some time lag (Angelstam et al. 1984). Although some populations follow this pattern, time series analyses give only limited evidence for four-year cycles and 63% of the populations in Fennoscandia were noncyclic (Newey et al. 2007).

Populations of mountain hares can be regulated by predation, from especially the red fox (Angerbjörn 1989; Marcström et al. 1989), parasites (Newey and Thirgood 2004), or starvation (Angerbjörn 1981). A controversial impact of parasitism and predation on *L. timidus* population dynamics has been recorded. For red fox, there are studies from both experiment, bag records, and islands that support the limitation of populations in Fennoscandia by predation, whereas predation is unlikely to be important in mountain hare populations in Scotland (Newey et al. 2007). This can be seen in Fig. 4 where the abundance of mountain hares increased strongly 1980 when the red fox population in Sweden was struck by sarcoptic mange (Danell and Hörnfeldt 1987). When the red fox population recovered 1990, the mountain hare population decreased. But there is an interesting case where the presence of lynx in Finland would release the mountain hares from the red fox predation (Elmhagen et al. 2010). By contrast among the Scottish populations, predation pressure is often lower due to human hunting. Here, parasites have instead been shown to impact body condition and fecundity leading to unstable population dynamics (Newey and Thirgood 2004), whereas in Fennoscandia parasitism does

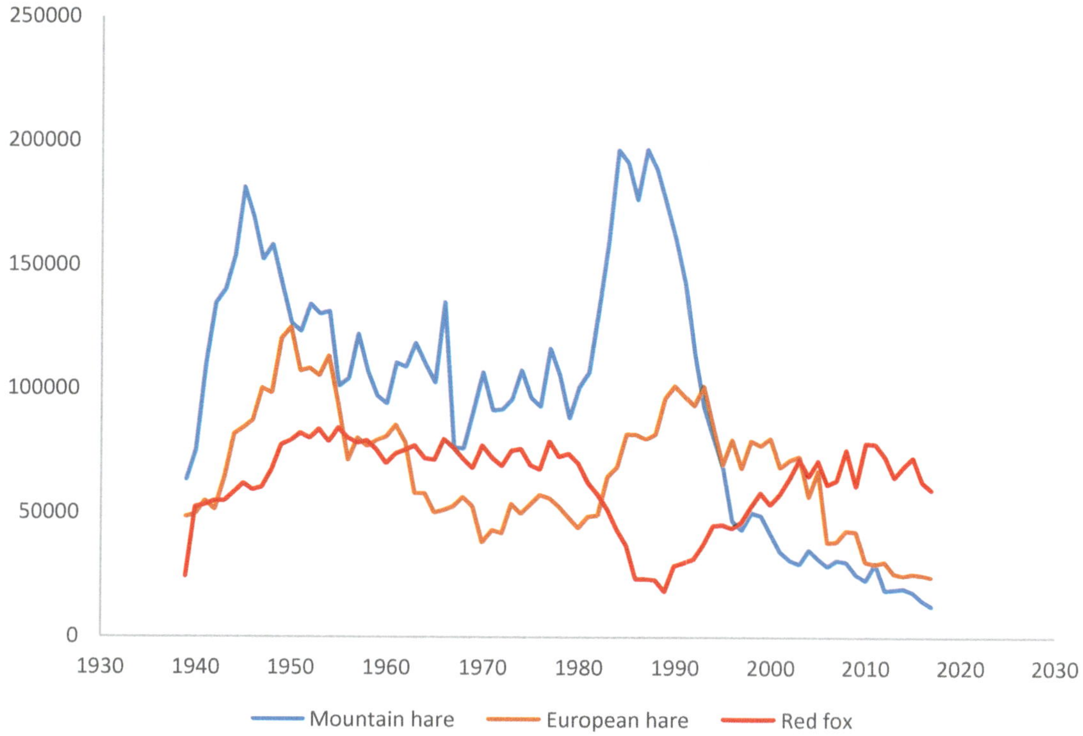

Fig. 4 Number of shot hares and foxes in Sweden from 1939 to 2017, according to bag statistics from the Swedish Hunting Organisation (www.viltdata.se). Note the period 1980–1990 when the red fox had sarcoptic mange

not seem to be of any importance to mountain hare population dynamics (Newey et al. 2007). On a general scale, it is not likely that mountain hares are limited by food; hares mostly die from other causes than starvation. But it is likely that all these factors interact with each other and with population density through effects on body condition and reproduction (Angerbjörn 1986b; Newey et al. 2007), with or without an effect of weather (Angerbjörn 1983).

There is a clear interspecific competition between mountain hare and European hare. The larger European hare is dominating the smaller mountain hare resulting in that the mountain hare in many areas retreats to less productive areas (Thulin 2003). There are also several diseases that can be spread from European hares, thereby increasing the negative effect from the European hare. Rabbits may also keep mountain hares from lower altitudes in Scotland (Flux 1970b). The diet of mountain hares and roe deer can be overlapping but the consequences of this are unclear (Hulbert and Andersen 2001).

Climate warming leads to changes in species' ranges, especially at high elevations, where the effects of climate warming are pronounced (Chen et al. 2011). Within the order Lagomorpha it has been shown that closely related species of similar size occurring at high elevations are likely to exhibit competitive interactions (Leach et al. 2015), and that habitat generalists are less likely to be threatened by extinction than habitat specialists (Verde Arregoitia et al. 2015). In northern Sweden, the mountain hare decreased at the same time as the European hare increased during a period of warming climate (Elmhagen et al. 2015). This suggests that interacting species at high elevations, such as the European hare and the Alpine mountain hare, are likely to be highly susceptible to future environmental change. Moreover, climate change may compromise specialists such as the Alpine mountain hare adapted to cold conditions, by competitive interactions with the European hare (Acevedo et al. 2012). This means that for the Alpine mountain hare, global warming might reinforce competition with the European hare and may compromise the co-existence between the two hare species (Acevedo et al. 2012).

Conservation Status

Although populations are stable across much of its geographic distribution with strong fluctuations occurring in northern Europe, there are declines in several areas, for example, Russia, large portions of Sweden (Fig. 4), the Alps, Scotland, and Ireland. Several possible causes for the decline of the mountain hare are under consideration, for example, disease, deforestation, agricultural intensification, predation, and interaction with other herbivores (Thulin 2003). There is evidence that in northern Europe areas of overlap between *L. timidus* and *L. europaeus* are not stable (Elmhagen et al. 2015; Jansson and Pehrson 2007; Thulin 2003), that interspecific competition between the two congeneric hare species is strong, and the European hare is invading the natural areas occupied by the mountain hare. This seems to be true not only for northern Europe but also for the Alps. Climatic variation along the elevation gradient in the Alps allows the coexistence of the two hare species (Fig. 5a). An evaluation of hunting bag statistics collected over 30 years in Grisons (Switzerland) showed that both hare species shifted their minimum elevations towards higher elevations (Fig. 5b-c; Schai-Braun et al. 2021). The study of the hunting bags of Grisons suggests that European hares are shifting their lower ranges faster to higher elevations than Alpine mountain hares (Alpine mountain hare: 28 m/decade, European hare: 31 m/decade). The available surface area decreases as animal species shift uphill. Thus, the Alpine mountain hare may lose habitat availability more rapidly than the European hare. The example from Grisons suggests that European hares living in the Alps may benefit from global climate change as they shift their lower range uphill. Therefore, the European hare might displace the Alpine mountain hare from its traditional range. There are indications that the only single factor capable of explaining the observed decline is interaction with the European hare (Thulin 2003). Thus, competitive exclusion or hybridization and introgression with European hares is a serious potential threat for the mountain hare. It seems that European hares and mountain hares are ecologically equivalent and interrelate in strong competition (Thulin 2003). Studies of food preferences show that allopatric mountain hares in Ireland eat much more grasses and herbs (Dingerkus and Montgomery 2001; Wolfe et al. 1996) than mountain hares that live in sympatry with European hares (Hewson 1962; Hulbert et al. 1996). Consequently, Wolfe et al. (1996) argued that competitive exclusion by the European hare restricts mountain hare distribution. Lind (1963) investigated where mountain hares have their forms, and discovered that in sympatric areas the forms are located in denser forest and further away from open fields than in allopatric areas. Moreover, Hewson (1976) noticed that European hares expand their range and habitat use when mountain hare densities are low. Yalden (1999) pointed out that in Bronze Age layers of Hartledale in England, the last remains of mountain hares were found together with the first remains of European hares. There are historical and recent indications that in southern and central Sweden, the population density and the geographical range of mountain hares are declining. The decline started simultaneously with the introduction of the European hare to Sweden by humans, and continues with the constant expansion of the European hare towards the north (Jansson and Pehrson 2007; Thulin 2003). This suggests that European hares have the potential to expand into mountain hare territory and that, after a time of sympatry due to range expansion by the European hare, the mountain hare disappears from optimal European hare habitat (Thulin 2003). This scenario has been supported by different projections and models (Bedson et al. 2021; Bisi et al. 2015; Caravaggi et al. 2017; La Morgia and Venturino 2017). In most hybridization events, female mountain hares mate with male European hares (Thulin 2003). Each time a female mountain hare mates with a male European hare, the local mountain hare population may lose a species-specific litter. Consequently, the observed decline in the range and population density of mountain hares may be a direct consequence of this unidirectional hybridization. The female hybrids will mate male

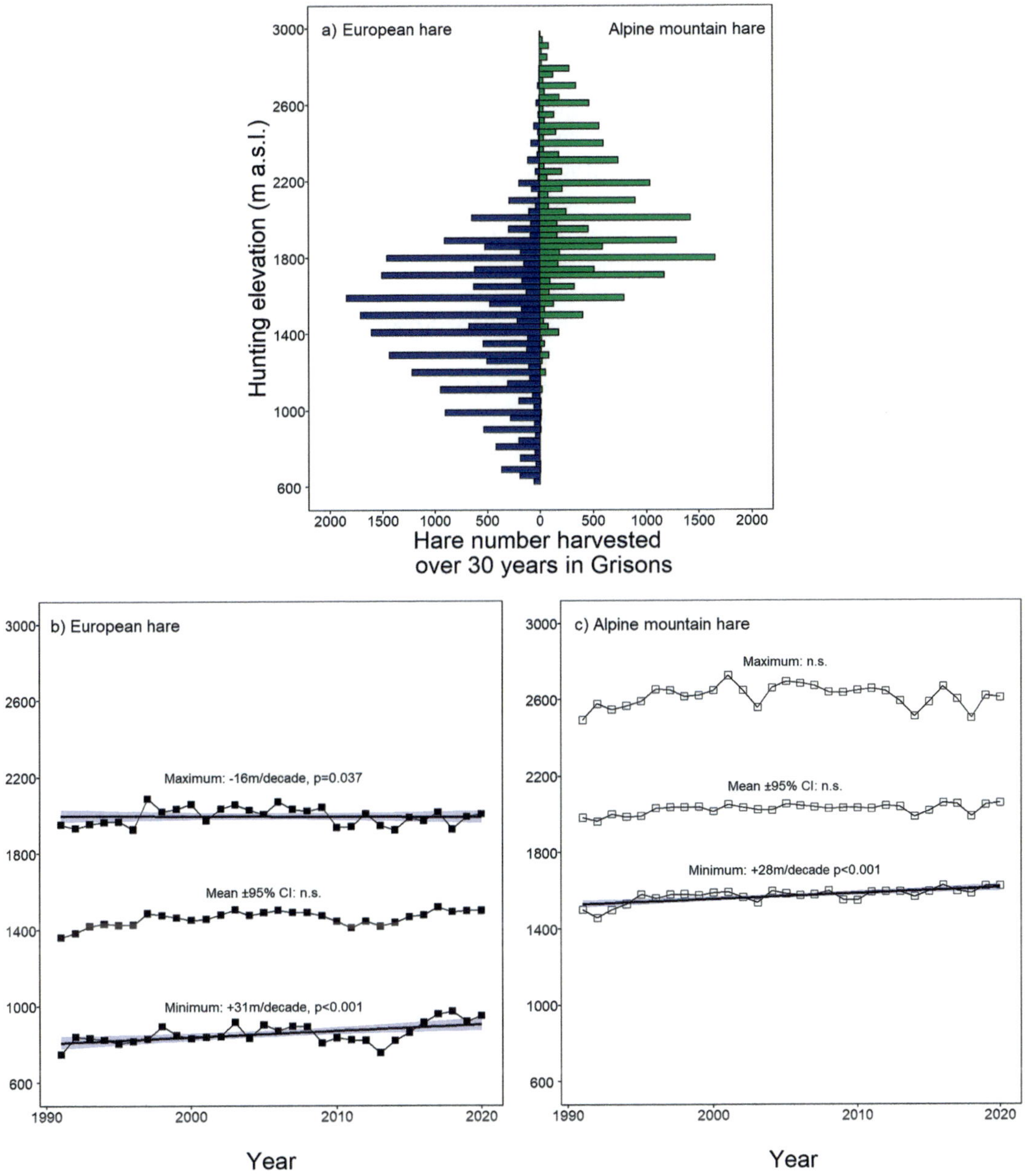

Fig. 5 (**a**) As indicated by hunting bag statistics from Grisons over 30 years, Alpine mountain hares and European hares live in sympatry in the Alps. Harvesting elevation, minimum (≤15th percentile), and maximum hunting elevation (≥85th percentile) of the (**b**) European hare and (**c**) Alpine mountain hare (Source: Schai-Braun et al. 2021)

European hares and the mtDNA from mountain hares will be transmitted to the European hare population, which conduct to the mitochondrial introgression. In areas where European hares currently live in sympatry with mountain hares, the proportion of European hares with mountain hare mtDNA is higher than in areas where both species lived in sympatry in the past (Thulin and Tegelström 2002). It seems that European hares with mountain hare mtDNA have lower fitness than European hares with species-specific mtDNA (Thulin and Tegelström 2002). This means that mountain hare mtDNA is incorporated into European hares in sympatric areas, but gradually disappears once the European hares become and remain allopatric. Accordingly, hybridization, introgression, and the subsequent disappearance of mountain hare genes might be regarded as refined competitive exclusion in which the traces of mountain hares are not only ecologically, but also genetically erased from the European hare (Thulin 2003). This might be one reason to the population decrease in areas with European hare sympatric distribution (Thulin 2003). Pathogens such as Caliciviruses and Tularemia can also be spread to areas through European hares. Nevertheless, the ancestral hybridization between the mountain hare in the hare species inhabiting in the Iberian Peninsula is bringing an extraordinary model study on species interaction, animal hybridization, reticulate evolution, and adaptive evolution (e.g., Melo-Ferreira et al. 2011; Melo-Ferreira et al. 2012; Melo-Ferreira et al. 2014).

Distribution of mountain hares in the Alps was highly predicted to decrease along with climate change (Rehnus et al. 2018). The abundance of mountain hare in Scandinavia was negatively associated to duration of snow cover and to abundance of predators (Pedersen et al. 2017).

The discovery of an introduced population of European hare to Ireland poses a serious threat to the endemic *L. t. hibernicus* (Reid et al. 2007). The range of the European hare has a wider niche breath than the Irish hare and has increased three times from 2005 to 2013 (Caravaggi et al. 2014). *L. t. hibernicus* shows a distinct ecological niche in Ireland, separated from the other *L. timidus* subspecies, suggesting that it might be an Evolutionarily Significant Unit (Caravaggi et al. 2017). Projections under global climate change suggest that, by 2070, bioclimatic space for invasive European hares in Ireland will expand (by 79%) but contract for endemic Irish hares (by 75%), further facilitating their replacement (Caravaggi et al. 2017). Hare coursing, where wild hares are caught and released in front running dogs, is banned throughout the UK but remains legal and widespread across the Republic of Ireland (Reid et al. 2007). Although the mortality in hare coursing is low, there are still wildlife welfare issues, especially since the Irish hare is under threat with a strong conservation perspective.

The mountain hare is listed under the Bern Convention, Appendix III as well as the EU Habitats Directive, Annex V.

IUCN Red List Classification: Least Concern (LC) (Smith and Johnston 2019).

Management

Mountain hare is a popular game species in most part of the distribution range. Release and restocking take place in many areas for hunting opportunities (Ahlgren et al. 2016) but often with little effects on the wild populations. However, this confers risks of spreading diseases, parasites, and also of genetic swamping. Supplemental feeding during winters can have positive effects on survival and reproduction. Red fox hunting is also a common action to increase the abundance of hares.

Hares may damage cereals, *Brassica* crops, fruit trees, and tree plantations, especially in winter (Allman 1946; Dinesman 1959; Wahlin 1950), and ring-bark pines (Borisenko 1983).

Mountain hares in the Alps in areas with human winter tourism showed increased levels of fecal glucocorticoid metabolites, an indicator of a stress response. To reduce the frequency of stress threats for mountain hares, management implications are to keep forests inhabited by mountain hares free of tourism infrastructure and retain undisturbed forest patches within skiing areas (Rehnus et al. 2014).

More species-rich pasture and provision of areas of cover, such as rushes, may arrest further declines of the Irish hare in Northern Ireland, or indeed promote numbers of hares, particularly in lowland areas (Dingerkus and Montgomery 2002).

Future Challenges for Research and Management

The abundance of mountain hares is decreasing all over the distribution area. This can be related to many different processes. A combination of global warming and habitat change are probably the main reasons (Elmhagen et al. 2015). This includes increased exposure to parasites and diseases, increase of generalistic predators, and agricultural practice. Regional declines in mountain hare populations suggest that investigations should be conducted to ascertain the causes. The current situation of the endemic Irish hare and the Alpine mountain hare demand particular research. As a species well adapted to cold climate and high altitudes, the mountain hare may constitute an interesting sentinel species to assess the impact of global warming in ecosystems and on biodiversity. Thus, specific long-term monitoring programs should be implemented on this species, in particular in the Alps and in some northern European regions.

Acknowledgments We are very grateful to John Flux who took an essential part in earlier drafts and to the editors of this volume.

References

Aarnio M (1983) Selection and quality of winter food of the mountain hare in southern Finland. Finn Game Res 41:57–65

Abe H, Ota K (1987) Fluctuation of a hare population in the Tomakomai experiment Forest. Res Bull Coll Exp Forestry, Hokkaido Uni 44:667–674

Acevedo P, Jiménez-Valverde A, Melo-Ferreira J, Real R, Alves PC (2012) Parapatric species and the implications for climate change studies: a case study on hares in Europe. Glob Change Biol 18:1509–1519. https://doi.org/10.1111/j.1365-2486.2012.02655.x

Adam KD (1982) Der Mensch im Eiszeitalter. Stuttg Beitr Naturk C 15:1–72

Ahlgren H, Norén K, Angerbjörn A, Lidén K (2016) Multiple prehistoric introductions of the mountain hare (*Lepus timidus*) on a remote island, as revealed by ancient DNA. J Biogeogr 43:1786–1796. https://doi.org/10.1111/jbi.12759

Allman D (1946) Observations on damage by hares at Clonegal Forest. Ir Forestry 3:92

Altuna J (1971) Fauna de Maiferos de los Yacimientos prehistóricos de Guipúzcoam, con cátalogo de los mamiferos cuaternarios del Cantábrico y del Pirineo Occidental. Munibe, San Sebastian 24:1–465

Alves PC, Ferrand N, Suchentrunk F, Harris DJ (2003) Ancient introgression of *Lepus timidus* mtDNA into *L. granatensis* and *L. europeaus* in the Iberian Peninsula. Mol Phylogenet Evol 27:70–80. https://doi.org/10.1016/S1055-7903(02)00417-7

Alves PC, Melo-Ferreira J, Freitas H, Boursot P (2008) The ubiquitous mountain hare mitochondria: multiple introgressive hybridization in hares, genus *Lepus*. Phil Trans R Soc Lond Ser B Biol Sci 363:2831–2839. https://doi.org/10.1098/rstb.2008.0053

Angelstam P, Lindström E, Widén P (1984) Role of predation in short-term fluctuations of some birds and mammals in Fennoscandia. Oecologia 62:199–208

Angerbjörn A (1981) Winter food as a limiting factor of dense mountain hare populations on islands, a comparative study. In: Myers K, MacInnes C (eds) Proceedings of the world lagomorph conference. University of Guelph, Gulph, pp 529–535

Angerbjörn A (1983) Proximate causes of mountain hare population crashes: a review. Finn Game Res 41:29–38

Angerbjörn A (1986a) Population dynamics of mountain hares *Lepus timidus* L on islands. PhD thesis, Stockholm University

Angerbjörn A (1986b) Reproduction of mountain hares (*Lepus timidus*) in relation to density and physical condition. J Zool (London) 208:559–568. https://doi.org/10.1111/j.1469-7998.1986.tb01523.x

Angerbjörn A (1989) Mountain hare populations on islands: effects of predation by red fox. Oecologia 81: 335–340. https://doi.org/10.1007/BF00377080

Angerbjörn A (1992) Mortality pattern in mountain hares (*Lepus timidus*) on islands. In: Bobek B, Perzanowski K, Reglin WL (eds) Global trends in wildlife management, vol 1. Swiat Press, Krakow, pp 87–93

Angerbjörn A, Flux JEC (1995) *Lepus timidus*. Mamm Species 495:1–11. https://doi.org/10.2307/3504302

Angerbjörn A, Pehrson Å (1986) Maternal investment in mountain hares under different feeding regimes. In: Angerbjörn A (ed) Population dynamics of mountain hares *Lepus timidus* L on islands. PhD thesis, Stockholm University, pp 46–68

Angerbjörn A, Pehrson Å (1987) Factors influencing winter food choice by mountain hares (*Lepus timidus* L.) on Swedish coastal islands. Can J Zool 65:2163–2167. https://doi.org/10.1139/z87-330

Angermann R (1967) Beiträge zur Kenntnis der Gattung *Lepus* (Lagomorpha, Leporidae) III. Zur Variabilität palaearktischer Schneehasen. Mitt Zool Mus Berlin 43:161–178

Ashrafzadeh MR, Djan M, Szendrei L, Paulauskas A, Scandura M, Bagi Z, Ilie DE, Kerdikoshvili N, Marek P, Soós N, Kusza S (2018) Large-scale mitochondrial DNA analysis reveals new light on the phylogeography of Central and Eastern-European Brown hare (*Lepus europaeus* Pallas, 1778). PLoS One 13:e0204653. https://doi.org/10.1371/journal.pone.0204653

Barrett-Hamilton G (1900) Exhibition of skins of the variable hare. Proc Zool Soc Lond 1900:87–92

Barrett-Hamilton G (1912) A history of British mammals. Gurney and Jackson, London

Baumann F (1949) Die freilebenden Säugetiere der Schweiz. Hans Huber, Bern

Bedson CP, Devenish C, Symeonakis E, Mallon D, Reid N, Edwin Harris W, Preziosi R (2021) Splitting hares: current and future ecological niches predicted as distinctly different for two congeneric lagomorphs. Acta Oecol 111:103742. https://doi.org/10.1016/j.actao.2021.103742

Berg C (1981) Endoparasitism in wild Norwegian hares (*Lepus timidus*). Parasitologiska Institutionen Åbo Akademi Information 16:88–89

Bergengren A (1969) On genetics, evolution and history of distribution of the heath-hare, a distinct population of the Arctic hare, *Lepus timidus* Lin. Viltrevy 6:381–460

Bille R (1974) Les animaux de montagne. Editions Danoel, Paris

Bisi F, Nodari M, Dos Santos Oliveira NM, Masseroni E, Preatoni D, Wauters LA, Tosi G, Martinoli A (2011) Space use patterns of mountain hare (*Lepus timidus*) on the Alps. Eur J Wildl Res 57:305–312

Bisi F, Nodari M, Dos Santos Oliveira NM, Ossi F, Masseroni E, Preatoni DG, Wauters LA, Martinoli A (2013) Habitat selection and activity patterns in alpine mountain hare (*Lepus timidus varronis*). Mamm Biol 78:28–33

Bisi F, Wauters LA, Preatoni DG, Martinoli A (2015) Interspecific competition mediated by climate change: which interaction between brown and mountain hare in the Alps? Mamm Biol 80:424–430. https://doi.org/10.1016/j.mambio.2015.06.002

Borg K, Höglund N, Notini G (1952) Nytt om haren. Svenska Jägar Meddelanden 17:1–90

Borisenko VA (1983) "Ringing" of the pine by the alpine hare. Biullentin Moskovskogo Obshchestra Ipsytatelai Prirody, Biologii 88:32–44

Bouche M (1989) Le lièvre variable dans le massif des Écrins. Contributions à l'étude écoéthologique de *Lepus timidus varronis*. Parc national des Écrins. France

Brummer-Korvenkontio M (1973) Arboviruses in Finland: V. Serological survey of antibodies against Inkoo virus (California group) in human, cow, reindeer, and wildlife sera. Am J Trop Med Hyg 22:654–661

Burgaz I (1970) The endoparasites of the varying hare (*Lepus timidus*) and the European hare (*Lepus europaeus*). Zool Rev (Stockholm) 32:61–66

Burgaz I (1973) A cross infection experiment with coccidia concerning mountain hares *Lepus timidus* and tame rabbits. Norw J Zool 21:327

Cabon-Raczynska K (1963) Occurrence of the blue hare, *Lepus timidus* Linnaeus, 1758 in Poland. Acta Theriol 6:303–305

Caravaggi A, Montgomery W, Reid N (2014) Range expansion and comparative habitat use of insular, congeneric lagomorphs: invasive European hares *Lepus europaeus* and endemic Irish hares *Lepus timidus hibernicus*. Biol Invasions 17:687–698. https://doi.org/10.1007/s10530-014-0759-1

Caravaggi A, Leach K, Santilli F, Rintala J, Helle P, Tiainen J, Bisi F, Martinoli A, Montgomery W, Reid N (2017) Niche overlap of mountain hare subspecies and the vulnerability of their ranges to invasion by the European hare; the (bad) luck of the Irish. Biol Invasions 19:655–674. https://doi.org/10.1007/s10530-016-1330-z

Chen I-C, Hill J, Ohlemüller R, Roy D, Thomas C (2011) Rapid range shifts of species associated with high levels of climate warming. Science 333:1024–1026

Collett R (1912) Norges pattedyr. Aschehoug & Co, Kristiania

Corbet GB (1980) The mammals of the Palaearctic region: a taxonom. Review. British Museum (Natural History), London

Corbet GB, Hill JE (1987) A world list of mammalian species, 2nd edn. British Musem (Natural History), London

Corbet GB, Southern HN (1977) The handbook of British mammals. Blackwell, London

Couturier M (1981) Le gibier des montagnes françaises. Arthaud, Grenoble

Dahl SK (1928) Über das Vorkommen des Scheehasen (*Lepus timidus* L) in der Krim. Buillentin Societe Naturales Crimee Simferapol 10:37–39

Dahl F (2005) Distinct seasonal habitat selection by annually sedentary mountain hares (*Lepus timidus*) in the boreal forest of Sweden. Eur J Wildl Res 51:163–169. https://doi.org/10.1007/s10344-005-0095-y

Dahl F, Willebrand T (2005) Natal dispersal, adult home ranges and site fidelity of mountain hares *Lepus timidus* in the boreal forest of Sweden. Wildl Biol 11:309–317. https://doi.org/10.2981/0909-6396(2005)11[309:ndahra]2.0.co;2

Danell K, Hörnfeldt B (1987) Numerical responses by populations of red fox and mountain hare during an outbreak of sarcoptic mange. Oecologia 73:533–536

Danilov P, Ivanter EB (1967) The pine marten in Karelia. Uchenye zapiski Petrozavodsk Universitet 15:179–197

Dinesman LG (1959) The effect of *Lepus timidus* on the regeneration of larch forests in central Yakutia. Biullentin Moskovskogo Obshchestra Ipsytatelai Prirody:17–25

Dingerkus SK, Montgomery W (2001) The diet and landclass affinities of the Irish hare *Lepus timidus hibernicus*. J Zool (London) 253:233–240

Dingerkus SK, Montgomery WI (2002) A review of the status and decline in abundance of the Irish Hare (*Lepus timidus hibernicus*) in Northern Ireland. Mammal Rev 32:1–11. https://doi.org/10.1046/j.1365-2907.2002.00098.x

Elmhagen B, Ludwig G, Rushton SP, Helle P, Lindén H (2010) Top predators, mesopredators and their prey: interference ecosystems along bioclimatic productivity gradients. J Anim Ecol 79:785–794. https://doi.org/10.1111/j.1365-2656.2010.01678.x

Elmhagen B, Eriksson O, Lindborg R et al (2015) Implications of climate and land-use change for landscape processes, biodiversity, ecosystem services, and governance. Ambio 44(Suppl 1):S1–S5. https://doi.org/10.1007/s13280-014-0596-6

Englund J (1965) Studies on food ecology of the red fox (*Vulpes vulpes*) in Sweden. Viltrevy 3:377–485

Fairley J (1974) Notes on the winter breeding condition of hares in the west of Ireland. Ir Nat J 18:17–19

Ferreira MS, Alves PC, Callahan CM, Giska I, Farelo L, Jenny H, Mills LS, Hackländer K, Good JM, Melo-Ferreira J (2020) Transcriptomic regulation of seasonal coat color change in hares. Ecol Evol 10:1180–1192. https://doi.org/10.1002/ece3.5956

Ferreira MS, Jones MR, Callahan CM, Farelo L, Tolesa Z, Suchentrunk F, Boursot P, Mills LS, Alves PC, Good JM, Melo-Ferreira J (2021) The legacy of recurrent introgression during the radiation of hares. Syst Biol 70:593–607. https://doi.org/10.1093/sysbio/syaa088

Flux JEC (1970a) Colour change of mountain hares (*Lepus timidus scoticus*) in north-east Scotland. J Zool (London) 162:345–358

Flux JEC (1970b) Life history of the mountain hare (*Lepus timidus scoticus*) in north-east Scotland. J Zool (London) 161:75–123. https://doi.org/10.1111/j.1469-7998.1970.tb02171.x

Gaiduk VE (1973) Reproduction of *Lepus europaeus* Pall. in Byelorussia. Vestn Zool 3:51–54

Gaiduk VE (1975) Influence of some abiotic factors upon moulting and pigmentation of hair cover in *Lepus timidus*. Zool Zhurnal 54:425–431

Galimov VR, Galimova EZ (1988) Hosts of the adult *Ixodes persulcatus* P. Soh. and their role in the circulation of tick-borne encephalitis. Meditsinskaia Parazitologiia i Parasitornye Bolezni 57:39–41

Gamboni A-S (1997) Comportement spatio-temporel d'une population de Lièvre variable (*Lepus timidus varronis*) au sud des Alpes. Diiploma thesis, Université de Neuchâtel.

Gautier A (1973) The fauna of the Palaeolithic site of Maisieres-Canal: fossil Mammalia. Institut Royal Sciences Naturelles de Belgique, Memoires 172:1–20

Genini Gamboni A-S, Bisi F, Masseroni E, Nodari M, Preatoni DG, Wauters LA, Martinoli A, Tosi G (2008) Home range dynamics of mountain hares (*Lepus timidus*) in the Swiss Alps. Hystrix 19:157–163. https://doi.org/10.4404/hystrix-19.2-4425

Giska I, Farelo L, Pimenta J, Seixas FA, Ferreira MS, Marques JP, Miranda I, Letty J, Jenny H, Hackländer K, Magnussen E, Melo-Ferreira J (2019) Introgression drives repeated evolution of winter coat color polymorphism in hares. PNAS 116:24150–24156. https://doi.org/10.1073/pnas.1910471116

Häkkinen I, Jokinen M (1974) On the winter ecology of the snow hare (*Lepus timidus*) in the outer archipelago. Suomen Riista 25:5–14

Hamill RM, Doyle D, Duke EJ (2006) Spatial patterns of genetic diversity across European subspecies of the mountain hare, *Lepus timidus* L. Heredity 97: 355–365. https://doi.org/10.1038/sj.hdy.6800880

Heller F (1960) Das Diluvialprofil in der Jungfernhöhle bei Tiefenellern, Landkreis Bamberg. Erlanger Geol Abh 34:3–17

Helminen M, Valanne K, Pirkola A, Sten J (1966) Summer feeding habits of the mountain hare (*Lepus timidus*) and of the brown hare (*L. europaeus*). Suomen Riista 18: 133–144

Hesford N, Fletcher K, Howarth D, Smith AA, Aebischer NJ, Baines D (2019) Spatial and temporal variation in mountain hare (*Lepus timidus*) abundance in relation to red grouse (*Lagopus lagopus scotica*) management in Scotland. Eur J Wildl Res 65. https://doi.org/10.1007/s10344-019-1273-7

Hesford N, Baines D, Smith AA, Ewald JA (2020) Distribution of mountain hares *Lepus timidus* in Scotland in 2016/2017 and changes relative to earlier surveys in 1995/1996 and 2006/2007. Wildl Biol 2020(2). https://doi.org/10.2981/wlb.00650

Hewson R (1958) Moults and winter whitening in the mountain hare *Lepus timidus scoticus* Hilzheimer. Proc Zool Soc Lond 131:99–108

Hewson R (1962) Food and feeding habits of the mountain hare *Lepus timidus scoticus*, Hilzheimer. Proc Zool Soc Lond 139:515–526

Hewson R (1968) Weights and growth rates in the mountain hare *Lepus timidus scoticus*. J Zool (London) 154: 249–262

Hewson R (1970) Variation in reproduction and shooting bags of mountain hares on two moors in north-east Scotland. J Appl Ecol 7:243–252

Hewson R (1976) A population study of mountain hares (*Lepus timidus*) in north-eastern Scotland from 1956–1969. J Anim Ecol 45:395–414. https://doi.org/10.2307/3881

Hewson R (1984) Mountain hare, *Lepus timidus*, bags and moor management. J Zool (London) 204:563–565

Hewson R (1990) Interactions between mountain hares (*Lepus timidus*) and other mammals. J Zool (London) 221:302–305

Hewson R, Hinge MDC (1990) Characteristics of the home range of mountain hares *Lepus timidus*. J Appl Ecol 27: 651–666

Hiltunen M, Kauhala K (2006) Selection of sapling stand habitats by the mountain hare (*Lepus timidus*) during winter. Mamm Biol 71:183–189. https://doi.org/10.1016/j.mambio.2005.12.004

Hiltunen M, Kauhala K, Lindén H (2004) Habitat use of mountain hare *Lepus timidus* in summer: the importance of different vegetation layers. Acta Theriol 49: 479–490. https://doi.org/10.1007/BF03192592

Hoffmann R, Smith A (2005) Order Lagomorpha. In: Wilson D, Reeder D (eds) Mammal species of the world: a taxonomic and geographic reference, 3rd edn. The Johns Hopkins University Press, Baltimore, pp 185–211

Höglund N (1957a) Fortplantningen hos skogsharen (*Lepus timidius timidus* Lin.). Viltrevy 1:267–282

Höglund N (1957b) Svenska Jägareförbundets viltmärkningar 1955 och 1956. Viltrevy 1:283–317

Höglund N, Viklund J (1953) On determining the sex of young hares. Suomen Riista 8:175–176

Holley A (1986) A hierarchy of hares: dominance status and access to oestrus does. Mammal Rev 16:181–186

Hulbert IAR, Andersen R (2001) Food competition between a large ruminant and a small hindgut fermentor: the case of the roe deer and mountain hare. Oecol 128:499–508

Hulbert IAR, Iason G, Racey PA (1996) Habitat utilization in a stratified upland landscape by two lagomorphs with different feeding strategies. J Appl Ecol 33:315–324

Iason G (1987) Seasonal reproduction in the mountain hare: ecological and physiological constraints. PhD thesis, University of Aberdeen

Iason G (1989a) Growth and mortality in mountain hares: the effect of sex and date of birth. Oecologia 81: 540–546

Iason G (1989b) Mortality of mountain hares in relation to body size and age. J Zool (London) 219:676–680

Iason G (1990) The effects of size, age and a cost of early breeding on reproduction in female mountain hares. Holarct Ecol 13:81–89

Iason G, Waterman PG (1988) Avoidance of plant phenolics by juvenile and reproducing female mountain hares in summer. Funct Ecol 2:433–440

Irvin A (1970) A note on the gastro-intestinal parasites of British hares (*Lepus europaeus* and *L. timidus*). J Zool (London) 162:544–546

Itamies J, Koskela P (1980) *Cynomyia mortuorumm yiasis* in *Lepus timidus*. Ann Entomol Fennici 46:24

Jackes AD, Watson A (1975) Winter whitening of Scottish mountain hares (*Lepus timidus scoticus*) in relation to daylength, temperature and snow-lie. J Zool (London) 178:403–409

Jaenson TGT, Petersson EH, Jaenson DGE, Kindberg J, Pettersson JH-O, Hjertqvist M, Medlock JM, Bengtsson H (2018) The importance of wildlife in the ecology and epidemiology of the TBE virus in Sweden: incidence of human TBE correlates with abundance of deer and hares. Parasit Vectors 11:477. https://doi.org/10.1186/s13071-018-3057-4

Jansson G, Pehrson Å (2007) The recent expansion of the brown hare (*Lepus europaeus*) in Sweden with possible implications to the mountain hare (*L. timidus*). Eur J Wildl Res 53:125–130. https://doi.org/10.1007/s10344-007-0086-2

Karlsen S (1983) Winter food preferences of the mountain hare in Norway. Finn Game Res 41:67–74

Kauhala K, Hiltunen M, Salonen T (2005a) Home ranges of mountain hares *Lepus timidus* in boreal forests of Finland. Wildl Biol 11:193–200. https://doi.org/10.2981/0909-6396(2005)11[193:hromhl]2.0.co;2

Kauhala K, Helle P, Hiltunen M (2005b) Population dynamics of mountain hare *Lepus timidus* populations in Finland. Wildl Biol 11:299–307. https://doi.org/10.2981/0909-6396(2005)11[299:PDOMHL]2.0.CO;2

Knipe A, Fowler PA, Ramsay S, Haydon DT, McNeilly AS, Thirgood S, Newey S (2013) The effects of population density on the breeding performance of mountain hare *Lepus timidus*. Wildl Biol 19:473–482. https://doi.org/10.2981/12-109

Koby F (1960) Sur l'extension maxima vers le sud-ouest de quelques représentants de la faune froide Würmienne. Anthropos 1. Supplement 1960:101–107

Kowalski E (1959) A catalogue of the Pleistocene mammals of Poland. Akademia Nauk, Warszawa

Kurtén B (2007) Pleistocene mammals of Europe, 1st edn. AldineTransaction, New Brunswick

La Morgia V, Venturino E (2017) Understanding hybridization and competition processes between hare species: implications for conservation and management on the basis of a mathematical model. Ecol Model 364:13–24. https://doi.org/10.1016/j.ecolmodel.2017.09.002

Lampio T (1952) An experiment in the control of disease in the snow hare. Pap Game Res, Helsinki 8:139–142

Leach K, Montgomery W, Reid N (2015) Biogeography, macroecology and species' traits mediate competitive interactions in the order Lagomorpha. Mammal Rev 45:88–102

Levänen R, Kunnasranta M, Pohjoismäki J (2018a) Mitochondrial DNA introgression at the northern edge of the brown hare (*Lepus europaeus*) range. Ann Zool Fenn 55:15–24. https://doi.org/10.5735/086.055.0103

Levänen R, Thulin C-G, Spong G, Pohjoismäki JLO (2018b) Widespread introgression of mountain hare genes into Fennoscandian brown hare populations. PLoS One 13:e0191790. https://doi.org/10.1371/journal.pone.0191790

Levänen R, Pohjoismäki JLO, Kunnasranta M (2019) Home ranges of semi-urban brown hares (*Lepus europaeus*) and mountain hares (*Lepus timidus*) at northern latitudes. Ann Zool Fenn 56:107. https://doi.org/10.5735/086.056.0110

Lind EA (1961) On the occurrence, daily activity, and spring moult of the snow hare (*Lepus timidus*) at Kilpisjärvi in Enontekio. Suomen Riista 14:106–111

Lind EA (1963) Observations on the mutual relationship between the snow hare (*Lepus timidus*) and the field hare (*L. europaeus*). Suomen Riista 16

Lindén H (1991) Analysis of game questionnaires: the validity of the alternative prey hypothesis in Finland. Suomen Riista 37:67–78

Lindlöf B, Lindström E, Pehrson Å (1974) On activity, habitat selection and diet of the mountain hare (*Lepus timidus* L) in winter. Viltrevy 9:27–43

Lockie JD (1961) The food of the pine marten *Martes martes* in west Rossshire, Scotland. Proc Zool Soc Lond 136:187–195

Lönnberg E (1905) On hybrids between *Lepus timidus* L. & *Lepus europaeus* Pall. from southern Sweden. Proc Zool Soc Lond 1:278–287

Luttschwager J (1956) Vorgeschichtliche Säugetierwelt in Schleswig-Holstein. Säugetierk Mitte 4:27–30

Malez M (1972) On the distribution of ice-age animals in the late Pleistocene of southeastern Europe. Radovi Jugoslavenske Akademija Znanosit i Umjetnosti 364: 133–180

Marcström V, Keith L, Engren E, Cary J, R. (1989) Demographic responses of arctic hares (*Lepus timidus*) to experimental reductions of red foxes (*Vulpes vulpes*) and martens (*Martes martes*). Can J Zool 67:658–668

Marques JP, Ferreira MS, Farelo L, Callahan CM, Hackländer K, Jenny H, Montgomery WI, Reid N, Good JM, Alves PC, Melo-Ferreira J (2017a) Mountain hare transcriptome and diagnostic markers as resources to monitor hybridization with European hares. Sci Data 4:170178. https://doi.org/10.1038/sdata.2017.178

Marques JP, Farelo L, Vilela J, Vanderpool D, Alves PC, Good JM, Boursot P, Melo-Ferreira J (2017b) Range expansion underlies historical introgressive hybridization in the Iberian hare. Sci Rep 7:40788. https://doi.org/10.1038/srep40788

Marques JP, Seixas FA, Farelo L, Callahan CM, Good JM, Montgomery WI, Reid N, Alves PC, Boursot P, Melo-Ferreira J (2020) An annotated draft genome of the mountain hare (*Lepus timidus*). Genome Biol Evol 12: 3656–3662. https://doi.org/10.1093/gbe/evz273

Mead-Briggs AR, Page R (1967) Ectoparasites from hares collected throughout the United Kingdom, January–March 1964. Entomo Month Maga 103:26–34

Melo-Ferreira J, Boursot P, Suchentrunk F, Ferrand N, Alves PC (2005) Invasion from the cold past: extensive introgression of mountain hare *(Lepus timidus)* mitochondrial DNA into three other hare species in northern Iberia. Mol Ecol 14:2459–2464. https://doi.org/10.1111/j.1365-294x.2005.02599.x

Melo-Ferreira J, Boursot P, Randi E, Kryukov A, Suchentrunk F, Ferrand N, Alves PC (2007) The rise and fall of the mountain hare (*Lepus timidus*) during Pleistocene glaciations: expansion and retreat with hybridization in the Iberian Peninsula. Mol Ecol 16: 605–618. https://doi.org/10.1111/j.1365-294x.2006.03166.x

Melo-Ferreira J, Alves PC, Rocha J, Ferrand N, Boursot P (2011) Interspecific X-chormosome and mitochondrial DNA introgression in the Iberian hare: selection or allele surfing? Evolution 65:1956–1968. https://doi.org/10.1111/j.1558-5646.2011.01261.x

Melo-Ferreira J, Boursot P, Carneiro M, Esteves PJ, Farelo L, Alves PC (2012) Recurrent introgression of mitochondrial DNA among hares (*Lepus* spp.) revealed by species-tree interference and coalescent simulations. Syst Biol 61:367–381. https://doi.org/10.1093/sysbio/syr114

Melo-Ferreira J, Farelo L, Freitas H, Suchentrunk F, Boursot P, Alves PC (2014) Home-loving boreal hare mitochondria survived several invasions in Iberia: the relative roles of recurrent hybridisation and allele surfing. Heredity 112:265–273. https://doi.org/10.1038/hdy.2013.102

Meneguz P, Rossi L (1990) Parassiti dell'apparato digerente della lepre variabile (*Lepus timidus*) dell'arco alpino. Parassitologia (Rome) 32:182–183

Miller G (1968) Evidence for selective feeding on fertilized plots by red grouse, hares, and rabbits. J Wild Manage 32:843–853

Mörner T (1999) Monitoring disease in wildlife-a review of diseases in the orders lagomorpha and rodentia in Sweden. Verhber Erkg Zootiere 39:255–262

Mostecky V (1964) Pleistozäne Säugetiere aus dem Steinbruch auf "Chlum" Mittelböhmen, unweit von Beroun. Sbornik Narodniho Muzea Praze 20:153–188

Mostecky V (1969) Late Pleistocene mammals from the "Chlupac cave" on Kobyla hill near Koneprusy (Bohemian Karst). Sbornik Narodniho Muzea Praze 25:1–55

Muroma E (1951) A list of parasites and diseases of the most important game animals in Finland. Suomen Riista 6:159–162

Myrberget S (1965) Returns of game species marked by the Norwegian hunting and fishing association, 1947–62. Pap Norw State Game Res Inst 21:1–24

Myrberget S (1972) Introduction of hares to Spitzbergen in 1930 and 1931. Fauna 25:126

Myrberget S (1983) Early reproduction in 1983 of mountain hares on Tranoy study area in northern Norway. Suomen Riista 30:35

Naumov SP, Shatalova SP (1974) Species indices of reproduction in mammals and peculiarities of their geographic variation (Leporidae taken as an example). Zool Ž 53:1856–1864

Newey S, Thirgood S (2004) Parasite-mediated reduction in fecundity of mountain hares. Proc R Soc Lond B Biol Sci 271:413–415

Newey S, Dahl F, Willebrand T, Thirgood S (2007) Unstable dynamics and population limitation in mountain hares. Biol Rev 82:527–549

Nodari M (2006) Ecological role of mountain hare (*Lepus timidus*) in the alpine ecosystems. PhD thesis, University of Insubria

Novikov GA, Airapetjantz AE, Pukinsky YB, Strelkov PP, Timofeeva EK (1970) The mammals of Leningrad region (in Russian). Leningradskogo Universitet, Leningrad, 360 pp

Nyholm ES (1968) Ecological observations on the snow hare (*Lepus timidus* L) on the islands of Krunnit and in Juusamo. Suomen Riista 20:15–31

Ognev SI (1940) Mammals of the USSR and adjacent countries. IV. Rodents. Izvestii Akademii Nauk SSSR, Moscow

Ohtaishi N, Nachiya N, Shibata Y (1976) Age determination of the hare from annual layers in the mandibular bone. Acta Theriol 21:168–171

Parovschikov VY (1961) The feeding habits of pine martens in the vicinity of Archangel. Zool Zhurnal 40:1112–1115

Pedersen S, Odden M, Pedersen HC (2017) Climate change induced molting mismatch? Mountain hare abundance reduced by duration of snow cover and predator abundance. Ecosphere 8:e01722. https://doi.org/10.1002/ecs2.1722

Pehrson Å (1983a) Caecotrophy in caged mountain hares (*Lepus timidus*). J Zool (London) 199:563–574. https://doi.org/10.1111/j.1469-7998.1983.tb05107.x

Pehrson Å (1983b) Digestibility and retension of food components in caged mountain hares *Lepus timidus* during the winter. Holarct Ecol 6:395–403

Pehrson Å (1984) Faecal nitrogen as an index of hare browse quality. Can J Zool 62:510–513. https://doi.org/10.1139/z84-075

Pehrson Å, Lindlöf B (1984) Impact of winter nutrition on reproduction in captive mountain hares (*Lepus timidus*) (Mammalia: Lagomorpha). J Zool (London) 204: 201–209

Picozzi N (1978) Dispersion, breeding and prey of the hen harrier *Circus cyaneus* in Glen Dye, Kincardineshire. Ibis 120:498–509

Pulliainen E (1981) A transect survey of small land carnivore and red fox populations on a subarctic fell in Finnish forest Lapland over 13 winters. Ann Zool Fenn 18:270–278

Pulliainen E (1983) The refuge theory and habitat selection in the mountain hare on a subarctic fell in Finnish forest Lapland. Finn Game Res 41:39–44

Raitis T, Ermala A (1972) Internal parasites in the snow hare (*Lepus timidus*). Suomen Riista 24:82–87

Rao SJ, Iason G, Hulbert IAR, Racey PA (2003) The effect of establishing native woodland on habitat selection and ranging of moorland mountain hares (*Lepus timidus*), a flexible forager. J Zool (London) 260:1–9

Rehnus M, Bollmann K (2016) Non-invasive genetic population density estimation of mountain hares (*Lepus timidus*) in the Alps: systematic or opportunistic sampling? Eur J Wildl Res 62:737–747. https://doi.org/10.1007/s10344-016-1053-6

Rehnus M, Palme R (2017) How genetic data improve the interpretation of results of faecal glucocorticoid metabolite measurements in a free-living population. PLoS One 12:e0183718. https://doi.org/10.1371/journal.pone.0183718

Rehnus M, Palme R, Filli F, Hackländer K (2010) Seasonal glucocorticoid secretion in mountain hares (*Lepus timidus*). Mammalia 74:347–350. https://doi.org/10.1515/MAMM.2010.032

Rehnus M, Marconi L, Hackländer K, Filli F (2013) Seasonal changes in habitat use and feeding strategy of the mountain hare (*Lepus timidus*) in the Central Alps. Hystrix 24:161–165. https://doi.org/10.4404/hystrix-24.2-4703

Rehnus M, Wehrle M, Palme R (2014) Mountain hares *Lepus timidus* and tourism: stress events and reactions. J Appl Ecol 51:6–12

Rehnus M, Braunisch V, Hackländer K, Jost L, Bollmann K (2016) The seasonal trade-off between food and cover in the Alpine mountain hare (*Lepus timidus*). Eur J Wildl Res 62:11–21

Rehnus M, Bollmann K, Schmatz DR, Hackländer K, Braunisch V (2018) Alpine glacial relict species losing out to climate change: the case of the fragmented mountain hare population (*Lepus timidus*) in the Alps. Glob Change Biol 24:3236–3253. https://doi.org/10.1111/gcb.14087

Rehnus M, Wehrle M, Obrist MK (2019) Vocalisation in the mountain hare: calls of a mostly silent species. Eur J Wildl Res 65. https://doi.org/10.1007/s10344-019-1331-1

Reid N, McDonald RA, Montgomery WI (2007) Factors associated with hare mortality during coursing. Anim Welf 16:427–434

Ritchie J (1929) The spread of the mountain hare in the Scottish lowlands and border country. Scott Nat 1929: 169–175

Robinson TJ, Elder F, Chapman JA (1983) Karyotypic conservation in the genus *Lepus* (order Lagomorpha). Can J Genetic Cytolo 25:540–544

Rosef O, Gondrosen B, Kapperud G, Underdal B (1983) Isolation and characterization of *Campylobacter jejuni* and *Campylobacter coli* from domestic and wild mammals in Norway. Appl Environ Microbiol 46:855–859

Rukovsky NN, Kupriyanov (1972) Some peculiarities of distribution and ecology of *Canis lupus* in the Onega peninsula. Zool Z51:1592–1596

Schai-Braun SC, Gander J, Jenny H, Hackländer K (2017) Is reproductive strategy of alpine mountain hares adapted to different elevations? Mamm Biol 85: 55–59. https://doi.org/10.1016/j.mambio.2017.02.008

Schai-Braun SC, Jenny H, Ruf T, Hackländer K (2021) Temperature increase and frost decrease driving upslope elevational range shifts in Alpine grouse and hares. Glob Change Biol 27(24):S. 6602–6614. https://doi.org/10.1111/gcb.15909

Schai-Braun SC, Posautz A, Alves PC, Hackländer K (2019) Gastrointestinal parasite infestation in the alpine mountain hare (*Lepus timidus varronis*): are abiotic environmental factors such as elevation, temperature and precipitation affecting prevalence of parasite species? Int J Parasitol 9:202–208. https://doi.org/10.1016/j.ijppaw.2019.05.009

Schai-Braun SC, Lapin K, Bernhardt K-G, Alves PC, Hackländer K (2020) Effect of landscape type, elevation, vegetation period, and taxonomic plant identification level on diet preferences of alpine mountain hares (*Lepus timidus varronis*). Eur J Wildl Res 66. https://doi.org/10.1007/s10344-020-01398-7

Schröder J, Soveri T, Suomalainen HA, Lindberg L-A, van der Loo W (1987) Hybrids between *Lepus timidus* and *Lepus europaeus* are rare although fertile. Hereditas 107:185–189

Seiskari P (1957) Winter home range of the snow hare. Suomen Riista 16:101–108

Seixas FA, Boursot P, Melo-Ferreira J (2018) The genomic impact of historical hybridization with massive mitochondrial DNA introgression. Genome Biol 19:91. https://doi.org/10.1186/s13059-018-1471-8

Sidorovich VE, Solovej IA, Sidorovich AA, Dyman AA (2008) Seasonal and annual variation in the diet of the raccoon dog *Nyctereutes procyonoides* in northern Belarus: the role of habitat type and family group. Acta Theriol 53:27–38. https://doi.org/10.1007/BF03194276

Slotta-Bachmayr L (1997) Biologie und Ökologie des Alpenschneehasen (*Lepus timidus varronis* Miller 1901): Verbreitung, Raumnutzung, Aktivität und Habitatwahl in den Hohen Tauern. PhD thesis, Paris Lodron Universität Salzburg

Smith AT, Johnston CH (2019) Lepus timidus. IUCN Red List Threatened Spec. 2019: e. T11791A45177198

Smith S, Sandoval-Castellanos E, Lagerholm VK, Napierala H, Sablin M, von Seth J, Fladerer FA, Germonpré M, Wojtal P, Miller R, Stewart JR, Dalén L (2017) Nonreceding hare lines: genetic continuity since the late Pleistocene in European mountain hares (*Lepus timidus*). Biol J Linnean Soc 120:891–908. https://doi.org/10.1093/biolinnean/blw009

Sokolov VS, Kryukova ME, Skurat LN (1987) Changes in the fine structure of sebacous glands in the specific inguinal gland of the arctic hare (*Lepus timidus*) during mating season. Dokl Akad Nauk SSSR 290:718–719

Sosnina EF, Novozhilova EN (1985) Lice of hares from the fauna of the USSR. Parozitologiya (Leningrad) 19: 290–295

Soveri T, Aarnio M (1983) Measuring body condition of hares during the winter. Finn Game Res 41:21–28

Sperber I (1974) Food utilization of *Lepus timidus* in the winter. Proc Intl Cong Game Biologists 11:175–177

Suchentrunk F, Polster K, Giacometti M, Ratti P, Thulin C-G, Ruhle C, Vasil'ev AG, Slotta-Bachmayr L (1999) Spatial partitioning of allozyme variability in European mountain hares (*Lepus timidus*): gene pool divergence across a disjunct distributional range? Z Säugetierkd 64:308–318

Sulkava S (1966) Feeding habits of the eagle owl (*Bubo bubo*) in Finland. Suomen Riista 18:145–156

Sulkava S, Huhtala K, Rajala P (1984) Diet and breeding success of the golden eagle in Finland 1958–82. Ann Zool Fenn 21:283–286

Thirgood S, Hewson R (1987) Shelter characteristics of mountain hare resting sites. Holarct Ecol 10:294–298

Thompson HV (1957) Mammal and bird pests on the land. Trans Intl Union Game Bio 3:57–59

Thulin C-G (2003) The distribution of mountain hares *Lepus timidus* in Europe: a challenge from brown hares *L. europaeus*? Mammal Rev 33:29–42. https://doi.org/10.1046/j.1365-2907.2003.00008.x

Thulin C-G, Tegelström H (2002) Biased geographical distribution of mitochondrial DNA that passed the species barrier from mountain hares to brown hares (genus *Lepus*): an effect of genetic incompatibility and mating behaviour? J Zool (London) 258:299–306. https://doi.org/10.1017/s0952836902001425

Thulin C-G, Jaarola M, Tegelström H (1997) The occurrence of mountain hare mitochondrial DNA in wild brown hares. Mol Ecol 6:463–467. https://doi.org/10.1046/j.1365-294x.1997.t01-1-00199.x

van der Merve M, Racey PA (1991) Body composition and reproduction in mountain hares (*Lepus timidus scoticus*) in north-East Scotland. J Zool (London) 225:676–682

Verde Arregoitia L, Leach K, Reid N, Fisher D (2015) Diversity, extinction, and threat status in lagomorphs. Ecography 38:1–11

Verkaik AJ, Hewson R (1986) Moult and rank in male mountain hares (*Lepus timidus*). J Zool (London) 207: 628–630

Vuolanto S (1972) Some aspects of the results of game marking with wing-tags in Finland in 1962–69. Suomen Riista 24:57–63

Wahlin B (1950) Hare damage and protection. Svensk Jakt 88:306–308

Walhovd H (1965) Age criteria of the mountain hare (*Lepus timidus* L) with analyses of age and sex ratios, body weights and growth in some Norwegian populations. Pap Norw State Game Res Inst 22:1–57

Waltari E, Cook JA (2005) Hares on ice: phylogeography and historical demographics of *Lepus arcticus*, *L. othus*, and *L. timidus* (Mammalia: Lagomorpha). Mol Ecol 14:3005–3016. https://doi.org/10.1111/j.1365-294X.2005.02625.x

Waltari E, Demboski JR, Klein DR, Cook JA (2004) A molecular perspective on the historical biogeography of the northern high latitudes. J Mammal 85:591–600. https://doi.org/10.1644/BER-101

Watson A, Hewson R (1973) Population densities of mountain hares (*Lepus timidus*) on western Scottish and Irish moors and on Scottish hills. J Zool (London) 170:151–159

Watson A, Wilson JD (2018) Seven decades of mountain hare counts show severe declines where high-yield recreational game bird hunting is practised. J Appl Ecol 55:2663–2672. https://doi.org/10.1111/1365-2664.13235

Wiley RH, Poston J (1996) Indirect mate choice, competition for mates, and coevolution of the sexes. Evolution 50:1372–1381. https://doi.org/10.1111/j.1558-5646.1996.tb03911.x

Wolfe A, Hayden TJ (1996) Home range sizes of Irish mountain hares on coastal grassland. Biol Environ Proc R Ir Acad 96B:141–146

Wolfe A, Whelan J, Hayden TJ (1996) The diet of the mountain hare (*Lepus timidus hibernicus*) on coastal grassland. J Zool (London) 240:804–810. https://doi.org/10.1111/j.1469-7998.1996.tb05327.x

Wu C, Wu J, Bunch TD, Li Q, Wang Y, Zhang Y-P (2005) Molecular phylogenetics and biogeography of *Lepus* in Eastern Asia based on mitochondrial DNA sequences. Mol Phylogenet Evol 37:45–61. https://doi.org/10.1016/j.ympev.2005.05.006

Yalden DW (1984) The status of the mountain hare, *Lepus timidus*, in the Peak District. Nat, Leeds 109:55–59

Yalden DW (1999) The history of British mammals. T. & A. D. Poyser, London

Zachos FE, Slimen HB, Hackländer K, Giacometti M, Suchentrunk F (2010) Regional genetic in situ differentiation despite phylogenetic heterogeneity in Alpine mountain hares. J Zool (London) 282:47–53. https://doi.org/10.1111/j.1469-7998.2010.00710.x

Zimova M, Hackländer K, Good JM, Melo-Ferreira J, Alves PC, Mills LS (2018) Function and underlying mechanisms of seasonal colour moulting in mammals and birds: what keeps them changing in a warming world? Biol Rev Camb Philos Soc 93:1478–1498. https://doi.org/10.1111/brv.12405

Index

K. Hackländer, P. C. Alves (eds.), *Primates and Lagomorpha*, Handbook of the Mammals of Europe,
https://doi.org/10.1007/978-3-030-34043-8

GPSR Compliance

The European Union's (EU) General Product Safety Regulation (GPSR) is a set of rules that requires consumer products to be safe and our obligations to ensure this.

If you have any concerns about our products, you can contact us on ProductSafety@springernature.com

In case Publisher is established outside the EU, the EU authorized representative is:

Springer Nature Customer Service Center GmbH
Europaplatz 3
69115 Heidelberg, Germany

Batch number: 10755744

Printed by Printforce, the Netherlands